NINE PINTS

NINE PINTS

A JOURNEY THROUGH
the MONEY, MEDICINE,
and MYSTERIES of BLOOD

ROSE GEORGE

METROPOLITAN BOOKS
HENRY HOLT AND COMPANY   NEW YORK

m

Metropolitan Books
Henry Holt and Company
Publishers since 1866
175 Fifth Avenue
New York, New York 10010
www.henryholt.com

Metropolitan Books® and m® are registered trademarks of
Macmillan Publishing Group, LLC.

Parts of this book originally appeared, in somewhat different form, in articles
in Longreads.com (chapter 3, "Janet and Percy") and in Mosaic Science
(chapter 6, "Rotting Pickles").

Library of Congress Cataloging-in-Publication data
Names: George, Rose, 1969– author.
Title: Nine pints : a journey through the money, medicine,
and mysteries of blood / Rose George.
Description: First edition. | New York : Metropolitan Books/ Henry Holt and
Company, 2018. | Includes bibliographical references and index.
Identifiers: LCCN 2018013647 | ISBN 9781627796378 (hardcover)
Subjects: LCSH: Blood.
Classification: LCC QP91 .G37 2018 | DDC 612.1/1—dc23
LC record available at https://lccn.loc.gov/2018013647

Our books may be purchased in bulk for promotional, educational, or business use. Please
contact your local bookseller or the Macmillan Corporate and Premium Sales Department at
(800) 221-7945, extension 5442, or by e-mail at MacmillanSpecialMarkets@macmillan.com.

First Edition 2018

Designed by Kelly S. Too

Printed in the United States of America

1 3 5 7 9 10 8 6 4 2

To the National Health Service

"Blood makes me feel so much better, and once I've had blood I want to play with my toys again."

—Owen Porter, 10

CONTENTS

NINE PINTS

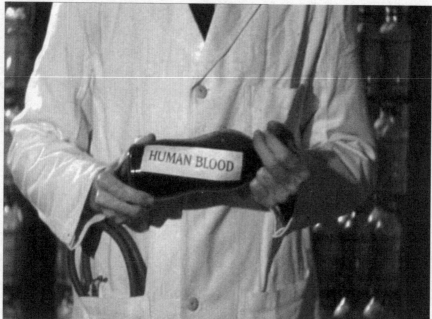

Ministry of Information trailer promoting blood donation, 1946

ONE

MY PINT

There is a TV but I watch my blood. It travels from a needle stuck in the crook of my right elbow, the arm with better veins, into a tube, down into the clear bag that is being hugged by a cradle that rocks then jerks, agitating its contents, stopping the clotting. Rock and wiggle. Rock, then wiggle.

I am giving away almost a pint, and it feels like it always does: soothing and calming. I watch the bag fill with this red rich liquid, which amounts to 13 percent of my blood supply.[1] I am comforted to know that 9 pints—8, now—of this stuff is moving around my body at any time at two to three miles per hour, taking oxygen to my organs and tissues, removing carbon dioxide, keeping my heart going, keeping me going.[2]

People have different rates of flow so the machine beeps with alarm when the output is too low. Today mine has been acceptable. Once, my veins were judged too small and I was turned away by the National Health Service Blood and Transplant (NHSBT), and I was insulted as if the rejection were moral, not medical. For a material that has been studied for thousands of years, blood still manages to run from rationality, even at walking pace.

Donating doesn't take long. I'm done in ten minutes. Female, A pos, time bled 11 a.m. Now I'm due to get thanked. Gratitude is the main theme here: the Wi-Fi password is "thank you." This is the main donor center in Leeds, my hometown and a city of three quarters of a million people. A bright, well-staffed place on one of the biggest shopping streets. Over the road at Red Hot restaurant, you can buy all you can eat from any cuisine in the world, all at once. One hundred dishes. Here, you can lie back and do not much—though clenching your buttocks helps keep your blood moving—and help three people, all at once. Give blood, and your donation can be separated by NHSBT, the public health agency that operates blood and organ transplant in England and Wales, into several lifesaving, life-enhancing gifts. By "gifts" they mean components such as red blood cells, platelets, plasma, and other useful fractions. Such details are available in NHSBT literature, as are phrases like "date bled." In the early days of the blood service, there were "bleeding couches." But now the straightforward language of biology has changed to one of altruism. It's all "donation" and "gift." The reality of it, that I am emitting a bodily fluid in public, is contained as much as possible, and not just in clear plastic bags.

Once decanted into its container, my blood is on its way to becoming something that even when given for free can be brokered and sold like ingots or wheat. It is also immediately much more perishable than it was in my veins: even when mixed with a storage medium, red blood cells have an official shelf life of between thirty-five and forty-nine days, depending on local laws.[3] They last longer than milk but not as long as cheese. This fragile but powerful substance can become a medicine, a lifesaver, and a commodity that is dearer than oil. Yet I give my blood freely because I know that my body will soon replace it and other people need it. I want nothing in return but a mint biscuit, a cup of tea, and a sticker that reads BE NICE TO ME. I GAVE BLOOD TODAY.

Every three seconds, somewhere in the world, a person receives a stranger's blood. Globally, 13,282 centers in 176 countries collect 110 million donations. The United States transfuses 16 million units of blood annually; the UK, 2.5 million. All of this blood is given to people when they have cancer or anemia or when they give birth; it can assist equally in trauma or chronic disease. Some accident victims can receive 60 units of blood; a liver transplant patient can use 100, or several bodies-full. A newborn can be saved with a teaspoonful. Read about the modern use of blood, and the word *precious* or *special* will appear alongside *health care resource*. Economists call the sale of organs and body parts a "repugnant market." But blood is different. The movement of blood—a body part, after all—is accepted unquestioned and common enough to be banal. But it is wondrous, still. It is as wondrous as blood.

Poor Odysseus. Deep in Hades, surrounded by ghosts and wraiths, and his mother won't speak to him. Not until she drinks the blood that Odysseus has taken from reluctant sheep. For Homer, blood had a power as fierce and invisible as electricity: a mouthful of blood, a switch flicked, and Anticlea could now speak to her son.[4] Of course Homer was awed by blood. There is nothing like it. It is stardust and the sea. The iron in our blood comes from the death of supernovas, like all iron on our planet.[5] This bright red liquid—brighter in the arteries, when it is transporting oxygen around the body from the heart, duller in the veins, when it is not—contains salt and water, like the sea we possibly came from.

We no longer sacrifice humans or beasts but the force of blood remains in language: blood feud, blood brothers, bloodlines. It remains in metaphor, where blood becomes an emotional state: my blood can be chilled, boiling, curdled. And its force remains in reality: most people associate the cheating cyclist Lance Armstrong with his abuse of erythropoietin (EPO), a hormone that stimulates the body to make more red blood cells. But I can't shake the image of him with a fridge full of his own blood, removed from his own body and ready to be transfused back into it.[6] A dose of fresh blood gives a cyclist enough strength—more red blood cells mean more oxygen—to push harder up the mountain, or an athlete to run faster around a track. The World Anti-Doping Agency includes blood on its list of prohibited substances, whether the transfusion be autologous (someone reinfusing his or her own blood), homologous (someone else's blood), or heterologous (blood from another species).[7]

The mythical Gorgon Medusa, with her head of snakes, showed the two-faced nature of blood best: the veins on her left side contained blood that was lethal, while the right side gave life. Transfusions can be two-faced, too. The right type of blood can save your life; the wrong one can kill you. I am calmed by the sight of my blood when it is being drawn or when I scratch it out from under my skin. I also curse it: along with stray menstrual tissue, it has wandered around my body for years to where it shouldn't be, so that I am riddled and glued by the adhesions of endometriosis, and every month they bleed too, in cacophonous sympathy.

We fear blood, still, despite our science and understanding, and we look to blood to tell us who we should fear. In 1144, the death of a young man named William of Norwich was attributed to Jews who had crucified him in order to use his blood as a sacrifice. This was the first documented case of what became known as blood libel, and it was enduring and lethal: for centuries, blood libel was used as a reason to massacre Jews and steal their property, across northern Europe, again and again.[8] In 2015, a Hamas leader in Gaza declared that Jews were still killing children and using their blood to knead into special Passover bread. The *Times of Israel* headlined this MEDIEVAL MINDSET.[9] I consider modern bans of gay men donating blood—the prohibitions are being relaxed now but persist—and see fear, not science. People with HIV are still jailed for not disclosing their status to sexual partners, long after HIV has become treatable to the point where it is not contagious. Chlamydia and hepatitis, now more life-threatening or disabling, get no such sanction. Blood is what artists still use to shock, although increasingly the menstrual kind.

Examine my blood with the right tools, and it can reveal who I am and what I was and what may become of me. I set up an alert to gather any mention of "a simple blood test," and it brings me news that my blood can be tested to show my biological or chronological age; whether I am likely to develop Alzheimer's, Parkinson's, or various types of cancer; whether surgery will give me delirium; whether my heart is failing; whether I am concussed. Most of these tests are still possibility and hope, or years away from being available. But already blood is a surveillance camera, the widest window with the best view into my past, present, and predictable future. Blood is one of the three main diagnostic tools of a doctor: the others are imaging and a physical exam.

Perhaps Hollywood describes blood best. In 1957, Frank Capra made a film for television. It was part of a series of educational films sponsored by Bell Laboratories; the year before, Capra had made *The Strange Case of the Cosmic Rays*, using puppets of Dostoyevsky and Dickens. In that context, his next film was perhaps normal. It was an extraordinary, partly animated portrayal of how blood moves around the body and what it does, its star a muscular cartoon he-man who, like the film, was named Hemo the Magnificent. He was blood, and he was magnificent with an attitude to match his muscles. "You men in white coats," Hemo said with disdain to the two human actors playing Dr. Research and A Writer, as Hemo stands surrounded by forest creatures (I didn't say it made sense), "you are not fit to tell my story."

> Humans think blood means disease, wounds, pain. These friends (the animals) they know me for what I really am: health, life. I'm the song of the lark, the blush on the cheek, the spring of the lamb. I am the precious sacrifice ancient man offered up to his gods, I am the sacred wine in the silver chalice. Down through the ages I am the price men pay for freedom. But to you scientists, I am a smear on a slide, a stain, a specimen, a sickness. My story is a song only poets should sing, not disease-lovers.[10]

I'm wooed by this, as much as his cooing animal chorus. (I'm less convinced by the analogy of vascular sphincters to railroad switchmen.) But my Yorkshire plain-speaking soul likes another description, from a consultant working for NHSBT. Blood, she said, is "the stuff that spurts out when you are not very well."[11]

The spleen is popular. Someone suggests the pancreas. Another offers, "The heart?" No one is sure, now that they have been asked. No one knows where blood is made.

The answer is: bones, mostly. Inside the bone, in the marrow, which most people probably think of as dog food, but which is the essence of us. "Gosh," says a hematologist when I tell him no one ever guesses this. "I wonder what they think the bone marrow does."

Perhaps they think bones are white and brittle, not vivid and vital. Perhaps they believe blood circulates ready-made, unchanging. Blood is always dying, always renewing, and rapidly: you can't yet grow back an arm, but you can lose plenty of blood and survive. The bone marrow produces two million red blood cells every second. It produces pluripotent stem cells that can become any cell, and red blood cells with no nucleus that can slink and slither through the tiniest of capillaries. Images of red blood cells show filled-in Cheerios or enticing pillows, so when I see a simulation of red blood cells, I want to jump into the picture and curl up in the middle of one. The American Society of Hematology prefers to liken them to doughnuts.[12]

Daily, the blood's thirty trillion red cells do a full circuit of the body, traveling about twelve thousand miles, three times the distance from my front door to Novosibirsk. The circulatory system of veins, arteries, and capillaries is about sixty thousand miles long, twice the circumference of the earth and more. Most of that is the capillaries, tiny blood vessels and holloways that reach nearly every cell in the body. In a resting human, the heart pumps a liter of blood every ten seconds, and it beats seventy-five times a minute. So does the heart of a sheep. A blue whale's heart, the size of an economy car, beats five times a minute (less on a deep dive); a shrew's, one thousand times.[13] The heart is busy, and so is blood. It has a lot to do. It carries oxygen to organs and tissues, as well as nutrients, heat, and hormones, the signals that regulate our functions, energy, sleep, mood. It carries out waste disposal, ridding the body of carbon dioxide and other unwanted matter. It clots when necessary. It fights infection and repels foreign invaders. It is a tissue and an organ at once. "The heart," a hematologist tells me, "is a pump for circulating our most important organ." Blood does all this—feeding station, temperature control, waste disposal, defender—and it never rests until you are dead.

Blood has fascinated humanity since it first spilled. Yet much about this "amiable juice," as Goethe called it,[14] remains remarkably mysterious. Take type. There are the four that you've heard of: the

ABO group of A, AB, B, and O. Then the rhesus factor that makes you positive or negative. All blood is categorized according to antigens, molecules found on the surface of red blood cells and on antibodies in plasma, the liquid part of blood. All blood cells have H antigens on their exterior, then A and B groups add A or B antigens or both. They are signals and markers: if a blood arrives in the body that has different molecules, it will be recognized and rejected. It is a highly effective alarm system. A and B will accept O because it has H antigens like they do. But an A type would reject a B type and vice versa. O negative has no A, B, or rhesus antigens, so anyone can have it. Every emergency department fridge will have O negative blood in it.

Get blood types wrong and the meeting of bloods will cause clots and clumping. The body can go into degrees of hemolytic shock, producing symptoms that range from itching to death.

In countries with good blood supplies, this is rare: the UK regards an incorrect blood transfusion as a "never event," something grave with serious consequences that was preventable. In 2015, never events included part of a chisel being left inside a patient, a fallopian tube removed instead of an appendix, and a B-positive person given A-positive blood, a mistake that was made clear by the patient getting chest pain and fever. The following year, there were three wrong transfusions (out of 2.5 million units transfused) but 264 near misses.[15] Globally, the chances of a transfusion-related infection are smaller than they have ever been. In poor countries, the prevalence of hepatitis B in blood is 0.3 percent; in a high-income nation, it's a tenth of that.[16]

The International Society of Blood Transfusion lists thirty-five blood group systems, and ABO is only one. Probably there are more like three hundred. Some more: Lutheran, Kell, Lewis, Duffy, Kidd, Diego, Dombrock, John Milton Hagen, Indian, and Globoside.[17] Most groups are named for the person who discovered them, so I'd quite like to meet Yt, Xg, and the cheerful Ok. One group is named Landsteiner-Wiener partly for Karl Landsteiner, the Austrian biologist who wondered why some blood mixed with other blood would clump and in 1901–3 discovered that not all blood was alike, that there were types and differences.[18] He later grouped blood into A, B, and C (later changed to O). It was an extraordinary discovery that won him the Nobel Prize, has enabled millions of people to be safely transfused with blood from perfect strangers, and I hope brought him more happiness than he publicly conveyed (photographs of him range from stern to terrifying). Perhaps his grimness came from his knowledge that he didn't know what blood types were for. We still don't.

Not that science hasn't been very busy. We can now change B-type blood to O type using an enzyme from a coffee bean (which clips off the B antigens from cells, leaving them nice and O). We have discovered that blood types can correlate with geography or ethnicity or a particular threat. Forty percent of Caucasians have type A blood, but only 27 percent of Asians.[19] The fact that O-type people are more susceptible to cholera was first noticed in 1977. During Peru's 1991 epidemic, people with O blood were eight times more likely to be hospitalized.[20] People from the Ganges delta, where cholera has always been endemic, have the lowest rate of O type anywhere. Recent research has shown that the cholera toxin thrives in intestinal cells derived from O-type stem cells, causing more severe infection.[21] In very bad news for A and AB men, a group of Turkish urologists recently found that their risk of erectile dysfunction was considerably increased compared to O-type men.[22] O-type people have a better chance against malaria while Bs come off worst. Ups and downs. These findings are inklings (a word nothing to do with writing but meaning "to utter in an undertone"). Why we have blood types, why they developed differently in different places at different times: we can still speak theories only in an undertone, not with certainty.

Yet in most countries, including mine, only mothers and patients and soldiers know what blood type they are. On a Portuguese warship, once, I stared in wonder at my escort's name badge, because under his name was his blood group. Pedro, A. I never got used to this. To know the sailors' blood types seemed wrong, as if I were reading on their name badges their latest sperm count or what their girlfriend best liked them to do in bed. It seemed prurient, invasive, like seeing inside them.

This is not rational. Common sense and blood sometimes repel each other. The Nazis, obsessed with the purity of blood, decided A was Aryan; B was inferior.[23] The Japanese even now believe that blood type involves far more than what antigens are on the outside of each blood cell. A types are perfectionist, kind, calm even in an emergency, and safe drivers; Bs are eccentric and selfish, but cheery. Os are both vigorous and cautious while ABs, obviously, are complicated.[24] A book on blood types in beautiful women was a bestseller, along with the author's follow-up, a book on blood types relating to lunch boxes.[25] Blood typing has serious consequences: people are denied jobs because of it and it is thought necessary to making a good dating choice. In 2011, when government minister Ryu Matsumoto resigned a week after taking office, having offended survivors of Fukushima and the earthquake, he blamed his blood type. "My blood is type B," he told reporters, "which means I can be irritable and impetuous [. . .] My wife called me earlier to point that out."[26] Blood type discrimination somehow nicely slots in with perceptions of inferior minorities: AB and B, more common in Taiwanese and Ainu people, are thought to be violent, backward, and cruel.

Cold War Americans thought blood type so important that they tattooed it onto adults and children. It would come in handy after the bomb dropped, when one physician predicted a city the size of Chicago would require nearly a million pints of donated blood.[27] In northern Indiana, wrote the historian Susan E. Lederer, using "a Burgess Vibratool instrument with thirty to fifty needles and an antiseptic ink, technicians tattooed the blood type and Rh factor on the chest of some 1,000 residents at the county fair."[28] Operation Tat-Type went on to tattoo children at five elementary schools, before the program was dropped because doctors didn't trust tattoos to be a fail-safe indication of blood type.[29] An editorial writer in Logan, Utah, reminisced about these "smudgy reminders" that can still be seen on middle-aged natives, though not read.[30]

The idea that blood is more than biology is not new. Nor is it resolved: blood is classified differently in different countries and even in the same country by different authorities. The UK exempts it from the Human Tissue Act, although it is a tissue, having cells. The United States thinks it a "biologic." The World Health Organization (WHO) added blood to its list of essential medicines (which even poor countries are advised to stock) only in 2013. In a lab in a London hospital, a man in a white coat moves away from his microscope and lets me see. It is nothing exciting for him, but it is the first time I have seen blood cells. The blood has been stained to be visible on the slide, so I can see clearly: the red cells, those biconcave discs, dumbbells and doughnuts. So vivid, though in a human body some are always dying and being replaced. I am my sixth version of myself, if most of my cells are replaced every seven years,[31] but I'm on my 143rd round of red blood cells, which live for about 115 days.[32] There is a popular philosophical question about identity and self, named Theseus's ship or Theseus's paradox: If all the planks were replaced in his vessel, was it still his ship? If I have replaced many of the cells I was born with, and none of the red blood I had at Christmas, am I still me?

When men first started to move blood between one body and another, they thought that they transmitted spirit with it. *Transfundere*, to pour from one vessel into another. A Mister Acton, writing in 1668 after the early transfusionists had published experiments, thought one experiment "most remarkable" because it was the "Transfusion of the Blood of a Mangie dog into a Sound one, to try whether the mange would be communicated with the blood." The Mangie dog was cured; "and the other who had received his blood, not become Mangie."[33] As blood was thought to cure spirit, not sickness, it was taken from suitable animals. Calves, lambs, mild and quiet creatures, were thought to transmit their sweet spirit to the frenzied and the troubled.

We can laugh at this, but we will be laughed at in turn. Our knowledge of blood is wide and unfinished.

Find yourself a blue coat first. There are plenty in the cupboard that look grubby but smell clean. Then sit on the bench provided and put your hair in a bonnet, something like a shower cap, and wrap your shoes in plastic. Follow the instructions above the basin and wash your hands thoroughly. No, more thoroughly. That's enough: they relaxed the restrictions a few years ago so you no longer have to wrap your beard or wear a snood. The pressure chamber now: this arrangement will be familiar to anyone who has been to a bank or traveled on a submarine. You step in and wait for one door to close before the other can be opened. The higher pressure of the air beyond keeps dust and bugs out. Now you have passed from the gray zone to the white zone. Now you are safe to walk onto the processing floor of the largest blood facility in Europe.

Today's bleed is 2706. Every day except Sunday, beginning just after lunch, one thousand gallons of blood arrive here from a several-hundred-mile radius, from many generous arms and veins. Filton is a small town in southwest England but also the name of this £60 million ($84.4 million) facility run by NHSBT, which processes a third of the blood donated in England and Wales. Getting here took months of asking. Nor I am allowed to identify or quote anyone. This is frustrating when the people I meet there are human and colorful beneath their white coats and plastic bonnets: along a corridor near the café there are photographs of staff in their leisure hours, and there is a diver, a knitter, a canicross competitor, and a newt collector. That's just one wall. But I understand why the NHSBT guards itself so tightly: the nation's blood is a vital and sensitive resource, and you wouldn't want just anyone coming to visit. Between the blood leaving me and entering someone else—a process known as "vein to vein"—an awful lot is done to it.

Blood arrives by vehicle and in bags. The plastic bag that I saw my blood running into is packed along with nine others into a blue cooler like a picnic bag. In the same bag are samples, three from each donation. All is put on a conveyor, and the donations go one way and the samples another. Testing is done at the same time as processing. All donations are tested, but Colindale, one of the three processing sites, doesn't have testing faculties, so Filton gets theirs too. Four thousand tests a day. Blood groups, obviously: ABO but also rhesus factor. Then, syphilis, HIV, hepatitis B, C, and E. First-time donors are also tested for human lymphotropic virus, which can cause leukemia. Particular donors, depending on where they have traveled or what they have done to their body with sharp implements—tattoos trigger a four-month deferral—can be tested for malaria, *Trypanosoma cruzi* (which causes Chagas' disease), West Nile virus, or cytomegalovirus. Anyone who has traveled to somewhere with Zika is currently deferred for twenty-eight days. For now. Tests can change. New infections come;

others die away. Zika has been known about since the 1970s, but it wasn't expected to cause trouble. White blood cells are removed from all donations—a process called leukodepletion—because it is in white cells that many infections travel, including the prions that cause variant Creutzfeldt-Jakob disease (vCJD), a vile and violent affliction that anyone growing up in the 1980s will visualize as piles of burning cattle and skeletal humans who fall when they walk because their brains are degenerating.

Safe. A blood system that is safe. Governments repeat this as if it can be true, not aspirational. But safety is a relative concept. Blood is a biological product and can never be safe because we can't plan for the next Zika or Ebola or HIV, until it comes.[34]

The processing space at Filton is vast, with a simple color scheme. The blue-coated staff and the red of the blood. There it is, hundreds of bags, hanging from hooks on what is called an overhead filtration device but what resembles a giant chicken rotisserie. It looks like a vampire's feast, and though the blood is safely contained in its bags, the color shouts through the sterility and bonnets and lab coats: this is not an inert material. Vivid, from the Latin. To live, to be living. Leukodepletion is being done by drip and filter: as the rows of bags rotate, the larger white blood cells are filtered out and the rest collects in the bottom of the bag.

Then anything can happen. Depending on need and logistics, a donation can become several useful products. Filton produces red blood cells, fresh frozen plasma (used for burn victims and to replace lost blood volume), platelets (used for clotting and by cancer patients), cryoprecipitate (also used to aid clotting), and leukodepleted whole blood, used for infants. Everything is organized and logged on a computer system, which obviously is called Pulse.

When Filton was being designed, NHSBT consulted the car industry. It was going to be an industrial process, and they wanted to have the most efficient production lines, like any industry. Like cars. The processing room was initially divided into three long lines, like a car production line. But it didn't work. This complicated business requires supervision, and supervisors were having to walk too far. Now the floor is set up in pods: small areas organized by a one-way flow system. A donation of leukodepleted whole blood comes in, it is processed by smart machines and dizzying technology, and it leaves at the other end of the pod as something else. If something goes wrong in a pod, a maximum of 96 units has to be quarantined, not the whole flow. All donations are centrifuged with considerable force, to separate plasma from the rest. Staff remember a hole in a bag—a manufacturing fault—and the power of the centrifuge throwing blood around the pod, with vigor. The humans got out of the way and off lightly: they changed their lab coats and got back to work.

The rest of the processing involves clever machinery and lots of tube connectors running between blood and bags. There is pressing and freezing and filtration and separation. It is technical and complicated and I must be starting to look the way I feel, because my guide asks kindly if he has fried my brain yet. It's not his fault. He does his best to translate. At various points I encounter equipment described as a giant condom or a snow globe. I am warned about "lumps and clumps." I'm invited to look for platelets in a processed bag and advised to spot "swirls," like a sprat in an ocean shoal of fish, seen from a plane. I see the sprats—I wouldn't from a plane, unless they were the size of a ship—and they are pretty. By the end of all this, I know that platelets look nothing like sprats, that if any donation is out of its allotted temperature for half an hour, it must be thrown away, and that anyone who complains that a unit of standard red blood cells costs £124.46 ($177.64) has no idea what a bargain they are getting. That is the price no matter what the blood group. Filton also processes blood groups so rare, they are kept frozen for ten years in an NHSBT facility in Liverpool. Plasma is cheaper, at only £28.75 ($40.39) a unit. Other products are costlier: some cryoprecipitate is £1,113.45 ($1,564.08) a unit.

In 2012, Filton flooded. Luckily, it was a Monday. No donations are taken on a Sunday, so nothing was being processed. A Friday night would have been awful. Other good times would have been during the Olympics, major football matches or sporting events, national holidays, Christmas, summer holidays, and Easter: these all badly affect blood donor numbers. The flooding could have been disastrous: all processed red blood cells must be kept in fridges at 39.2 degrees Fahrenheit until the test samples have cleared and they can be released from quarantine.

Plasma is frozen from fresh—and called fresh frozen plasma—and kept that way until it is used.

That is, male plasma. Away from the pods and the processing, my guide shows me some cages. They are filled with bags of plasma that should be yellow but isn't always. My plasma, probably, is green. I'm female and menopausal and taking hormone replacement therapy, and all those factors—along with the contraceptive pill—turn women's plasma green. Off and odd-looking. They don't know why, and it doesn't really matter, as all female plasma is discarded anyway, since NHSBT introduced a policy of "male donor preference" in 2003. So many female donors take hormones of one sort or another, it's not worth all the screening, not when NHSBT has enough plasma without it.

There are other conditions as well as being female that can turn plasma into a discard. My guide searches through the bags before he finds one with triumph. This is a nice one. I peer at it. Look, he says. Fat. I see globs of fat floating among the golden plasma. Staff are required to inform a donor if his or her blood is alarmingly fatty, even when the bag's appearance may just be the result of a fat-laden pre-donation meal. Obesity threatens life like HIV does. Perhaps now HIV is treatable, fat is worse. None of the NHSBT videos about Filton mention the discards. I don't mind if I become medical waste: the rest of my blood is useful. Discards used to be incinerated, but that was expensive as well as unsettlingly Old Testament. Burned blood. Now it gets "alternative treatment," which my guide thinks is a form of fancy landfill.

The blood products that survive discard can be processed in twenty-four hours. By six p.m. each day, donations set off for their destinations. Filton serves ninety hospitals with regular deliveries but sometimes the hospitals need irregular ones. In that event blood can go by taxi, a bag on a front seat like a passenger. At even odder hours, a fleet of volunteer bikers might deliver blood. I find a captivating Pathé film from 1967 about a Volunteer Emergency Service, which involved young motorcyclists zooming around London transporting blood as a charitable enterprise. This was probably to do with Father Bill Shergold, known to his flock as Farv, and an East End priest who thought bikers—generally despised and feared by the public—could be modern knights, upholding ideals of courage, courtesy, and chivalry. When Farv was seventy, and retired, he was approached by Wrangler to star in its advertising. Shergold asked his rector if that was acceptable, and the rector replied, "Of course you must do it. Good for the Church to be seen doing ordinary, rather silly things." Ordinary and rather silly, too, was the idea of doctors at Plymouth Hospital in the 1970s that blood samples could be delivered by carrier pigeon. The idea and the pigeons did not take off.[35]

I can't discover what happened to the Volunteer Emergency Service, but blood bikers are thriving and vital. The Nationwide Association of Blood Bikes and its regional chapters deliver thousands of blood bags a year, for nothing. They also deliver breast milk, spinal fluid, surgical instruments, and fecal matter for fecal transplants. Their slogan is "Saving Lives and Money," and, as their publicist writes on the Blood Bikes site, "despite many of us being middle aged and a bit flabby," they are doing more and more deliveries and getting more recognition. As this is Britain, the recognition often consists of "free hot drinks in certain cafes [and] nods and waves from police and paramedics."

Normally, blood that has been bled—Filton uses more corporeal terminology than the donor center, with its donations and gifts—arrives all day from one p.m. until eleven p.m. The quickest turnaround is blood that can be issued by one p.m. the next day. Donations from first-time donors take longer: they have to go through the testing twice, for extra safety. For this reason, blood bankers like two types of donors: young ones and ones that come back. They always sound desperate for both. A nation-state needs 1 to 3 percent of its population to give blood to maintain an adequate blood supply,[36] the higher the better. The UK needs two hundred thousand new donors every year. This should be blood from voluntary non-remunerated donations because this is the type of blood supply that the WHO thinks is safest. People who aren't paid for blood don't generally lie about their health. But of 172 countries surveyed by the WHO, 80 reported that only 1 percent of the population was donating. That's not enough. In Africa, the WHO judges that most countries in the region do not have enough donors for a safe or adequate blood supply.

Seventy-one countries get more than half of their blood from

"family replacement" systems (where patients are encouraged to provide blood given by relatives) or paid blood sellers. I have grown up in
a country with one of the best and safest blood supplies in the world.
I have been spoiled. The scientific wizardry of Filton; the efficient
blood donation and delivery system. That is not how much of the
world gets its blood.

It is easy to wander the corridors of Delhi's major hospitals, unmolested
by staff or security. The corridors of Safdarjung Hospital, one of the
city's largest, are usually full and noisy, and even my obviously foreign
face provokes no interest. People have their own troubles.

 In law, India's blood supply is rigorously monitored and a voluntary system. This fiction is as flimsy as paper. Countries that can find
enough donors or sellers to meet their blood requirements are in a
minority. In fact, India relies heavily on a family replacement system.
Patients who require blood must supply blood from a relative or friend.
Upstairs, at the hospital blood donor center, I meet a young man who
is donating blood for his pregnant wife. It is his first time donating,
and he says he's not nervous, but when the needle goes in he bares his
teeth. With this suffering, he will get a blood credit of one unit for his
wife to use if need be. This is how the United States' blood supply began
and why we talk of blood banks. Bernard Fantus, then director of therapeutics at Chicago's Cook County Hospital, invented the concept of
blood banking. Blood storage was already being done by then, but it
was Fantus who thought up debt and repayment of blood, who saw
blood as a product to be transacted, not a gift. He was straightforward
about this, saying in 1937 that "just as one cannot draw money from
a bank unless one has deposited some, so the blood preservation department cannot supply blood unless as much comes in as goes out."[37]
    Fantus's reach continues today. The US system, a network of 786
blood collection centers ranging from a large Red Cross facility to a one-
room community blood bank down the street, still operates according
to principles familiar to bankers. Blood that Americans think they are
giving for the good of their local community may then in fact be traded
all over the country. There is even a spot market or clearinghouse, where

purchasers requiring urgent blood resupply are obliged to take less popular blood types as part of the transaction. Your order of O negative comes with a side order of unwanted AB. Worldwide, money and blood mix more often than not.[38]

In India, family replacement, a benign phrase for a transaction, is impossible for migrants, the alone, the ones without networks. So other networks step in. I meet a man who tells me he has a WhatsApp group with his friends. They are all fans of a particular Tamil film star, and on that basis they have decided to give one another blood. There are now Facebook and WhatsApp groups trying to match blood givers and receivers. In the waiting area of the blood donation center at Safdarjung, I find a group of people who have formed an informal blood camp simply because they are friends, and kind.

Their patriarch, an elderly man given the honorific Baba-ji, has been an unpaid hospital volunteer for years, bringing milk to the poorer patients. He arrived at seven thirty a.m. every morning, and through this work came to realize that some people needed blood as well as milk. Safdarjung has a reputation for amputations, and each amputation surgery requires at least 2 units of blood. Sometimes 4. They are expected to replace that, but how? "These people are from outside," says Baba-ji (real name: Mr. Kayast). "From Orissa, Nepal, Jharkhand." They have an accident, it's poorly treated, so they come to the city but alone. No friends, no blood.

Every Sunday morning, then, Baba-ji and others come and give blood to strangers. Eleven of them have donated today, and each will ensure their blood goes to a particular patient. The last time Baba-ji donated—he is now too old, as Safdarjung's age limit for donors is fifty-five—it was for a man in bed number thirty-nine. An amputation. A new national blood policy wants family replacement to be phased out, but it will be difficult. Data released by the National AIDS Control Organization (NACO) of India in 2016 showed that hospitals had demanded blood from ten million Indians over the previous five years to compensate for transfusions given to their relatives or friends.[39]

NACO would prefer that hospitals relied on all-voluntary blood donor sessions, known in India as blood camps. I set up my own, inadvertently, when I visit Rotary Blood Bank, one of the few independent voluntary-based blood banks, in outer Delhi. The interview was going badly. My hosts were polite, but not illuminating. I was offered a tour of the building, and we entered the donating room, which was empty. The equipment looked clean enough, my frustration was mounting— the woman had just told me that no one sold blood in India, which is nonsense—so I thought giving away millions of white and red blood cells might pass the time. My hosts looked astonished, then delighted. I signed a form, stretched out my right arm, and that was all it took. Now the staff chatted about their families and their work with warmth and openness. Blood spilled, ice broken.

Visiting British journalists are not going to fill the hole in India's blood supply. They shouldn't accept my blood anyway: like anyone else who was born in Britain or living there before 1996, I am a global pariah. Until a cure is found for variant Creutzfeldt-Jakob disease, whose infectious prion particles can live for years undetected in our bodies, my blood is considered unsafe anywhere outside my island.

Delightful impromptu replacements for family replacement, like Baba-ji's group, will not plug the gaps either. But God and gods may. "Guru movement and political party adherents," wrote Jacob Copeman, a social anthropologist who studied religion and blood in north India, "vie to donate the most blood in a kind of national league of virtuous beneficence." In 2005, the Dera Sacha Sauda devotional order of north India held the Guinness world record for the most blood donated in a single day: sixty-five bathtubs full, 12,002,450 milliliters.[40] Indian medical professionals don't like massive blood camps. One told Copeman they were "blood massacres" because of all the blood that was wasted due to lax medical standards and an excess of "quality not sufficient" units.[41] Nor do they like the family replacement system. They want an anonymous, volunteer-based system because that is the safest. The trouble is, Indians don't agree with them. You give blood under pressure and only for your kin, not to be nice to strangers. Giving blood brings weakness because it is a life force. Losing it can damage a man's virility and a woman's fertility. When Jawaharlal Nehru was photographed giving blood in 1942, he was criticized for damaging his health, which was an important national treasure.[42]

India's Supreme Court banned the sale of blood in 1996.[43] It has also banned untouchability. Both bans are equally flexibly interpreted and both banned activities flourish happily. In 2008, for example, police acting on a tip-off raided a series of squalid tin sheds near Gorakhpur, Madhya Pradesh, and found blood slaves.[44] As Scott Carney reported in *The Red Market*, poor migrant men were kept in sheds by a local dairy farmer, Pappu Yadhav, and persistently bled to the point of death. Police found five sheds and freed seventeen men, who had been bled twice a week. Some had been imprisoned for two and a half years. Hemoglobin levels in a normal adult male should be 14 to 18 grams per deciliter of blood. The blood slaves had 4 grams.

Yadhav's setup was extreme but not unique. In 2017, Lucknow police arrested Mohd Arif, known as Shibbu, a former artisan who became a blood broker, buying blood from "donors," storing it in an ordinary kitchen fridge, and selling it to blood banks and hospitals. He paid agents 500 rupees ($7) to find sellers, paid sellers 1,000 rupees per blood sale, and sold the blood on for 4,000 ($60).[45] Hang around outside any hospital and look like you need something, and you will be offered blood for sale. Indian journalists have no trouble finding men such as Pramod, interviewed by Nikhil M. Babu, who offered a troop of "clean boys" who sold blood for 4,000 rupees ($60) a dose.[46] In poor farming regions such as Bundelkhand in Uttar Pradesh, where thirty-five hundred farmers have committed suicide in the last five years because of poverty and failed crops, blood is the new crop. It is illegal for hospitals to buy it, but they do, for a pittance. One farmer interviewed by Reuters sold two bottles and was given 1,200 rupees ($18.45).[47]

When an activist named Chetan Kothari requested data on HIV infections from India's National AIDS Control Organization, 2,234 people reported that they had been infected from transfusions.[48] Self-reporting is problematic: some people with HIV may prefer to blame the blood supply rather than drug use or illicit sex. But getting HIV through a blood transfusion in India is three thousand times more likely than in the United States,[49] where the risk is 0.01 percent, as in other high-income countries.[50]

A hospital as large as Safdarjung hopefully does its screening properly, although the reception booth at the blood bank on the ground floor is ramshackle. Baba-ji and his eight friends and relatives have done their bit. They have their glow. They have given blood for nothing to strangers, as perfect a gift as is possible, and narrowed a small bit of the gaping rift in India's blood supply. Now India just needs to find millions more people like them, and so does every other country like India.

Persuading people to give up their blood is not easy. Blood is a scarce and valuable resource. Why should anyone give it up for free? Every blood system will use incentives, even the voluntary non-remunerated ones. In the United States, government regulations forbid any incentive that can be exchanged for cash or be transferred or sold to someone else. Sports and event tickets can be given away, so they are out, but a discounted hotel room is permitted.[51] Even so, blood banks have offered pizza, football tickets, ice cream (in a Ben & Jerry's–sponsored initiative called A Pint for a Pint), haircuts, a day's gym membership.[52] There are other, more coercive ways to get blood. In 2015, Judge Marvin Wiggins, a circuit judge in Alabama, told petitioners in his court to give blood if they didn't have any money to pay fines. For those who refused and couldn't pay, "the sheriff has enough handcuffs." A professor of medical ethics described this as "wrong in about 3,000 ways."[53] Wiggins, as judges are supposed to do, was basing his ruling on precedence, as Susan E. Lederer writes in *Flesh and Blood*. "In 1940, a Chicago judge ordered Thomas Donohue to contribute blood to the blood bank at Cook County Hospital in lieu of the alimony he owed his wife. (His wife did not welcome his 'alcoholic transfusions.')" After the attack on Pearl Harbor, the mayor of Honolulu demanded blood from traffic violators, and in Worcester, Massachusetts, Lederer writes, "sightseeing motorists attempting to view the damage from a 1953 tornado learned that 'a pint of blood is your admission fee down here.'"[54] They paid.

Much thinking is done about how to attract donors. To entice, blood authorities need to understand what motivates someone to give away a considerable amount of their blood supply. Altruism, that most obvious motivator (sociologists actually talk of a "warm glow" effect), turns out to be more complicated than we think: countries that have non-remunerated volunteer donor systems can have wildly different rates of donations. In Luxembourg only 14 percent of people give blood; over the border in France, it's 44 percent. Nor do rates of blood donors match rates of other types of volunteering, which you'd expect if blood donors were just nice people wanting to do good. Instead, as economist Kieran Healey found, what changed donor rates was what kind of institution was collecting the blood. The Red Cross was the most popular. State authorities, such as NHSBT, were the next best. Germany saw donor rates drop when it allowed payments for blood. The lower rates persisted even when the agencies that paid for blood went bust.[55]

I have a key ring given me long ago when I donated blood that has my blood type on it—A+, the second commonest—but recently NHSBT has moved away from trinkets. Donors, it found, simply want a thank-you. Eighty-three percent of first-time donors remembered staff thanking them. Emotions work better than key rings. In 2012, Stockholm's blood bank Blodcentralen began sending text messages to people who had donated, telling them that their blood had been used, and for whom. It was a powerful way to humanize what is, after all, an anonymous system, where your blood disappears into a bag, into a back room, into a van, and away. The WHO classes the texts and other technological outreach as "mHealth." Blodcentralen's communications manager Karolina Blom Wiberg is more passionate. "Donors, including me, love getting them. They hit you right in the gut when you think that someone has in this instant been helped by my blood."[56]

NHSBT decided to try out the texting. "We borrowed it with pride," says Mike Stredder, head of Donor Services at NHSBT. They weren't as detailed: NHSBT data allowed them only to tell donors when their blood had been delivered to a hospital, not when it was used. (My last two have gone to hospitals in Derby and Sheffield: you're welcome, cancer patients, new mothers, trauma victims, and anemic people of Derby and Sheffield.) The response was so positive, says Stredder, "we cut short the trial and just rolled it out because I've never seen such overwhelming positive response to anything I've ever done in my career, in such a short period of time."

Donors were reporting that they were going to be donors for life now. Social media comments were overwhelmingly good. In short, NHSBT probably should have done it a long time ago. Also, it's cheap: it costs only 3 pence (4.2 cents) to send each text, or £35,000 ($49,165) for the whole donor base.[57] As efforts to retain donors go, it's the most successful for decades. Retention is what blood authorities dream of. Perfect donors are the ones who come back. They require less testing; they can be relied upon. They may even turn up during football finals or Wimbledon. Getting new donors is another matter, and particularly the young kind. The donor base is aging, and younger people are giving less blood. Between 2005 and 2015, new donors decreased by nearly a quarter. NHSBT has launched youth-friendly campaigns such as Missing Type, where the letters A, B, and O went missing from various organizations. The prime minister was suddenly living in Dwning Street; people were searching on Ggle and reading the *Dily Mirrr*. It was smart and accessible, and more successful than, for example, a program in Australia, when the local blood service attempted outreach in universities. Too many underweight young women were coming in and fainting, and then their friends started fainting in sympathy: the outreach retreated.

"We need black blood."

I say, "Pardon?," not because I didn't hear but because I want to hear this again, because I can't believe Stredder has said it. Blood and race have such a sensitive history. In the early days of US blood banking, blood bags were labeled as N for Negro or AA for African American, and hospitals as august as Johns Hopkins refused to transfuse Caucasian patients with Negro blood. Lemuel Diggs, who ran the blood bank at Memphis's John Gaston Hospital, had stored different "colored" blood on different refrigerator shelves, openly. Southern states routinely segregated "white" and "black" blood. Most admitted their policies had no scientific basis. The Red Cross, which refused donations from African Americans for plasma collection, called its reasoning "a matter of tradition and sentiment rather than of science."[58] The US War Department directed that "for reasons which are not biologically convincing but which are commonly recognized as psychologically important in America, it is not deemed advisable to collect and mix Caucasian and Negro blood indiscriminately for later administration to members of the military forces."[59] Nazis did the same, refusing non-Aryan blood and dying of their wounds as a result. In the United States, blood segregation stopped only in 1972, when Louisiana finally repealed its blood label laws (along with the segregation of water fountains, bathrooms, trains, dance halls, and marriage).[60]

But Stredder is emphatic. Black blood. Not even blood from BAME, an ungainly acronym standing for Black, Asian, and Minority Ethnic. They made appeals for more BAME blood in the past, Stredder tells me, but now they need to be more "robust." Antigens and antibodies commoner in black people's blood make it more appropriate for transfusions for sickle-cell anemia, for example, a genetic condition that produces deformed red blood cells (shaped like sickles), leading to anemia, fatigue, and awful pain. Sickle-cell patients can require several transfusions a month, and they can develop immune reactions to "foreign" antigens. The closer the cross match of blood to their own variety of antigens, the better. Black donors are more likely to have the Ro subtype (a version of the rhesus group), and Ro blood is more likely to be used to treat sickle-cell anemia. But only 2 percent of the British population has it.

At Tooting Blood Donor Centre in London, I meet part of the 2 percent. I'd asked to meet a young donor, and Azeez had been suggested, as Azeez had offered his blood, again and again, because it is one of the rarest kinds. It is Ro but also U negative, a combination found only in black people, usually those of African heritage. He was unusual because of his blood but also because of who he was. Neither of us knew what the other looked like, but I had an advantage: if a young, black man walked into the donor center, then it would be him, because donors like him are only slightly less rare than his blood. Numbers of young donors are falling: half of British donors are now over forty-five. Also, Azeez is Muslim. Attracting Muslim donors is another challenge: as one non-donor reported to a NHSBT survey, he didn't know whether blood donation was permissible, and he was an imam. (Most scholars think it is, as long as money does not change hands and it is done in good faith.) Azeez didn't need to forensically explore the scriptures: in his mind giving blood was obviously a deed of charity and one of the five pillars of Islam.

He has persuaded some of his friends to donate. This is the most successful way to attract new donors: they'll come if they know someone who gives blood or if someone in their family has required blood. The UK needs between 1.6 and 1.7 million units a year. That is 6,000 units a day, but NHSBT likes to keep a buffer, and has 30,000 to 40,000 units available on any one day.[61] All these numbers make the UK an exception. When the WHO compiled its latest global status report on blood safety with data from 180 countries, it concluded that 112.5 million donations had been made in the reporting year (2013). High-income countries collected almost half of those donations, despite having only 19 percent of the global population. Low-income and lower-middle-income countries, where nearly half the world's people live, collected only 27 percent.[62] NHSBT needs to find 200,000 new donors every year to maintain a safe blood supply. When one of the best blood systems in the world struggles, what chance does everyone else have?

In the reception area at Tooting Blood Donor Centre, I read a book. Around me are the signs of the ordinary: the noises of the donor center; the cheap supermarket cordial; the chatter of the receptionists about somethings and nothings. All of it makes the place seem unthreatening and soothing. It detracts from the truth about blood donation and transfusion: that it is something that hasn't happened for very long and that it involves a substance that still defies science's attempts to understand it. If we could understand how to stop people bleeding, we wouldn't need so much blood. We can create synthetic hearts. We can build organs on 3-D printers. But despite decades of research and many millions of dollars, we have yet to make blood. What it can do, how it does it, what it can carry, how it works: all these things have some answers but not all. The only sure thing about blood is our enduring fascination with it and its answering mystery.

The book is leather-bound, or fake-leather-bound, and embossed with gold lettering. Both these things expressed the fact that it is worth reading. A prized thing. That is because it is full of life and death. It is the thank-you book, from recipients to donors. I read its pages and think that if I weren't a blood donor already, this might make me one. The recipients haven't written much, but their words express their surprise that someone could do something as extraordinary as give away a part of their body to them and want nothing back.

There are two questions on each of the sheets in the blood donor book. The first is "What has this transfusion meant to you?," then "What would you like to say to your blood donor?" One writes that she now has a better lifestyle, as if a blood transfusion has improved her clothes shopping (she means she can get out of bed). One is thankful that the transfusion "has improved my blood and changed the color from pink to red."

But the man who best describes the power of blood, this baffling, splendid substance, writes in an elderly hand. He gives no sense of what blood has done for his health or why he needed it, but his economy of words is rich. Under the question "What has this transfusion meant to you?" he has answered, only but sufficiently, "Good."

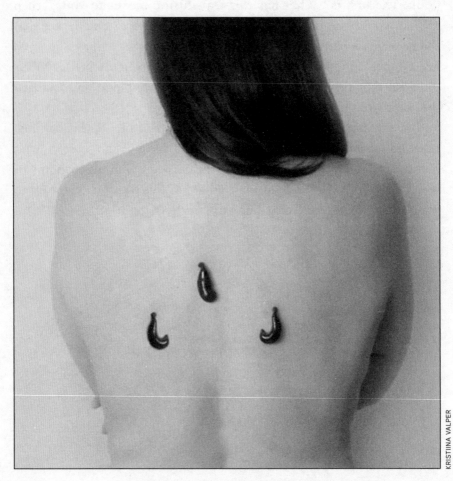

Hirudotherapy in Tallinn, Estonia

TWO

THAT MOST SINGULAR
AND VALUABLE REPTILE[1]

Six seconds. Perhaps ten. Twelve, if it is cautious or dopey. After that, the jaws will activate, the hundreds of teeth will engage, the leech will begin to eat, and its meal is your blood. Are you wading through a tropical pond in fierce humidity? Have you returned to your guesthouse to find with horror a passenger on your leg? Are you Humphrey Bogart, tugging the *African Queen* and Katharine Hepburn through a brown river, cursing the "filthy little devils" that cling to you? Possibly (except the last). But you are equally likely to be in a sterile room of a modern hospital, tended to by nurses who attach these bloodsucking animals to you without a shiver. You accept them equally calmly, because it has been explained to you that these leeches may save your breast, or your finger, or your ear, or your life.

Less than half a mile from the busy M4 motorway, in the southwest of Wales, there is a walled entrance off a road whose Welsh name I can't pronounce, and a small sign saying BIOPHARM. A long and winding driveway passes sheds of unclear purpose and ends in a small yard beyond an imposing cream-colored manor house. The view is unexpected: I can hear the motorway, a distant roar, yet here is a quiet landscape of green fields stretching away with no apparent end. The nation's only leech production business looks like a health farm. Which I suppose it is.

Biopharm was founded by Roy Sawyer, an American zoologist transplanted to Wales. Its name is as broad and generic as its business is specific: it is one of fewer than half a dozen suppliers of medicinal leeches in the world. My hosts are Sawyer's daughter Bethany, a young woman with a mild Welsh accent and a stern demeanor who is Biopharm's manager, and her colleague Carl Peters-Bond, a fair man who is what grandmothers would call "cuddly." They may revise that when they learn that one of his job titles is "leech-growth technician." Carl is also a practiced Biopharm guide, as well as a retriever of visitors who, given how little signage there is, must inevitably wander into the narrow corridor and wonder if they are in fact in buildings that house a globally renowned supply of medicinal leeches or in someone's garage.

The reception room is furnished with a large table, a library of leech-related and zoological books, and several ceramic jars in a cabinet, identified by their ornate lettering spelling out LEECHES. These jars were used by apothecaries who sold leeches and sometimes rented them in the high times of the leech industry, from the beginning of the nineteenth century to the early twentieth, a period known as "leech mania." The mania was for bloodletting, or "the taking of blood from a person with therapeutic intent," in one description, which had been used for a few thousand years to cure ailments ranging from headaches to near hangings.[2] There was hardly any disease or condition for which bloodletting was not thought useful; it was even used to treat severe bleeding. For most of human history, we have preferred to remove blood, not add it. By the end of the nineteenth century people still believed so powerfully in the force of bloodletting that they "were in the habit of coming to be bled at their own request, just as they now apply to have their teeth drawn."[3] Bleeding was done by opening veins with lancets and knives (and earlier with stones, fish teeth, or whatever could cut[4]), but a leech was kinder, sucking its blood from the capillaries, not veins, and providing its own natural anesthetic.[5] By the twentieth century, bloodletting and leeching had fallen from favor with the rise of better surgery and medicine and germ theory. By then the leech had also been harvested almost to extinction from its natural habitat in most of Britain and across much of Europe. Although leech fans are always hopeful of finding undiscovered populations, the known habitats of the British medicinal leech, now, are wetlands near Dungeness, some ponds in the New Forest, and this building near Llanelli.[6]

Coffee is served and seats taken and then we wait for three student nurses arriving from Swansea, who are coming for a leech training day and who are lost. Meanwhile: chat. Carl has worked at Biopharm for twenty-four years. He used to work with fish. "Just aquariums, like." Then he was approached by the leech people on the basis that leeches need tanks too, and he has been at Biopharm ever since. His Welsh accent soothes as Welsh accents tend to do, even when he is talking about creatures that can make you bleed for ten hours and look like slugs and are slimy.

The leech is not a slug. Nor is it a bug, reptile, or insect. Sometimes it is not slimy either. The leech is an animal belonging to the phylum Annelida, a zoological category that includes over fifteen thousand species of segmented bristle worms and 650 species of leeches in the subclass Hirudinea.[7] Not all leeches suck blood and not all bloodsucking leeches seek the blood of humans. Many have evolved to have impressively specialized food sources: one desert variety lives in camel's noses; another feeds on bats. Some eat hamsters and frogs. The Giant Amazon leech, which can grow to 45 centimeters (17.7 inches) long, feeds by inserting a proboscis—like a long, 10-centimeter (4-inch) straw—into its prey.[8]

The leeches that I have driven several hundred miles to encounter are freshwater, bloodsucking, multisegmented annelid worms with ten stomachs, thirty-two brains, nine pairs of testicles, and several hundred teeth that leave a distinct bite mark.[9] Depending on the era you live in, this resembles either a wound made by a circular saw or a Mercedes-Benz logo. Biopharm breeds both *Hirudo verbana* and *Hirudo medicinalis*. Until recently, *verbana* was thought to be the same species as *medicinalis*, and both were called the "European medicinal leech." Now we know they are genetically distinct. The *medicinalis* are northern European by origin; the *verbana* are more Mediterranean.[10] Biopharm also breeds *Hirudinaria manillensis*, named for its native habitat of Manila but also called the Asian medicinal leech or the buffalo, after its habit of dining on bovines. The European is bred to suck the blood of humans; the Asian is for veterinary use. The buffalo leech is fatter, hungrier, and less picky, used to piercing through hairy cattle legs or bellies. Europeans, more refined as bloodsuckers go, avoid stubble, perfume, hair products, and peculiar skin smells.

Both varieties have two characteristics in common: they inject their host with a local anesthetic so that they are rarely noticed until they have tucked in. Because of this, a leech bite will usually feel like a vague sensation, not a nip or scratch. Once their teeth are engaged, they emit the best anticoagulants known to exist, so that their blood meal keeps flowing long after they have stopped feeding, often for up to ten hours. The leech is in many ways a simple animal, but its anesthetic and anticoagulant have yet to be bettered by science. Roy Sawyer has often called the medicinal leech a "living pharmacy." Only eight compounds in leech saliva have been identified, and there are probably hundreds that are useful. One, discovered by John Berry Haycraft in 1884, was later distilled into what became known as hirudin,[11] vastly more efficient as an anticoagulant than man-made heparin, the next best blood thinner. Another is a "potent inhibitor of collagen mediated platelet adhesion and activation."[12] This was isolated after researchers noticed a strange thing: the leech bite made blood flow for hours, but not because of the hirudin. Instead, another substance—which Sawyer named "calin," Welsh for "heart"—was stopping platelets doing what they were supposed to do, which is to aggregate and form a clot.[13] On one patent database, there are ten entries for Biopharm covering antithrombins, hyaluronidase, inhibitors of fibrin cross-linking, protease inhibitors, and heparin-containing formulations.[14] Not only is the leech a medicinal treasure chest, but its Mercedes-Benz bite is spectacularly efficient, the tripartite shape much less damaging than a scalpel incision, which can damage tissue. Apart from the bloodsucking issue, it seems to me that the leech is one of the more polite parasites. All in all, it is an astonishing creature, but I still don't want to pick one up.

The student nurses have arrived with a whoosh of vigor and enthusiasm. They take a moment to enthuse over the musk turtle housed in a tank, a female named Jimmy that to me just looks lonely, then the leech jars and images. In my notebook, I write, "They are almost entirely undisgusted." And later, "They are going to be really good nurses." Then the introductory Biopharm video can begin. It starts with a twenty-year-old scene from the Elizabethan-era *Blackadder*: Edmund Black-adder is in love and wants a cure. He consults a doctor and the doctor prescribes leeches. The joke is that in Blackadder's day the cure is almost always leeches. Doctors were called leeches, but because of an etymological coincidence, not for their fondness for prescribing worms. (Leeches were named for the Old English word *laece*, meaning "worm," and derived from Middle Dutch; doctors were also called *laece*, but derived from the Old Frisian *laki*, meaning "a physician.")[15] This name for physicians lasted until the Renaissance and bequeathed to us such wonderful book titles as *Leeching, Wortcunning, and Starcraft of Early England*.

The leech is an ancient companion, and the idea of using it to treat maladies occurred to humans thousands of years ago. The body's ailments were thought to be due to too much blood, among other things. Along with fleams and lances, the leech was an essential tool in the bloodletter's armamentarium. Babylonians wrote of a striped blood-sucking worm that became "thick with blood" but also described the leech as a daughter of Gula, the goddess of healing. Dhanvantari, the Hindu god of healing, medicine, and Ayurveda, is usually pictured with a jar of leeches in one of his four arms. A wall painting in the tomb of the Egyptian scribe Userhat, thought to date to three thousand years ago, pictures a figure applying leeches.[16] The first written record of leeching was probably in the *Alexipharmica*, a listing of poisons and their antidotes done in hexameter—as all medical literature should be written—by Nicander of Colophon, thought to have lived in the second or third century BCE, but there are leech references in Sanskrit, Persian, Chinese, and Arabic literature.[17] The Chinese scholar Wang Chong, wrote Robert Kirk and Neil Pemberton in their fine book *Leech*, told of a king who accidentally swallowed a leech with his meal but said nothing, not wanting to embarrass his hosts. "Later, the king found himself cured of his chronic affliction. This was, Wang Chong explains, a happy effect of the leech having drawn blood from the site of illness within the king's body."[18] (Actually a leech in the throat can swell and suffocate you.)

Bloodletting suited humoral medicine, the prevailing dogma for thousands of years. This held that the body was made up of four humors or liquids. The medical historian Hermann Glasscheib called them "the four juices." A human was only a vessel for these life juices, which were yellow bile, black bile, white phlegm, and red blood. A cold, for example, was due to an excess of the white juice, which was then expelled through the nose and mouth. If someone turned yellowish, there was too much yellow juice. "The body had three doors through which it could evacuate nocive matter," wrote Glasscheib. "Through the skin in the form of sweat, through the kidneys as urine and through the bowels as feces. But since there were four juices there must also be four exits. The doctors invented this fourth door in the shape of bloodletting." Hippocrates, Galen, Paracelsus: all the medical celebrities believed in the power of relieving the body of blood. In the *Canon of Medicine*, the fourteenth-century Persian polymath Ibn Sina (Westernized as Avicenna) devoted considerable attention to bloodletting. It was "a general evacuation," good for all sorts, medicine that was preventative and curative at the same time. It was useful when "the blood is so superabundant that a disease is about to develop [or when] disease is already present."[19] Different blood vessels served different purposes. Bleeding the veins between the eyebrows was good for long-standing headache, cutting the veins under the tongue—only lengthways, otherwise it was difficult to stanch—was useful for angina or tonsillar abscess. Opening the sciatic vein relieved podagra and elephantiasis; menstrual problems were alleviated by cutting the saphenous vein in the leg. The same vein was good for emptying blood from other organs. He did not recommend that blood be removed from humans aged less than fourteen or more than seventy. Young adults could increase their tolerance for adult-level exsanguination with "small blood-abstractions."[20] It's easy to scorn this misplaced precision from our privileged position in the twenty-first century. My tonsils were removed as a child because it was standard procedure. I bled for hours into one side of the pillow and then, when the nurse turned it over phlegmatically, the other. Tonsillectomies are now considered old-fashioned and are rarely performed, only forty years later.

Bloodletting was as unquestioned as Band-Aids. Sometimes, it was a job requirement: monks had to be bled several times a year in bleeding houses called *seyneys* or *flebotomaria*, either as a sort of general body maintenance or for a more intriguing reason. Monks were supposed to be celibate, and chronic and enforced celibacy was thought to entail a dangerous buildup of semen (*retentio semenis*), which could lead to blood poisoning.[21] Bloodletting avoided blood poisoning.[22] Monks didn't appear to mind and treated bloodletting as a holiday: they were relieved of choir duty and work, a fire was lit in the infirmary, and they got to eat meat.[23]

By Avicenna's time, the job of bloodletting could be done by medical men but also by barbers. They were used to sharp instruments, and a papal decree forbidding monks from performing medical tasks meant monastery barbers began to diversify into doing small acts of surgery. This practice spread, so that barbers became instead barber-surgeons and formed a guild. The first barber-surgeon on the registry of the Worshipful Company of Barbers was recorded in 1312.[24] The bleeding barber is the reason modern barbers display red and white striped poles: the pole was a stick for the patient to grip; the white stripes were the bandages, the red stripes the blood. The ball on the top was probably a deformation of the blood-gathering bowl.

Liber Albus: The White Book of the City of London, a 1419 rule book published by Lord Mayor Richard Whitington (better known as Dick, and for having a cat), provided instructions for all aspects of city life. As well as looking darkly upon foreigners, who could not be hostelers or sell meat, and forbidding the baking of bread made from bran, the rule book pronounced that no barber "shall be so bold or so daring, as to put blood in their windows openly or in view of folks; but let them have it carried privily unto the Thames, under pain of paying two shillings unto the use of the Sheriffs."[25] Barbers were bloodletters until surgery was established as a profession; there were tussles, and King George II finally put an end to the rivalry by setting up two separate guilds in 1745.[26] After this, surgeons did surgery and barbers did what barbers do now, but both could pull teeth.

For two and a half thousand years, if you scratched at any account of illness, bloodletting would come out.

The video has moved on from *Blackadder*. We are approaching the nineteenth century, when leeching became leech mania. Leeching was accepted enough already that it was applied to royalty. When the Prince Regent of England fell ill in 1816, he was given thirty-six leeches in one go.[27] In 1825, Emperor Alexander of Russia caught a fever in the Crimea. The empress urged him to submit to leeching, but he "rejected the proposition with great obstinacy and violence."[28] Only when he worsened did he accept leeches on his head. He died anyway.

It was another ruler who indirectly began the reign of the leechers. A surgeon in Napoleon's army, François-Joseph-Victor Broussais, became "the most sanguinary physician in history."[29] The Napoleonic Wars, wrote Robert Kirk and Neil Pemberton, meant that "civilian surgeons had been absorbed into the military leaving few with the skills to perform bloodletting via the lancet. Broussais's brilliance was to present an entirely new system of medicine that sounded modern yet was grounded in a simple, familiar and apparently safe therapy."[30] Broussais's theory, based on his finding traces of blood in the digestive system on autopsies, was that all illnesses arose from inflammation of the guts. He called these inflammations "phlegmasies" and believed they could all be relieved by bleeding. So could head colds, syphilis, menstruation, flu, cholera, and gout. But he knew that bleeding was dangerous: people would often be bled until the point of "syncope," or near death. All those opened veins got infected. Broussais thought the leech was a much better idea: even copious leeching did not kill as often as vein opening, and leeches were in abundance. Leeching was particularly useful for trauma, "when, for example, a wheel has passed over the body."[31] Inflammation could also be reduced by applying leeches to the anus. "You will in an instant remove a phlegmasie from six inches to a foot wide."[32] (He does not describe this in any more detail, leaving me to wonder what a foot-wide inflammation out of the anus looks like. Briefly.) A toddler needed only one or three leeches, and a woman, fifteen. A grown man should be given sixty in one application. Broussais was a star; his lectures were so notorious, the minister of war (or police, depending which source you believe) once had to close the lecture-hall doors to keep out besieging hordes, and his theories were revered. At the beginning of the nineteenth century, France produced enough native leeches to export them. By 1833, it had to import

41.6 million leeches.[33] French doctors by now ordered their patients to be leeched even before they met them, no matter what the trouble. A worm prophylactic.

Where were all these leeches coming from? When they were abundant in European marshes and ponds, leech gatherers would walk bare-legged through ponds to harvest them. The Wellcome Collection in London has a bucolic engraving of Yorkshire women gathering leeches to put them in small barrels, and Wordsworth wrote a poem about a Lakeland leech gatherer, who "roamed from Pond to Pond and moor to moor / Housing, with God's good help, by choice or chance / And in this way he gained an honest maintenance."[34] It was an honest occupation but not poetic or bucolic. In France, it was known as "blood fishing."

> You would suddenly see a young woman soften and sway, as if she was drunk or dizzy, sometimes slump into the pond, her legs in the mud but her head in the clouds. Her companions knew what this flagging meant: a weakness caused by the insatiable vampirism of leeches. So they would quickly get the stunned girl out of the mud to free her from the slimy parasites.[35]

The usual tonic was strong red wine; it revived and was thought to replace the blood the leech harvester had lost from the animals stuck all over her bare legs, which she then removed with hot ash or salt. The occupation paid badly and was confined to the poorest and most desperate, and it was doomed: by mid-century, the native western European leech was getting scarce. Europeans tried to breed leeches, but it was tricky. The first to succeed was M. Béchade of the Gironde in France, who in 1835 invented a revolting method of sending horses, donkeys, and cows into ponds for leeches to feed on.[36] When the animals showed an understandable aversion to being cut open and sent into the ponds, they were strapped into a box, wheeled into the ponds, and bled anyway. Elderly horses were often chosen for this fate. It was, wrote Claude Seignolle, like condemning the old horse—who had given years of loyal service—to "two deaths," and the first was the more horrible of the two.[37]

But there were fortunes to be made. Béchade did so well his company continues to trade, as Biopharm's main competitor Ricarimpex. Leech farming was profitable enough that crooks and frauds abounded. In 1856, the French ministry of agriculture condemned the practice of fattening up leeches for sale with old blood (usually from abattoirs) and noted that it contravened Articles 1 and 2 of the 1851 Penal Code. Inspectors were sent to pharmacies to do random leech checks, taking worms, weighing them, then putting them in saline and squeezing them, before weighing them again to see if they had been plumped up with animal blood.[38] The increasing rarity of the native leech was reflected in its price: a thousand animals had cost 5 francs ($1); now it was 20 francs and more in winter.

The era of leech import-export and smuggling began. Hungary, Russia, Portugal: they had leech populations to spare, and all gave up their native leeches for profit, and lots of profit. Shipping magnates loved the leech trade, transporting leeches across seas and oceans from Germany, Russia, Hungary, and Portugal to the United States and Brazil. Containers could be tins or pots or glass or cases, but they had to be sturdy, because leeches were canny. Many a ship arrived in port with leeches all over the deck. Brazil's mania was fueled by the leeching of the Brazilian emperor Pedro I and his wife Leopoldina as well as various Portuguese royals. Slaves, usually young boys from Angola, Mozambique, and the Congo, were trained in leeching and other barber skills, and this made them more valuable. In 1844, as Roy Sawyer writes in an exhaustively researched paper on the Portuguese and Brazilian leech trade, a girl trained in domestic duties sold for 220,000 milréis ($98,300). A slave talented at leeching sold for three times as much. Leeches, meanwhile, sold for 200 milréis ($89) apiece. "In other words," wrote Sawyer, "at one point the life of a barber slave was worth as little as 500 leeches, and that of a domestic girl for less than 175 leeches. This reflected more on the high price of leeches than on the low price of slaves."[39]

Critics of leech mania were surprisingly few. Or perhaps they didn't survive long enough to record their opposition. An early objector was Lord Byron, whose objections were satisfyingly lyrical. In 1825, parts of a letter written by Lord Byron's doctor Francis Bruno appeared in the *Times* and gave an account of the poet's death the year before. The poet, who had joined the Greek insurrection against the Ottoman Empire, had passed "a very gay day" in Missolonghi when he fell ill. A first attempt at bleeding was canceled because of a commotion in the lord's bowels. Two days later, after pain in his forehead, seven leeches were applied to his temples, and they took two pounds of blood. "I perceived," wrote Bruno, "that his Lordship had a very great aversion to bloodletting."[40] "Have you no other remedy than bleeding?," the patient asked. "There are many more die of the lancet than the lance."[41] This is admirable wordplay from a man on his deathbed, but it didn't stop the doctors: they thought blood should out, and they got it. "Come," said Byron toward his end, "you are, I see, a damned set of butchers. Take away as much blood as you will but have done with it."[42] They did, and he died.

In 1827, a doctor named Joseph-Marie Audin-Rouvière had published a "No more leeches!" anti-bleeding polemic. It was a "murderous system," wrote Audin-Rouvière, whose prestige was "almost inexplicable." He described the visit of a typical doctor. He does not consult his patient or ask him about symptoms but, from the threshold, cries, "Leeches! Leeches!"

"How many?"
"Sixty, eighty."
"But the sick man has no strength, he is eighty years old."
"The leeches will give him strength!"

He tells of a Dr. Frappart who, during the course of one patient's sickness, applied eighteen hundred leeches (so that the course of the sickness probably ended in death); and of the case of Monsieur Martainville, a newspaper editor, whose gouty fingers received five hundred leeches. "Everybody knows that M. Martainville still has gout." If each leech took an ounce of blood, wrote Audin-Rouvière, then a patient could lose twelve pounds of blood. A Broussais leeching could take 80 percent of a patient's blood volume, putting them at risk for the most severe category of traumatic hemorrhage, one that usually ends in death. Audin-Rouvière hoped that justice would be done to Broussais, but it wasn't.[43] In the decades after his death in 1838, leech mania subsided, to the gratitude of leech gatherers and leeches. An obituary of this "immense celebrity" mourned the loss of the great man to medicine. Broussais did leave something of a legacy: paisley was probably inspired by his leech mania (it was actually Persian), and his questioning of humoral medicine was useful, though his conclusions could be murderous. This "ardent defender of inflammation and leeches," as an obituary writer described him,[44] had also overseen an animal hunted to extinction in many countries and lied about how many patients had survived his theories.[45] But there are hospitals named for him, in France and Italy, for Broussais the most bloody and bleeding.[46] There is no memorial for the millions of leeches his bizarre theory wasted or the patients his theory killed.

The video is over. My nerves wake up along with my disgust mechanism, because next is the tour of the tanks, and leech wrangling. For an animal that biologists describe as rather simple, the leech needs complicated handling. Biopharm's leech raising is done over three large rooms, each kept at a different temperature. The further in we go, the further along the path of the leech to becoming a hospital device, the colder it gets. All the tanks and equipment are built to exact specifications, most of it devised by Carl. It is the engineering and the precision that keep him at Biopharm, not the leeches. Everything here, he says with pride, is bespoke.

The first room is kept at 78.8 degrees Fahrenheit. It gives the pleasurable jolt that a winter walker gets entering a tropical hothouse, a sudden wash of heat. I take a photo though the sight is just dozens of tanks draped in white muslin. Carl notices. "You can take a picture of the room but not of the tanks." Breeding leeches is a sensitive process of feeding and starving and warming and cooling, and leeches can be spooked even by the noise of a smartphone click. The tanks are where leeches are born, by the happy meeting of any two of them: leeches are hermaphrodites and very flexible. Carl lifts a corner of muslin covering a tank and picks one up. It's a European and surprisingly beautiful, its belly striped with iridescent gold and green. Even Carl, the sober engineer, admits, "The colors are quite nice. If you see anyone else's leeches, they're not as nice as ours. I select them for color."

Elsewhere are the buffalo leeches. They are kept for animal use: cats can be leeched because of polycythemia vera, a condition of excess hemoglobin in the blood. Dogs are often leeched to relieve swollen or infected ears, a problem particularly common in French bulldogs. Carl thinks Biopharm supplies leeches for 90 percent of French bulldogs with cauliflower ears, as aural hematomas are nicknamed. I like that statistic, but I am not sure it can be backed up. Carl says the buffalo costs an arm and a leg, an appropriately physical metaphor, but like the pharmacists' leeches of old, rented, squeezed, and re-rented, buffaloes can be reused. Also, buffaloes get hungrier more quickly than medicinal leeches so they can be used more frequently: even after feeding to satiety on a cat, they will be ready to eat again in six to eight weeks. The European medicinal leech spends a year digesting one meal.

The menu at Biopharm is always black pudding. In the two years it takes to raise a European leech for medicinal use, it is fed sheep's blood served in sausage casing every six months. Biopharm used to feed its residents with cow's blood, which was more successful. The leeches ate it more readily, and one cow held the blood volume of ten sheep. But bovine spongiform encephalopathy (BSE) has ruled out cow blood, for leeches and humans.

Carl points out an immobile leech on the bottom of the tank. "That's what they do in the wild. When they feed, because they have a huge reserve of blood, they'll bury in the mud or moss." He describes the leech as a sort of oil tanker: all its reproductive organs are on the front where the cab would be. "The central organs are on its side. It's got two hearts, one on each side. The bulk of it is storage." A fed leech can swell to up to five times its body weight. A small leech can expand eightfold. Carl sticks his finger in the water and a leech immediately appears. "He's sniffing around now." Actually, it's more of a tasting: Carl thinks they sense the sugars and oils in the skin. He picks one up but isn't bitten. "I'm not very attractive to leeches." A bigger problem is leeches biting each other. They can digest at different rates. "Maybe one leech has shrunk down to three hundred milligrams and it's in a tank with a leech that is three or four grams." That is a recipe for murder: a big hungry leech will eat from a small hungry leech, and sometimes the biting can be fatal. The best method for peace among leeches is to adjust the temperature so they are half asleep and half awake. The safest leech is a spaced-out leech.

Biopharm also experiments with tank size to give leeches the optimal amount of exercise. Carl is tank builder, leech grower, and personal trainer: leeches have to be exercised twice a day. It's not complicated, as training programs go. "I'll go and pick one up and put it at the other end of the tank." It will swim, and it can lose weight quite quickly. Sometimes it gets more exercise than Carl bargained for. Their most annoying talent, he says, is for escape, even from Biopharm's tanks. He has often arrived home to find some attached to his ankles. "I'm usually surprised if I don't find ten leeches in the footwell of my car. They stick to your shoe and then they dry out." He says this, and we all look at our feet.

Leeches can shift. In a race between a slug and a leech, who would win? It depends on the conditions and climate, says Bethany Sawyer, but leeches can move faster than expected. "A lot of people assume that the leech's relative is a slug because they're black and look like that. So we have people who ring up wanting them for weird and quirky photo shoots." Once, it was a fashion student. Often, it is someone who wants to do a "wacky" promotion of various things, events, places. "And they get the leeches thinking they are going to be really slow and they're going to have all the opportunities in the world to take the perfect picture, they're going to be able to set them down, to tell them to sit and stay.

"And that's it," says Carl. "They turn the lights up and whoooomph! Off they go." How far can they travel? "Anywhere it's damp. So anywhere in Wales, really."

When they swim, they come fast and beautifully. On land, leeches move by suction: they suck with the front sucker, then the rear, and that is their locomotion. It is an efficient but not elegant movement. (It is nothing like earthworm locomotion, which is done by peristalsis-style burrowing, in waves.) But in water, they are different. They are sinuous. "By flattening and manipulating their bodies into wavelike patterns," write Kirk and Pemberton, "leeches are capable of swimming at speed and with an elegance few other creatures can rival." Leonardo da Vinci drew leeches in his notebooks in an attempt to understand the physics of their movement.[47] The motion is dorsoventral, as done by whales, dolphins, and eels: up and down, not side to side (think a butterfly stroke, but done by an Olympic athlete, with grace and power).

It doesn't matter how good a swimmer a Biopharm leech is. It will be packaged in gel and sent to a hospital pharmacy, and sooner or later—its work done—it will be killed. In 2004, the US Food and Drug Administration (FDA) gave *Hirudo medicinalis* an unclassified status as a marketable medical device.[48] Single-use only: all leeches employed in hospital settings must be exterminated with alcohol solution once they have fed and dropped off. This seems ungrateful, but a filled leech is a biohazard. Leeches can transfer blood from one person to another. "They're worse than that," says Carl. "They're a needle that can walk." Biopharm sells a special euthanasia kit called Nosda to dispatch the leeches humanely. This includes the alcohol required, various pots, and, with misplaced kindness, "leech-friendly forceps."[49]

The leeches in the cold room are almost hospital-ready. They have had four feeds in their lifetime and been starved for six months. If he's lucky, says Carl, he can get a leech from birth to a hospital pharmacy in two years. But usually it's about three. The starving is because a hungry leech, when applied to a human, is an efficient leech. We are not allowed into the final room, as it is bathed in UV light to make the leech as sterile as possible. Nor do we see the packing: leeches make their onward journey in a proprietary polymer gel. There is skullduggery in leeching: when I ask Carl if there is any corporate spying, he won't answer, except to say, "We don't need to. No one has a yield like ours." Ninety percent of the leeches born at Biopharm grow up to be walking needles. It helps that they are flexible, with a tolerance of temperatures from 23 degrees Fahrenheit up to 104. If it's hotter, they travel with ice chips. They have to arrive in good order: they have work to do.

And one of them struck the slave of the high priest and cut his right ear. But Jesus said, no more of this, and he touched his ear and healed him.

—Luke 22:50–51

Amputated ears are harder to fix than the Bible maintains. They are filled with tiny blood vessels, so when they are torn off—the medical term is "avulsed"—it is difficult to reattach them. It is a tiny tapestry torn in half, but the tapestry is made of hairs whose diameter varies from 0.3 to 0.7 millimeters (a human hair is actually thicker). Every tiny thread must be reattached and must work. It is fiendishly complicated. As three doctors wrote in one paper, "Cases of successful microvascular reattachment of totally amputated ears have been conspicuous by their absence." This paper appeared in 1987 and caused a sensation. You wouldn't know why from the title, which was "Microsurgical Reattachment of Totally Amputated Ears." Nor would you know why from the images. They are graphic, showing an avulsed ear, then a reattached one. They do not show the reason the paper became as renowned as one of its authors, pediatric surgeon Joseph Upton.[50] The reason was leeches.

In 1985, a three-year-old boy from Massachusetts named Guy Condelli had his ear bitten off by the family dog.[51] He was taken to Boston's Children's Hospital, where his surgeons included Joseph Upton. The surgeons proceeded as they usually did with amputated ears: the detached ear was examined under a microscope in the operating room. Several sets of blood vessels were identified, measuring 0.2 to 0.5 millimeters in diameter. "It was impossible," wrote the paper's authors, "to distinguish arteries from veins." The boy was given a general anesthetic and the ear was stitched back to its rightful place. But it began to turn blue. The blood was being pumped into the area by the arteries, which are more robust and quicker to recover, but the veins weren't working and the blood couldn't be pumped away again. The blood was stuck, dark and ominous through the skin. During his operation, the boy had been given 5,000 units of heparin, a powerful anticoagulant, to loosen the congestion. It didn't work. By day five, as the images show, the ear looked black. The child was in trouble.

Joseph Upton had worked as an army surgeon during the war in Vietnam. He had heard about leeches and maggots being used. "I started calling round the country to my friends," he told a reporter, "trying to find some hungry leeches."[52] This was unlikely to be successful. In Carl Peters-Bond's words, American leeches "are rubbish."

The Asian medicinal leech is 25 percent less effective than the European, in Carl's estimation, but the American variety is twice as bad. Its anticoagulant doesn't work as well, and Upton needed the most powerful anticoagulant he could get. So Upton needed a European medicinal leech and the United States didn't have any. Nor were leeches licensed to be used as medical devices. Upton eventually found Biopharm, founded only the year before by Roy Sawyer, and ordered some of its products. I ask Bethany how they got to Boston. "Flown. The pilot took them." Imagine this: a pilot flying over the Atlantic who has to worry about storms and turbulence and keeping three hundred people alive, and he also has a box of leeches behind his seat. That would be a fidgeting pilot.

The leeches arrived safely and were placed directly on the congested tissue. Upton described the procedure in his article. "When they were engorged, they would fall off the ear. New ones were applied when discoloration occurred. Following initial application, the color immediately improved." Or, as Upton told a reporter more plainly, "The ear perked right up." It perked, it pinked, it was saved.

For a paper that was describing a revolutionary and extremely successful procedure, it is oddly reserved, beyond the usual dispassion of scientific journal writing. Graphic pictures of ripped-off ears are shown but not the leeches that saved them. There is no triumphal trumpeting of the first successful use of leeches in decades. Instead, the authors write glumly, their use was "not new." Nothing to see, no big deal that multisegmented annelid worms that most humans find revolting had just been let into the most sterile environment possible and performed a revolutionary act of anticoagulation on a three-year-old boy who would now have two working ears.

In a way, this self-undermining was justified. A pair of Slovenian surgeons had rehabilitated the medicinal leech in the 1960s.[53] But leeches hadn't been used in an American or British operating room for decades. Thirty years after Guy Condelli's operation, the leech occupies a peculiar place in modern life. To the general public, it is simply disgusting. They think leeching is "evil quackery," says Bethany Sawyer, and that it belongs in the Middle Ages along with pestilence, boils, and Blackadder. In 2016, the Olympic swimmer Michael Phelps was revealed to be a fan of "cupping," a technique popular across the world for centuries.[54] Cupping applies cups—glass, usually, but cow horn will do—to the skin. A flame is lit, then extinguished, and a vacuum created inside the cup is supposed to draw blood into the tissues and provoke an anti-inflammatory response. It's meant to improve blood flow. The sight of dark red circular marks on Michael Phelps caused derision. One science writer tweeted: "What next, leeches?" A *New Yorker* writer, wanting to convey that someone thought something was pointless, wrote, "It would be as useful as applying leeches to a head wound."[55] Guy Condelli's ear—a head wound—shows that this is wrong. But the leech is still a symbol of the ignorant and old ways, when a woman was known to be hysterical because her womb wandered around her body, and the application of half a mouse to a wart was thought sensible.[56]

Unless you know better. Unless you are a plastic surgeon. When the surgeon Iain Whitaker did a telephone survey of all the sixty-two plastic surgery units across the UK in 2002, 80 percent of the fifty that replied had used leeches postoperatively in the salvage of compromised free flaps or digital replants within the last five years. Three units had used leeches more than sixteen times a year; fifteen had used them up to five times.[57] There are abundant leech papers in journals of plastic surgery, maxillofacial surgery, and microsurgery detailing proper leech procedure. Leeches are judged to be effective in the salvage of various essential body parts such as fingers, ears, nipples, nasal tips, and penises. (Leeches were not used in the famous penis reconstruction of John Wayne Bobbitt but were apparently on standby.)[58]

By now, leeching is most commonly used in flap surgery, the transfer of a living piece of tissue from one part of the body to help another part, used in breast reconstruction, open fractures, large wounds, and improving cleft palates. Unlike a graft, the flap comes with its blood vessels attached, and the incoming blood vessels must be attached to the existing ones, an intricate and infernal weaving. Attaching tiny blood vessels calls for a microsurgeon, and if the blood vessels get congested, the microsurgeon calls for a leech. It is unquestioned practice now. Biopharm sends out leeches every day. Today's packages are going to Cyprus and Finland.

Things have changed since Upton's fly-by-night leech delivery. Both *Hirudo medicinalis* and *Hirudo verbana* are now endangered species, listed in Appendix II of the Convention on International Trade in Endangered Species of Wild Fauna and Flora.[59] To export a leech you need a permit from CITES, and a permit can take six weeks to organize. This seems odd, in a building where leeches are breeding like, well, leeches. "Ours aren't endangered at all," says Carl. "We've got plenty."

"We had an instance," says Bethany, "where a Saudi doctor rang and said, look, this five-year-old boy is going to lose his foot, we really, really need them. It was frantic, it was just constant work between us speaking to the CITES authorities, and speaking to the airline trying to get all the air freight organized. We are two and something hours away from Heathrow. Our courier has gone away with everything and we had to speak to customs and say, can we get it done today rather than in two or three weeks. They actually held the Saudi Airlines flight for two hours for our courier to get up there." As further proof that the boy was probably the well-connected kind of Saudi, the leeches were carried by the wife of one of the owners of Saudi Airlines.

CITES is a frustration. "I'm just waiting for somebody from CITES to cut his arm off," says Carl, "and they say, you have to wait for the permit." Airlines are another trouble. There aren't many medical devices that are alive, so while leeches are medical devices, airlines treat them as livestock, and they don't like carrying livestock. Some airlines are better than others at keeping their hold temperatures cool in the summer. Boiled leeches are always a possibility.

It's time for the handling. Carl doesn't have any appropriate Europeans, so we are invited to pick up a buffalo. He fetches a leech that looks big and black and, though I know better, a lot like a slug. He says, "This species is slimy. It's super, super slimy." Is that relish in his voice? He invites the first junior nurse to put her hands out. He will put his hands underneath in case the shock makes her drop it, but she is not shocked. She is entirely sanguine, in the English sense (the French have decided it means fiery and excitable). This is a nurse who won't balk at bedpans. Carl gives an audio tour of the leech while it moves on her hands, sniffing and looking for a bite. "This is the biting end. That's the hold-ing end. The reproductive organs are there." He tells us the color means that the leech is pregnant, and the nurse nearly drops it. "Don't worry. They're as tough as old boots."

Will it splat?

"No. And they take quite a long time to bite." An attached leech can bite in six seconds; a handled leech is slower. "You'd be surprised how much time you've got." He estimates twelve to fifteen seconds, although he no longer needs to count. "I can see where the muscles are, when it's about to bite. I can see exactly." He also knows which leech it is, and who its parents were: this one has two stripes, and his and her parents had one each. I ask him if he gives leeches names and his expression says that is a very stupid question. "No. But they defi-nitely have traits. Some are more aggressive than others, some are faster than others. Some are really placid."

The second nurse takes the leech. She squeals. "Don't worry," says Carl. "The head is nowhere near. Chill out!" I ask her what it feels like and she says, "Like there's a leech on my hand." One more nurse, one more period of fewer than fifteen seconds of leech holding, and then it's my turn.

I have blanked out what it felt like. My recording tells me that I said, "It's all right, actually." Then, immediately, "You can have it back now." A picture shows my face screwed up in a classic expression of disgust. I remember something that felt not like slime but like nothing else. Cool. Alive. Something I didn't want on my hands. I have no idea how I'd tolerate them on a ravaged ear, a reconstructed breast, a torn-off finger. I have no idea how I could be persuaded.

After months of reading papers on leeches, I notice something. Leech therapy "was well received by virtually all," wrote Whitaker in his survey of plastic surgery units, "with only a small number of units reporting patient noncompliance within the last five years."[60] In a jour-nal of head and neck surgery: "No patients had leech therapy stopped because of inability to tolerate the treatment."[61] A patient information leaflet on leech therapy: "The nurse will explain leech therapy to you and make sure you understand the process before applying the first leech."[62] By all these accounts, persuading a patient to undergo leeching is no more problematic than offering a needle. But the leech is not a needle, despite Carl's analogy. It is a bloodsucking parasitic creature that has suckers and teeth. It is disgusting. Or is it?

There are good reasons that humans find things disgusting. For disgust theorists—there are such people, and they are great—it's because things and creatures that disgust are things and creatures that are dangerous. A discarded hair that can transmit disease is more disgusting than one attached to a head. A caterpillar, which is unlikely to infest you, is less repellent than a worm, which might. This biological determinism is not fixed: what is found disgusting varies according to age, geography, and status. The disgustologist Dr. Val Curtis found that Indians were disgusted by urine, sweat, menstrual blood, cut hair, childbirth, vomit, mice and rats, lower castes, and decaying waste.[63] The Dutch, meanwhile, were repelled by shit, stickiness, and fishmongers' hands, as well as cats and dogs. When the author William Miller explored the disgust of Americans, he found them to be repelled by "feces, bodily fluids, pustules, rotting wastes, severed limbs, pubic hair, sexual fluids, graveyards, slaughter houses, compost, carrion, slugs, maggots, bloodsucking parasites and deformity."[64]

Disgust is why the leech has been perverted into a symbol for malignity, parasitism, evil, and corruption, so that its secondary dictionary definition after "parasitic or predatory annelid worm" is "a person who extorts profit from or lives off others." It is why Adolf Hitler and Nazi propaganda equated Jews with leeches, with pronouncements such as the Jew "whines for the favor of 'His Majesty' and misuses it like a leech fastened upon the nations."[65] It is why Dr. Peter Mark Roget, in his thesaurus, gave the leech five entries, and one is bane, subcategory "troublemaker," along with parasite, threadworm, tapeworm. I love Roget's thesaurus, but this seems unfair. I accept that the medicinal leech is not benign, however useful: as it lacks enzymes to digest blood by itself, it relies on bacteria in its gut to do the work of digestion. Patients' wounds have subsequently been infected by *Aeromonas hydrophila*, other members of the *Aeromonas* genus, and *Vibrio fluvialis*, so antibiotics are given prophylactically as routine. But the leech will not infest your guts like a tapeworm nor trouble you like a thread-worm. A single leech will not kill you, though several hundred may. I'm not sure, given all the use and abuse that humans have made of leeches, who is the parasite and who the prey.

It's true that leeching presents risk. Getting the leech to bite where it is wanted is not straightforward, and the literature is full of helpful suggestions of how to coax it into the needed position. A group of Mumbai doctors wrote to the *Journal of Plastic, Reconstructive & Aesthetic Surgery* in 2009 recommending applying the leech enclosed in a syringe tube, a method with "an obvious aesthetic advantage."[66] Cutting a square in the center of a piece of gauze and guiding the leech to it is another option. My favorite suggestion appears in an article in the *Lancet* of 1849 titled "Leeches Drunk Will Bite Till Sober." Instructions: "Put the leeches that you are going to use in some warm porter, and directly they kick about in it, take them out, hold them in a cloth, and they will bite instantly, without fail, even if they have been before tried for some time without any success."[67]

Leeches move more than other medical devices. In the literature, this is known as leech migration and it causes countless problems, because they can migrate both inside and outside the body. Some leeches have migrated to a patient's throat; others to the bronchus, air passages leading to the lungs. In the past, leechers attempted to train their leeches by sewing a string to them. In a treatise on the leech, the nineteenth-century physician James Rawlins Johnson thought leeching useful for "phrenitis," an ailment believed to derive from a retention of the menses, and, according to Zacutus Lusitanus, a sixteenth-century Portuguese physician, best treated by fastening four leeches to a piece of thread and introducing them as closely as possible to the uterus. "Lusitanus is so warm an advocate for their employment, that he declares there is no disease but will become mild under this mode of treatment, and particularly should the leeches be applied to the vessels of the anus."[68] There have been as many proposals to counter leech migration as there have been papers on repairing free flaps. But the best way to get a leech to bite accurately and to stay put is to have it watched. The people who must do the watching are nurses.

For months, I can't find nurses in the leeching literature. They must be there, but their voices are silent. It is a puzzle. Without nurses, who would apply the leech and soothe the patient, and watch for leech migration and kill the helpful creatures when they are finished? And how do nurses overcome the disgust mechanism they must surely feel, because they are human, without conveying it to the patient? Then, late one night, I find a paper from the *British Journal of Nursing*, "Nurses' Experience of Leech Therapy in Plastic and Reconstructive Surgery." The authors are Alison Reynolds and Colm OBoyle.[69] The study area was a thirty-bed plastic surgical ward in a major Dublin teaching hospital staffed by twenty-six nurses, five consultant plastic surgeons, and eight hospital doctors. The ward uses leeches about once every three months.

Reynolds was no leech specialist at first. She had encountered them as a student nurse, rotating around hospital departments and working on the plastic surgery ward. Then, in the early 2000s, leeching was more common because of the times. Ireland was rich. Eastern Europeans— mostly Poles—came to work in factories but without enough English to understand instructions. Poor comprehension met heavy machinery and traumatic injuries abounded. I ask Reynolds to recount her first leeching, but it was too long ago and there have been too many instances since then. She thinks she was on nights and was told, "You have someone here who's for leeching," and her reaction was what my reaction would be: "How am I meant to do that?" There was no official policy to give guidance. Nothing was written down. You asked your colleagues, and you took the things that looked like slugs, and you did your best.

The paper came about because she needed to write a thesis for a master's degree. She discussed her work in plastic surgery nursing with OBoyle, a midwife who was her thesis supervisor, and when she got to the fact that in plastic surgery they used leeches, he said, "Stop!" It had to be leeches. But then she did what students are meant to do and looked at the literature. "And there was literally nothing." Irish emphasis. Literally. Nothing. One paper written by a nurse in the late '90s. Plenty of papers by surgeons about flaps and veins. But nothing from nurses about what having someone for leeching meant, nor how they felt about it when they set off for the patient's room clutching a tub of annelids. She panicked and went to OBoyle. "I've got no foundation!" But he said, "Your foundation is that there's nothing out there."

She interviewed seven plastic surgery nurses. Every one disliked using leeches. "They might visibly squirm and distort their faces as they described the leeches and their need to be close to and manipulate them." They described them as "the black slug," "bloodsucking, slimy bugs," and "creepy crawlies." Other things they didn't like: that the leech, that "dirty walking needle," went against all their training: hygiene control was paramount, yet a live creature full of potentially toxic blood was moving around their hygienic hospital ward. Reynolds uses the anthropologist Mary Douglas's famous definition of dirt as matter out of place. "The recasting of such matter (parasite) that is conceptually out of place [. . .] would appear to be a challenge." Having someone for leeching means extra care but no extra staff, or money. Plastic surgery nursing does not count as specialized, unlike oncology or midwifery. This is ludicrous, and Reynolds is indignant at it. Leeching is specialized. Not everyone can do it. It's hard.

Migration, for example. I tell Reynolds that most surgeons gloss over this. If she were less polite, she'd snort, but her derision comes through anyway. "That's typical surgeons. Migration is a big problem and anyone who says it isn't clearly isn't doing the therapy themselves." Leeches move. They always move. "They'd be anywhere. Anywhere. Especially when you're on nights, the lights are down, we've got a little nightlight, but you're walking in watching your step because you could walk on one." She has found leeches up curtains, on the radiator, on the floor, in the bathroom, in the shower. "You try to go in every fifteen minutes or half an hour but if it's fallen off, it can fall anywhere, and then they just slide around, so you have trails of blood."

It sounds so gothic and shocking. Why do patients agree to it? Because they have to. Only patients whose transplantation has failed are offered leech therapy, and the leech is the last resort. "It's not like 'I don't need those tablets, I don't need them' and there are no repercussions." Your repercussions are a lost finger or ear, or one reconstructed breast and one void. "You thought you were going to come in, you were going to have the surgery and you were going to go home and be yourself again, that this was going to be your identity." No one offered leech therapy, says Reynolds, has ever refused. I press her. Even so, how do you persuade someone to submit to it? She says the surgeons come around first and suggest it. All official, like, with their white coats and terminology. "But I know that [the patient] is sitting there and I know they're thinking, I'm going to ask the nurse later about this. The consultant will go through the medical stuff. But the itty-bitty things? They'll ask the nurses." She's always right. The consultant leaves, and the patient says, "I didn't really know what he meant, does he mean the little slug thing?" Reynolds always answered truthfully. "But I would always give my encouragement because I've seen this work. It might not, but at least we can say we've done everything."

There are differing levels of acceptance. She delights me with a tale of a farmer. "He loved it. He thought this was amazing. He said, 'Put them on, I'll watch them, don't worry, when that fella's finished, I'll ring the bell.' He was watching them thinking it was the most fasci-nating thing ever. He'd say, that one did well now didn't it?" His treat-ment was successful and he phoned all his friends to tell them about it. "But he was a farmer. A different kind of outlook."

The Biopharm video includes a news report about a woman named Michelle Fuller, from Bradford. She was young, she had children, and she had mouth cancer. Later, running across Yorkshire moorland with a new acquaintance, I'm surprised at her lack of surprise when I say that I'm writing about leeches. It's because Michelle was her cousin. She tells me Michelle tried the leeches because she was sure they would save her life. No doubt about it. The leeches were placed on her new tongue, which had been constructed from a flap, four times a day for ten days. "I have never been squeamish," Michelle told the local paper. "And I just said you have to do what you have to do."[70] The *Daily Mail* wrote that "bloodsucking leeches saved a woman from cancer," but they didn't.[71] She died eight months later, aged thirty-three, a few weeks before her planned wedding, and I salute her.[72]

The farmer and Michelle and other broad-minded patients: they're the easy ones. "Some people," says Reynolds, "just don't want to know. 'Don't tell me, I'm not going to look, come in, come out, do what you need to do, but I'm not an active participant.'" The face and breasts are the hardest. Women undergoing breast construction are "already

emotionally distressed and they find it harder." Delayed reconstruction therapy, for example: these women have had chemotherapy and
radiotherapy, and only then a reconstruction. Already they have waited
months, and they have the surgery and then one breast fails. "That's
so huge for a woman. It's part of your identity, it's who you are, and
it's not going well and then you have leeches stuck on your boobs as
well."

Some of the nurses Reynolds interviewed can't bear leeches, either,
but because they are nurses, and wonderful, they suck it up. In one
newspaper article titled "A Sucker's Born Every Minute," an unnamed
doctor at Cedars Medical Centre reported an interesting technique for
getting nurses used to the idea. "'We first ask them if they like animals,"
the physician says. "Then we work up to leeches."[73] Or the more junior
nurses do trade-offs with more veteran ones who don't mind the creatures. Reynolds knows better but, despite her knowledge and expertise, thinks leeches are "slugs. One hundred percent slugs."

They are also trickier to dispose of than a sharp needle into a bin.
From Biopharm, I'd got the sense of a dispatch method that was clean
and humane, if animal murder can be humane. No, says Reynolds. It's
nothing like that. Wild *Hirudo medicinalis* or *verbana* may live for
twenty-seven years,[74] but these are service animals with a human-
dictated life span. Its job done, the useful leech is placed in a plastic
pot—like the urine sample ones—and sprayed with alcohol solution.
Rather than it being anesthetized and dying painlessly as I'd imagined,
there is an explosion. "They're so big," says Reynolds. "And you spray
them with seventy percent alcohol, and they burst open with all the
blood."

You will not find many vivid news reports about patients who have
successfully undergone surgery and overcome venous congestion of a
free flap. They are quiet, the successfully leeched, and there is no public triumph. Perhaps they need to forget. Perhaps they can't square the
instinctive horror of the leech with its powerful ability to heal. Perhaps
they cannot summon gratitude for these slug-like creatures. They
should.

In the nineteenth century, a man named Thomas Erskine was convinced that being bled by two leeches had saved his life. Despite the
abundant use of leeches, gratitude to them was rarely expressed. But
Erskine—who served as Lord Chancellor in the most optimistic Ministry of All the Talents—named his leeches Home and Cline, after two
eminent Victorian surgeons, and made them pets.[75] Another man who
gave leeches their due was George Merryweather, a family doctor in
the Yorkshire port of Whitby, famous for being a setting in *Dracula*
but also where Dr. Merryweather devised the Tempest Prognosticator.
In a long submission to the Whitby Literary and Philosophical Society
in 1851, Merryweather set out his startling claim: that a leech could
feel weather, and that many leeches could predict it.[76] This barometer
ability of the leech had been reported anecdotally before. The eighteenth-
century poet William Cowper wrote to his cousin of a storm and of "a
leech in a bottle that foretells all these prodigies and convulsions of
nature. [. . .] No change of weather surprises him, and that in point of
the earliest and most accurate intelligence, he is worth all the barom-
eters in the world."

Merryweather read of this and went further. Rather than one leech,
his barometer featured twelve, each in a white glass bottle, pint-size.
He took care that they could see one another, because he thought
leeches could get lonely. He called them his "jury of philosophical coun-
sellors" and "my little comrades." In his writing, a kindness percolates. He is apologetic in his submission that he must predict yet another
storm, though the sky is blue. "I am sorry to interfere with your engage-
ments this beautiful weather," he writes in February 1850. "I do not
trouble you with a little blustering of the wind." But he is rare in that
his kindness extends to leeches, these animals thought to be only useful or to be abused.

He thought of calling his device the Atmospheric, Electro-Magnetic
Telegraph, conducted by Animal Instinct, but somehow decided that
Tempest Prognosticator was more straightforward for foreigners to
understand. He suggested that it be placed in stations around the coast
to serve all shipping, along with life boats, and that mariners be issued
a book of tempest signals, to interpret better. The prognosticator
appeared at the Great Exhibition of 1851, and Lloyds of London did

its own tests.[77] But the leech machine failed: the mechanical barometer, using a sealed liquid, was adopted instead, partly because it didn't require maintenance, feeding, or the changing of water every few months. Merryweather's comrades were retired, and the leech returned to its role as an unloved worm.

Leeches are called upon less often now, because surgery is better. "We're a bit flat really," says Carl, when I ask how business is. The young nurses at Biopharm thought their hospital used to keep them on the burn unit, but they don't anymore. Surgeons are defter at reattaching veins and getting drainage, but only with clean cuts and decent-size veins. "Where there is trauma," says Carl, "someone is burnt or something gets caught in a chain or a belt: that's when they can't stretch the veins to get them together. So they attach as many arteries and veins as they can and use leeches to borrow a bit of time." In America, Biopharm's leeches are used frequently for scalp injuries. "Bears," says Carl. "They tear the scalp off. It's one of the few things when there's very little that they can do surgically. They attach the scalp and then plaster it with leeches. It works very well." Canning factories used to be good for business. Carl remembers one case where a canning employee in Finland had two fingers and a thumb ripped off. "I think they used something like eight hundred leeches over five days. I don't know how they used so many." Degloving cases, where the skin is ripped off, have also dropped.

But in disasters overwhelmed surgeons can't afford to spend ten or twenty-five hours of surgery on each patient, and leeches can save time. After the San Francisco earthquake in 1985, Roy Sawyer wrote in a biographical account that Biopharm had requests for hundreds of leeches. The same after the terrorist attacks of 7/7 in London.[78] Wars often require leeches: according to Carl, the Royal Centre for Defence Medicine in Birmingham keeps two hundred leeches in its pharmacy. "They have a really big stock," says Carl, because they have people coming in with no leg, no foot. It's not a case of putting one or two on, they've got to put fifty on." Regular hospitals used to keep a leech buffer—half a dozen or so, in the pharmacy—but most don't now.

July has been quieter than expected. In summer, people do DIY, and, Carl says with meaning, "They cut the hedge . . ." I ask if they attend surgery conferences to increase trade. "When we can afford it." It's a rare chance for them to meet surgeons and doctors: usually they only deal with the pharmacist, who orders the leeches in and sometimes not expertly. "We'll get a pharmacist ringing in at six in the morning and saying, can I have six leeches, and we'll say, no you don't want six because by the time the courier gets to you, you'll be needing more."

Surely there's an untapped leeching population in the rugby and boxing worlds where cauliflower ears and contusions abound? For now they just get the odd human kickboxer from a local club. They are walk-ins, popping in for leeches when required, no need to book. Carl can't remember what the boxers were charged, but the usual price is £9.50 to £10 ($13.35 to $14.07) a leech. Whichever way you look at it—compare it to the painstaking two-year breeding process, or calculate it against three or more visits to a casualty department for a cauliflower ear, or a lost finger or breast or scalp—that is a bargain.

What about diversifying into hirudotherapy? The general public already thinks leeching is quackery, say Bethany and Carl. Or that it's something to do with maggots. But when I bring up hirudotherapy, their reaction is dour and sour and no wonder. The website of the British Association of Hirudotherapy is accompanied by images of women with leeches hanging out of their mouths or in their genitalia. In the curious minds of hirudotherapists, leeches applied correctly will treat diseases caused by "insufficient micro-circulation." These include blood defects, joint disorders, neuroses, boils, hemorrhoids, varicose veins, asthma, heart attack, stroke, depression, infertility, memory disorders, diabetes, hair loss, and detached retinas. The actress Demi Moore told talk show hosts that leeches had fed on her to "detoxify my blood," saying they released a cleansing enzyme. This is scientific nonsense: if the leech's decongestant ability was not localized, its host would bleed to death. The only thing hirudotherapy claims it can't treat, says Carl, is death. But science is immaterial to hirudotherapists, as are laws. I ask Carl where hirudotherapists get their leeches from if it's not Biopharm. He says nothing for a while, then relents. "A pond somewhere. A lot of them come from Russia without CITES. All the leeches that come in are smuggled, in a suitcase. They get their relatives to bring them in." It's easy to check: CITES regulates imports and exports. In 2017, Romania was allowed to export forty thousand *Hirudo medicinalis*; Turkey was allowed a quota of 200 kilograms of *Hirudo verbana*. Russia had no leech quota for 2016–17.[79] Carl says Russia exports two hundred leeches every few years. "But the hirudotherapists are using five hundred a week."

Perhaps a few things on that long hirudotherapy shopping list of conditions are not entirely lunatic. Carl has heard that they're good for tendonitis. Their saliva has some kind of anti-inflammatory effect. "But there's no exact science as to why it works." In 2003, the surgeon Richard Fiddian-Green wrote a letter to the *British Medical Journal*. He remembered doing a ward round at St. Mary's Hospital in London with a senior consultant and encountering a patient with extremely painful pericarditis (a swelling of the fluid-filled sac enclosing the heart). It is usually treated with anti-inflammatory drugs. But the consultant thought differently. "The most effective treatment for pericarditis I have ever seen," he told us, "was three leeches applied to the precordium."[80]

Anecdotes abound of leech application helping to relieve arthritis or other inflammation. I'd like to ask Roy Sawyer about it, but Bethany says he doesn't do interviews anymore. She nods in the direction of the cream mansion house, where he is. I start to think of Jane Eyre and attics but I tell them that I picture him as an Indiana Jones figure, striding through swamps for science, his movement purposeful and his legs bare and ready. They don't stop laughing for a good while. "Not quite," says Bethany. Her parents are planning a holiday, she says. "My mother thinks it's going to be a relaxing holiday, but he's already got plans to find a swamp somewhere." Neither she nor Carl looks like they relish this idea. When Sawyer is out leech hunting, he sends them regular e-mails of his progress. Images of dissected leech gonads in their morning in-boxes are not uncommon.

Sawyer is going swamping because he may be retired from giving interviews but not from learning about leeches. There is so much to discover about these multisegmented annelid worms, this bane and troublemaker and parasite, this bloodsucking creature that straddles old and new medicine as serenely as it moves through water or attaches itself to Carl's trouser leg. It has given us much, this wee black slug, but Roy Sawyer is convinced it has more. "Secretions from bloodsucking animals," he told a reporter, "could be to cardiovascular diseases what penicillin was to infectious disease in the past." He listed the possibilities, and his belief was as high as the rain forest. "Blood clotting, digestion, connective tissue, disease, pain, inhibition of enzymes, anti-inflammation. You name it, the leech has it."[81]

Janet Vaughan on a medical mission, India, 1950

THREE

JANET AND PERCY

She was a name on a plaque and a face on a wall. I ate beneath her portrait for three years and gave it little attention except to notice that the artist had made her look square. There were other portraits of women to hold my attention on the walls of Somerville, my Oxford college: Indira Gandhi, who left without a degree, and Dorothy Hodgkin, a Nobel Prize winner in chemistry. Vera Brittain, Iris Murdoch, Dorothy L. Sayers. In a room where we had our French-language classes, behind glass that was rumored to be bulletproof, there was also a bust of Margaret Thatcher, a former chemistry undergraduate. Somerville was one of only two women's colleges at the University of Oxford while I was there, from 1988 to 1992, and the walls were crowded with strong, notable women. (The college now admits men.)

The plaque was on the exterior wall of my first-year student residence, a building named after Dame Janet Maria Vaughan, the woman in the portrait and principal of Somerville between 1945 and 1967.[1] She was alive when I was an undergraduate and, according to obituaries, was known for always dressing in tweeds and for going to the Bodleian Library even in her late eighties, inadvertently annoying other readers when her hearing aid hummed and whistled. (She turned it off when asked.) But when I arrived at Somerville and was assigned a room in Vaughan, I thought only with some relief that everyone would finally be able to spell my Welsh third name—Vaughan, too—that is usually a puzzle even to English speakers. The name did not make me think back to the three surgical procedures I have had, or to my birth, where bags of someone's blood may have hung from hooks to help or save me. I did not look up from my dinner plate at Janet Vaughan's portrait and thank her for her role in helping to make blood transfusion standard medical practice. I should have.

In any developed country with good health care and blood to give, someone receives blood every two seconds, more or less.[2] This is an American figure, but it can be applied widely. Yet the system of widespread donation of blood by anonymous volunteers, and its transfusion into people who need it, dates back not even a century. In England, where I live, the National Health Service Blood and Transplant (NHSBT) began as the Blood Transfusion Service in 1946; the National Health Service was founded only two years later.[3] Last year NHSBT collected nearly 2 million units of blood; in the United States, where 36,000 units of red blood cells are used daily, the figure is 13.6 million.[4] Globally, 112.5 million donations of blood are made annually.[5] But in 1920s Oxford, when Janet Vaughan was a medical sciences undergraduate at Somerville, the mass donation, storage, and delivery of blood was unthinkable.

Vaughan was born to privilege and story. She was descended from a noble line of physicians: William Vaughan ministered to William and Mary of Orange; Henry Halford Vaughan was physician-in-ordinary to Georges III and IV, and to William IV and Queen Victoria. Vaughan's grandmother Adeline Maria Jackson was "one of the seven famous Pattle sisters, famous for their great beauty and for the wild ways of their father James Pattle of the Bengal Civil Service, 'the biggest liar in India.'" Even when he had drunk himself to death, Pattle was trouble. His corpse, due to be transported to England in a barrel of rum—not an outlandish storage device, given its preservative properties—exploded during the night while stored outside his widow's room. "There was a loud explosion," wrote Vaughan in an unpublished autobiography titled "Jogging Along," "[and] she rushed out and found Pattle her husband menacing her in death as he had menaced her in life."[6]

Janet's mother, Madge Symonds, was also a beauty. She had grown up the daughter of John Addington Symonds, a poet and Renaissance scholar who spent much of his married life exploring homosexual relationships. Accounts of his life sound suitably glamorous—houses in Perugia, Venice, and Davos—but I guess that behind that glitter was distress for his wife and daughters, obliged to tolerate his affairs with young men. When Madge married William Wyamar Vaughan, she exchanged the dazzle for a suburban house in Bristol, where Janet was born, then headmaster's residences at Giggleswick, Wellington College, and Rugby. She was, wrote her daughter, "a caged butterfly or hummingbird," and the cage was her life as a headmaster's wife. At Wellington, she made sure to eat chicken off the bone with her fingers to shock the butler. She must have got out sometimes, enough to make great friends with her husband's cousin Virginia Woolf. Madge is Sally Seton in *Mrs. Dalloway*, who "sat on the floor with her arms around her knees, smoking a cigarette. [. . .] It was an extraordinary beauty of the kind she most admired . . . a sort of abandonment, as if she could say anything, do anything."[7] (Bloomsbury enthusiasts think Madge was Virginia's first love.)

The Vaughans were connected but not wealthy. Janet was given an indifferent education by governess, while her two brothers were sent to good schools. At fifteen, she was enrolled at a school where the headmistress believed in training girls only to be well-read wives. She thought Janet "too stupid to be educated." She was almost too dead to be educated: it was the First World War, the school was evacuated to "four empty houses at Great Malvern," and she came close to dying of pneumonia. Of the war, she remembered cold and never-satisfied hunger, both at home and at school. Even so, she ignored her head teacher's dismissal of her brain and read voraciously. She took the entrance exam for Oxford, a test known formally as responsions and informally as the Smalls (at Cambridge it was the Little-Go),[8] and failed twice. Just before her third little go, her mother booked them both in at the

Mitre hotel in Oxford "and firmly ordered a bottle of claret every evening."[9] She passed, and was accepted at Somerville College, one of four women's colleges at Oxford, to study medical sciences.

Janet arrived at Somerville in January 1919 with nothing more than "a little ladylike botany."[10] For her first year, she wasn't a proper Oxford student: until 1920, the university allowed women only to study, not graduate.[11] Janet found physics a mystery; she had never heard of acid and alkali. "I was a public danger," she told the journalist Polly Toynbee, when she was interviewed as one of six women featured in the BBC series *Women of Our Century*. "[I was] . . . in the lab handling phosphorus and I knew nothing about it at all."[12] She learned fast. When her exam results landed on the mat at home in Rugby, Janet was even more surprised by the first-class degree than her family, who were bewildered by this daughter too stupid to be educated, now with an education pedigree better than anyone else's, and one that hardly any women had been granted. She was especially surprised because during her final oral examination, the senior examiner shook his head at her, saying, "And to think that a B.A. of Oxford should spell vomiting with two *t*'s." (She was probably dyslexic.)

Janet, who spelled vomiting with two *t*'s but had a First from Oxford, set about becoming a physician. There were rotations in various surgical "firms" (teams of doctors led by a senior consultant). She studied under Bill Williams, a man who threw scalpels at frustrating pupils in the operating theater but who taught her valuable things about wounds and burns that would later be useful in war. For obstetrics, she was sent into the slums. "Terrible poverty," she told Toynbee. She encountered "a woman with no bed except newspapers."[13] She saw lines of children sitting up in bed with rheumatic hearts, who would die because there was no National Health Service and good health cost money. She saw that poverty is deadly. "How anyone could do medicine in those days," she wrote, "and not become a socialist I find hard to understand. What I hated most was people's acceptance: 'Yes, I have had seven children and buried six, it was God's will.' I hated God's will with a burning hatred."[14]

Her brothers were conservatives, but these experiences turned their sister "firmly and forever" into a socialist. The slums introduced her to politics—the best kind of politics, the kind that saves people—but also to blood, her lifelong scientific interest. Anemia accompanied poverty, because good food with a good iron content was costly. The treatment for anemia was arsenic, which to Janet seemed ethically wrong and medically inefficient (it didn't work). So she did what Oxford had trained her to do and read the literature. She read of George Minot, an American hematologist who had successfully treated anemic patients with raw liver and later won the Nobel Prize.[15] This treatment made more sense than arsenic, and she wanted to try it. She was a pathologist by now: her mother had died, and she thought doing lab research rather than hospital rounds would enable her to care better for her widowed father. She did some surreptitious research, arranging with a doctor friend on the appropriate ward to treat his anemic patients with raw liver. "I did the blood counts, the house physician kept a straight face on ward rounds when the senior physician demonstrated to students the magnificent effects of his prescribed treatment with arsenic, and the patients got well."[16]

She thought a concentrated liver extract would work even better. She approached her professor of medicine, a man who did not see a woman who was young and think both those things to be handicaps. He said she could test raw liver extract, but first on dogs. "He gave me some money and said I could work in Harrington's lab, the great chemist, but that I must go and collect the mincing machines of my friends and the pails of my friends because the hospital hadn't got any mincing machines. So I went round and collected mincing machines, I collected Virginia Woolf's mincing machine, and I minced liver with Minot's book lying on the table telling me how to do it and I produced some filthy-looking stuff at the end of it."[17] This became a scene in Virginia Woolf's *A Room of One's Own*: Janet in a kitchen with mincers, Minot's paper propped open like a recipe book, a parody and inspirational illustration that women can have other occupations than "the perennial interests of domesticity."[18]

The extract was fed to one dog, which sickened, then another. The same. Janet said, no more dogs, the extract was too precious. She went home and took the extract herself. "The next morning when I came back to the hospital there were all the professors of medicine, chemistry, surgery, waiting on the doorstep to see if I was still alive." It was fed to a patient, "a nice old laboring man" who was dying of pernicious anemia.[19] He survived, and a senior professor took all the credit for the miraculous new treatment. Janet had other things to do. Her father had remarried, which meant she was released from her weekend visits to Rugby, where she had acted as hostess for him, so she won a Rockefeller Scholarship to Harvard, becoming the only female student there. She wasn't allowed to work with patients, being a woman, so she ended up with pigeons, using them to do groundbreaking research on vitamin B_{12} in blood that wasn't fully acknowledged for fifty years. She called them her Bloody Pigeons.[20]

How I love the brisk nervelessness of this woman. Some of it comes from privilege. But much of it is her own, as much as her fictional room was. She had the confidence to make fissures in patriarchal concrete, but also the confidence to get married, because she wanted to. She returned from Harvard to marry David Gourlay, who ran the Wayfarers Travel Agency, and they moved into 33 Gordon Square in Bloomsbury, above the business. She kept her name, "not for any feminist reasons, but because I had already published several papers, and it seemed a pity to lose my medical identity." Which is a good feminist reason.[21]

At the London Hospital, where she then worked, no one spoke to her in the passageways. Her reputation with blood diseases was known, but if a doctor wanted advice about a patient, he sent a note and she sent one back. She was forbidden from bedside consultations. At lunch, she and Dorothy Russell, later an eminent professor of morbid anatomy, sat with the secretaries. When George Minot came to London on his way to Sweden to fetch his Nobel Prize, he was hosted by the Gourlays. Minot was invited to a formal hospital function, but Janet was not. "I brought him to the door in my car to meet Professors and Physicians [and] I was merely told what time to pick him up, though it was well known that I was acting as his London hostess."[22]

She doesn't sound angry about this, perhaps because she is writing when many of the worst fights seem to have been won, when she helped to fight them. She is angrier about the anemic patients she treated with liver who then said, "Don't give me any more of that medicine, doctor. It makes me hungry and I can't afford [to eat]."[23] She taught her patients to fight authorities to get extra milk, for the extra iron it would give them. She taught her students that to practice medicine, they must learn to deal with the public assistance board, with bureaucracy, as well as with the hospital dispensary. I like her clear fulmination, which persisted sixty years after the young trainee doctor set out into the slums. I wish she were here to fulminate against the sly dismantling of our welfare state and the National Health Service. She wouldn't stand for it, as we should not.

The Gourlays built a cottage in the country and called it Plover's Field. They had two daughters. Just before the birth of her first child in 1934, Janet published *The Anaemias*, still considered a pioneering hematology textbook. She was invited to set up a department of pathology at Hammersmith Hospital, and she did, and it thrived. She was frequently consulted on obscure blood diseases and drove around London with her car "always full of interesting specimens."[24] She was busy and happy.

But this was the 1930s, and war was coming. First, it came to Spain. Her Bloomsbury friends went to fight, and Vanessa Bell's son Julian was killed. Vaughan began to work with the Spanish Medical Aid Committee and joined the Communist Party but soon lapsed. She said that no one seemed to notice she had left.[25] She sold possessions to raise money for Basque children; she spoke on soapboxes at street corners. The time was tense and unclear. In 1937, the Committee of Imperial Defence had calculated that sixty days of bombardment of London would produce more than a million casualties.[26] At the beginning of the 1938 Munich conference, medical staff were told that should the peace negotiations fail, they should expect either thirty-seven thousand or fifty-seven thousand casualties in London that weekend.[27] Whatever the figure, Janet knew from the Spanish conflict that "we shall want some blood. We shall want a great deal of blood."[28]

———

Maybe Medea was the first. The fearsome witch, as written by Ovid, who cut the throat of the aging Aeson, father of Jason, and drained his blood, then refilled him with herbs and potions, and revived him. To revive, to revitalize, to give back life. She wasn't particularly scientific about it, walking around incanting with her streaming hair, nor did she use blood. Her potion was made from "roots dug from a Thessalian valley," but also "hoar-frost collected by night under the moon, the wings and flesh of a vile screech-owl, and the slavering foam of a sacrificed were-wolf."[29]

But the principle was established, though shakily and magically: that in the same way that the loss of blood could drain a creature of life, a suitable replacement of fluid could bring it back. For two thousand years, humanity definitely liked to let out blood but didn't think of putting it back anywhere, except through the mouth: the drinking of blood to acquire strength and life force is as old as the (Roman) hills. There were obvious obstacles: blood freed from the vein quickly clots, as it is supposed to. Getting blood back into a vein is a difficult skill that even modern phlebotomists and physicians sometimes lack. And before blood could be moved in and out of the body, its movement around the body had to be understood.

William Harvey, physician to kings, came to understand this, probably in 1616, by cutting open scores of animals to see what their hearts and blood were doing.[30] Charles I let Harvey experiment on royal deer, but their hearts were too quick. He could not catch their rhythm or secret. He moved on to the slow-blooded, the cold animals and the dying ones, whose hearts beat slowly, who let him see. He understood that the body operated a one-way system: the heart pushing the blood out through ventricles to the arteries, and the blood returning by the veins, the whole system far more ingenious than most urban planners could devise, because of flaps on the veins that compel the blood to go only one way. (An earlier Italian anatomist called these flaps *ostiole*, little doors.)[31] Harvey published his findings in 1628, including a dedication that manages to be wondrous and wonderfully oleaginous, in which he called the animal's heart "the basis of its life, its chief member, the sun of its microcosm . . . [from which] all power arises and all grace stems." Just like the king, who is "the basis of his kingdoms [. . .] the heart of the state."[32] Oily it may seem, but Harvey's diplomacy and skill enabled the king's personal doctor to survive his master's execution. His findings weren't accepted. In 1680, John Aubrey wrote that he had heard Harvey say "that, after his Booke of the *Circulation of the Blood* came out, that he fell mightily in his Practice, and that 'twas beleeved by the vulgar that he was crack-brained."[33]

Yet his work began the era of transfusion. Thousands of dogs, horses, lambs, and hens were cut open, their blood drained, replaced, their lives expendable. It became a fashionable quest for the eminent. Sir Christopher Wren was carrying out experiments in 1659. He was, wrote a historian of the Royal Society, "the first author of the Noble Anatomical Experiment of Injecting Liquors into the Veins of Animals. [. . .] By this Operation divers Creatures were immediately purg'd, vomit'd, intoxicated, kill'd or reviv'd according to the quality of the Liquor injected."[34]

The attempt to transfuse a man became a race between a few men: Richard Lower was the front-runner in England, backed by the Royal Academy of Science, and Jean-Baptiste Denis, doctor to King Louis XIV, his French competitor. Lower worked with dogs. The descriptions and illustrations of his experiments are gruesome: dogs spread-eagled, their veins or arteries cut, the "emittent" dog connected to the receiver by quills, the experiment judged to be over when the emittent dog began "to cry, and faint, and fall into Convulsions, and at last dye by his side."[35] This was the beginning of the age of pitiless vivisection (which hasn't ended yet). Another physician, Mr. Thomas Coxe, who experimented with transfusing pigeons, was in 1665 "particularly desired to try the changing of dogs' skins" and this was not judged to be remarkable.[36]

The transfer of animal blood—a xenotransfusion—was chosen because blood was believed to contain character that could be transmitted. A lamb would give a man its "mild and laudable" nature, a calf the same. Denis began with lambs' blood, infusing a young boy who was wasting away, who survived and thrived and grew fat, "a subject of amazement." There was an older man next, who was purely an experiment, "having no considerable indisposition," who also survived.[37] His most famous case was his final one: two transfusions of Antoine Mauroy, a madman, wife beater, and former valet to nobility. To calm Mauroy's "phrensy," Denis chose the blood of a calf. Despite science knowing nothing about blood types or incompatibility or much about the nature of blood, Mauroy's body at first did not react adversely. The second transfusion was larger, and Denis described, without knowing, hemolytic shock: "As soon as the blood began to enter into his veins, he felt the like heat along his Arm . . . his pulse rose presently, and soon after we observed a plentiful sweat all over his face." He vomited up bacon and fat, but the next morning woke calm. "He made a great glass of Urine, of a colour as black, as if it had been mixed with the blood of chimneys."[38] (It was not soot but his dead cells killed by the foreign blood.)

Mauroy survived long enough for Denis to claim that he was the first to successfully transfuse blood into a human. Mauroy soon died; his wife was probably executed for his murder and Denis was disgraced. Lower, meanwhile, also picked a madman: a bachelor of divinity, Arthur Coga, whom Samuel Pepys described as "cracked a little in his head." Coga was of decent family—his brother was master of Pembroke College—but he was a drinker, and he was promised 20 shillings for his trouble,[39] which consisted of 12 ounces of sheep's blood let into his veins in just over a minute.[40] Coga survived but transfusion didn't. There were too many deaths and disasters, as well as what the surgeon Geoffrey Langdon Keynes described as fears that "terrible results, such as the growth of horns, would follow the transfusion of an animal's blood into a human being."[41] Thomas Shadwell, in his Restoration satire *The Virtuoso*, skewers the practice as quackery, having his transfusionist character Sir Nicholas Gimcrack report on his patient:

> From being Maniacal or raging mad, [he] became wholly Ovine or Sheepish; he bleated perpetually, and chew'd the Cud; he had Wool growing on him in great quantities, and a Northamptonshire Sheep's Tail did soon emerge or arise from his Anus or humane Fundament.

In the face of derision, Gimcrack protests: his patient has written him a letter, he says, and sent the good doctor some of his own wool. "I shall shortly have a Flock of 'em," he says, "I'll make all my Clothes of 'em, 'tis finer than Beaver."

Transfusion was banned in France and abandoned in England until James Blundell, an obstetrician working in the early nineteenth century, transfused ten patients. Two were already dead and stayed dead. Three died and five survived. He dealt with the trouble of clotting blood by working quickly, and he refused to use any animal blood but human because

> What then was to be done on an emergency? A dog, it is true, might have come when you whistled, but the animal is small; a calf, or sheep, might, to some, have appeared fitter for the purpose, but then it could not run up stairs.[42]

Blundell knew about dogs, having started his experiments on them first. After he had transfused five dogs with human blood, one died on the table, two or three lived a few hours then died, and one survived for five days and died. Obviously, he concluded, "the blood of one genus of animal cannot in large quantities be substituted indifferently for the blood of another, without occasioning the most fatal results." The "blood of the brute" should not be transfused into men. Or women: he was a noted obstetrician, and had spent decades in practice in Edinburgh, where he watched women give birth and bleed to death from it, frequently. (One of the section headings in his *Principles and Practice of Obstetricy* is "After-management of floodings.") Bleeding is expected in childbirth, even today, and is mostly caused by the detachment of the placenta and by cuts and tears. Five in one hundred women lose more than a pint of blood in the first twenty-four hours, a condition known as postpartum hemorrhage.[43] Most of those five in one hundred women survive, but postpartum hemorrhage still kills 127,000 women worldwide. Blundell's women? Most should have died, according to the death rates of the day. And some did. Some, such as the women who were cases number 5 and 6, and who had been dying of blood loss, recovered. Case number 5, a woman dying of uterine hemorrhage, received 14 ounces of blood. After she had taken in 6 ounces, she pronounced herself "as strong as a bull."[44]

Blundell considered transfusion to be "of so much importance to mankind, that, [. . .] I seize with pleasure the opportunity of treating the topic." But despite his efforts and experiments—he devised an extraordinary device called an Impellor—transfusion was put aside. It was too risky and the unknowns too large, until Karl Landsteiner discovered that blood came in groups and that they should not always mix. Then the urine the color of soot, the convulsions, the deaths made sense: all blood looked alike, but it wasn't. Thirteen years after Landsteiner's discovery, science was given its ideal experimental setting to explore blood more. It was given a world war.

October 16, 1914. Biarritz, southwestern France. Corporal Henri Legrain, of the Forty-fifth Infantry, lies in a hospital bed. He arrived from the front, bleeding out, and he has not stopped bleeding. In other beds are other men, dying of blood loss, one by one. In the same ward is Isidore Colas, "a small, brave Breton," who in October had been fighting in his artillery regiment in the Marne Valley when his leg was wounded by a shell. He is in recovery, now, in l'Hôpital Biarritz, when a doctor asks for his blood. It is a sensible request: transfusion required proximity, because you needed to do it quickly to beat the blood clotting. Colas was in the next bed along. The newspapers called Colas brave not just because he was a recuperating patient but because the doctors could not put him to sleep while they cut down into his vein with a scalpel. Colas "listened without hesitation, without expressing any emotion whatsoever," and when the time came, he gave his arm, connected to the other man by a silver tube, and his blood for two hours, and only the water pouring from his brow showed what he was enduring.[45] The results were spectacular. "I saw [Legrain] regain colour," said one doctor, "little by little, and come back to life."[46] Legrain was so revived, he leaned over to his donor, the little Breton, and kissed him on both cheeks (because he was French, and because he was alive). Both men lived a long life, and the modern era of transfusion was launched.

From 1915, Major Lawrence Bruce Robertson, a Canadian surgeon, began using a technique of indirect blood transfusion that he had learned in civilian life at a Toronto hospital. This involved withdrawing blood, transporting it in a syringe, and then transfusing. This method freed him to use transfusion more than otherwise: there was no need to find a soldier, cut open his arm, keep him close. He saw how dramatic a change blood could bring in a soldier in shock. "The change from a pallid, sometimes semi-conscious patient with a rapid flickering pulse to a comparatively healthy looking conscious and comfortable patient with a slower and fuller pulse is dramatic evidence of the value of transfused blood." Another doctor wrote more lyrically that men who seemed lifeless were given blood and "it was like putting a half dead flower in water on a hot day."[47] Another young surgeon, an American captain named Oswald Robertson, used recent developments in blood storage and pioneered the use of blood mixed with sodium citrate, then stored in glass bottles on ice. Robertson called this process "a blood dump,"[48] but it was also the first disembodied blood transplant and the world's first blood bank.

By 1918, base hospitals and casualty stations on the Western Front were transfusing 50 to 100 pints of blood to an average of fifty wounded personnel daily. In the context of millions of wounded men, that's not much blood. By the next war, blood transfusion had been fully accepted by the British military, which planned in prewar years to set up an efficient blood supply to its forces and operated it extremely effectively. Field Transfusion Unit trucks carried refrigerators containing 1,100 pints of fresh whole blood, and other units carried plasma, "slung underside in containers which in the ordinary way hold trench mortar shells; four bottles and two transfusion sets took the place of three shells."[49] Transfusion was so routine that the Ministry of Information's official account of wartime blood use, *Life Blood*, had a chapter called "The Tenth Man's Chance." At the battle of El Alamein, one in ten men received blood, three bottles each.[50]

In the Wellcome Library in London, I find a propaganda film released by the Ministry of Information in 1941, a time when neither "propaganda" nor "Ministry of Information" sounded sinister. The film is called *Blood Transfusion* and is narrated by accents that now sound cut-glass and royal but then were normal on-screen and on the wireless.[51] The film tells us with appropriate images that blood transfusion was widely used in the First World War on the Western Front, then the setting changes to the living room of a house on Talfourd Road in the southern London borough of Camberwell. A scene from 1921 is reenacted: a telephone rings. It is black and Bakelite, and answered by Percy Lane Oliver, a middle-ranking civil servant of middle age who is playing himself and who was about to become historic. Oliver, son of a Cornwall lighthouse keeper but a Londoner since childhood, worked for Camberwell council. He was an ardent volunteer and had been awarded an OBE (Order of the British Empire) in 1918 for running four refugee hostels. In 1921, he was forty-three years old, married to Ethel Grace, and honorary secretary to the Camberwell Division of the British Red Cross. He wore glasses, was balding in a way that seemed that he had always been like that, and had a face that fitted the name Percy.

The call came from King's College Hospital, a mile and a half away, and the caller wanted blood. As the film showed, and as the story goes, Oliver immediately found three other volunteers from among his Red Cross colleagues in case his blood group wasn't the one required, and all four set off for the hospital. Nurse Linstead, one of the Red Cross employees, was chosen to give "a pint of the best" and so became the country's first voluntary blood donor.[52] Within a few weeks, the Olivers had organized twenty-two volunteers ready to give blood if needed, and so began a system of voluntary blood donation that continues today.[53]

That is the official history. The reality is less cinematic. The Camberwell Division had supplied blood donors several times before that historic call, according to medical historian Kim Pelis.[54] Nurse Linstead was not the first voluntary blood donor: even when transfusion was not routine, during the nineteenth century, husbands gave blood for their wives in childbirth. The First World War had popularized blood transfusion, but it wasn't a revolution. By the end of the war, even the best field hospitals were transfusing only fifty patients a day: hardly anything, among the appalling numbers of wounded. The blood to treat bleeding soldiers came from bleeding soldiers. There was no difficulty in procuring donors, wrote Major General W. G. MacPherson in his medical history of the Great War: "The spirit of comradeship among the troops gave a plentiful supply." This consisted of "lightly wounded men, dental patients, and men suffering from sprains, flat feet and minor injuries."[55] Soldier donors weren't paid but they were soon offered three weeks' leave in England, a powerful incentive.[56] Harvey Cushing reported that when volunteers were sought for transfusion experiments, and "Blighty leave" given as inducement, they came "like trout to a fly."[57]

But the wartime spirit of comradeship did not survive the transition to peace, and the notion of an organized system of blood donation faltered. The medical profession applied its Semmelweis reflex, a refusal to accept change, named after Ignaz Semmelweis, who realized that doctors delivering babies after performing autopsies were lethally unhygienic but was scorned for decades. When it came to storing blood, "the feeling in England," wrote Victor Horsley Riddell, "is that this is carrying change too far."[58] Surgeons and doctors stuck to what they knew: blood should be used fresh if it was used at all. Fresh blood meant having the donor come to the patient, slice open a vein—the term was "cutting down"—and then convey the blood either by connecting the two veins (direct transfusion) or by using a syringe or pump to transfer the blood (indirect transfusion). Most donors expected money for their blood. There was a register of paid blood donors kept in Liverpool that was made available to all local hospitals. In Bradford, hospitals paid £10 ($36) a donation to donors who cleared the Wassermann test, the standard screening method for syphilis.

Volunteers were an option, but they were less available or reliable. Medical staff often had relied on an informal circle of potential donors. Patients used the blood of friends and relatives. You used whoever you could, or whoever was nearest. A hospital doctor in Edinburgh, wrote Dr. Alastair Masson, used junior doctors or medical students. "My House Surgeon, Dr. Carmichael," or "Mr. Handyman, a healthy and powerful young student."[59] Another resorted to relatives, though this didn't always work. Once, when no relative could be found, the doctor asked a student to give of his blood. "However, he was about to sit his finals a couple of days later so it was thought advisable to take only a little. In the end, the patient was given 600 ml blood taken from one nurse, two residents, three students and the writer." Many disapproved of this practice: a letter writer to the *Lancet* deplored the exploitation of the medical student, already "a hard-worked person, little able to give a pint and a half of his blood." What was to be done? Doctors were sometimes seen walking the streets looking for donors, offering cash. People in public service such as the police and firemen were considered good targets for giving blood for nothing, although one eminent surgeon thought this a bad idea. "The policeman's lot is said not to be a happy one, and it would be putting rather a severe strain on his already superhuman benevolence to expect him to give his blood to all who need it." In Evanston, Illinois, firemen were asked to donate blood because the local police chief complained that his men—avid blood donors—were looking anemic.[60]

The donor pool was also reduced because doctors wouldn't consider half the population. Women, the same surgeon believed, would present the "disability of nervousness." Also, our veins are smaller. In the United States, male donors were preferred because doctors couldn't bear to cut into a woman's arm.[61] Women were no good.[62]

Overseas, payment was also usual. At the Second International Blood Transfusion Congress in Paris, it was reported that Parisians were paid on a sliding scale: 100 francs for the first 200 grams and 50 francs for each 100 grams thereafter.[63] A correspondent wrote to the *British Medical Journal* about a Frenchman who had given 257 liters of blood that year and was still selling.[64] By the 1920s, hospitals in New York were paying $100 a pint.[65] American newspapers reported young women who funded college with blood donations. Blood transfusions were now being used to treat more than thirty maladies, and selling blood was one of the few sustainable industries in the Depression. Hospitals allegedly tried to ensure that their sellers were healthy. But "since the profit is considerable," wrote the *New York Times*, "there is a temptation to make sales as frequent as possible." Despite some interesting measures to protect donors' health—a Massachusetts law dictated that donors get a pint of whiskey as well as $25—doctors complained that the donor was often in greater need than the recipient.[66] The money created a blood sale infrastructure, with middlemen, professional sellers, and fierce competition. In a magazine account of his time as a professional blood seller in 1929 New York, Charles Nemo (Nobody) described his life living in a blood sellers' boardinghouse, watched over by a middleman. A fellow blood seller traveled around the United States looking for the best markets for his blood. Baltimore was no good as hospitals limited sellers to a quart of blood a year. Philadelphia was more promising: there was a shortage of blood donors "after the police force became tired of being heroes for a day by volunteering blood."[67] Now and then the press published objections to the notion of selling blood. At New York's Flower Hospital in the early 1920s, women medical students—but not the men—gave their blood for nothing when the price of blood became exorbitant.[68] Blood altruism was praised in the press, but the sellers were dominant: even in the 1930s they were powerful enough to form a union.

The Olivers wanted something different. This mixture of paid and co-opted donors—known as "on-the-hoof"—was inefficient. They believed strongly in the voluntary ideal, and Percy Oliver didn't see why this couldn't apply to blood. Countries that paid donors attracted drug takers and promiscuous people. Paying for blood attracted "a very different class of person." The conviction that unpaid blood was better was shared by Geoffrey Langdon Keynes, who had been converted to the power of blood transfusion during the Great War, and who later wrote a textbook on it that made him almost as famous as his economist brother John Maynard. "It was not difficult to find paid donors," he wrote. "But it was not so simple to obtain in this way individuals whose Wassermann reaction was likely to remain permanently negative." After Nurse Linstead gave her pint of blood to King's College Hospital, the Olivers decided to do something radical. They would set up a register of reliable blood donors who would never ask for payment. Percy Oliver noted their names on a database, which in 1921 meant index cards, and each card would list contact details, plus the donor's blood group and health history. Donors would be screened ahead of donating and their blood group noted. A telephone would be manned day and night. Hospitals would call for blood; the Olivers would have it brought to them, in the shape of a person. In return, doctors had to follow their rules. "The needle method of extraction alone is to be used. Opening the vein, cutting down upon it, or levering it up, is forbidden."[69] It made more sense: a less invasive needle meant a donor could be reused, and a less painful procedure meant they would more readily volunteer in the first place.

They began to set their scheme in motion, using their home as an office. (Percy Oliver continued to work for the council.) It was named

the London Blood Transfusion Service, and "The Service" by its volunteers. It may have consisted of some index cards and a phone, but
the Olivers' operation was the world's first voluntary blood panel and
the beginning of a shift to a model of altruistic blood donation in Britain
that has endured one hundred years. In the first year, only four donors
were signed up and they were called upon only once.[70] The next year,
the donors numbered thirteen.[71] In August 1922, a woman whose husband was dying at Guy's Hospital was "reduced to stopping strangers
in the street to ask them to give their blood." One of those strangers
belonged to the Camberwell Division of the Red Cross; the woman in
trouble contacted Mr. Oliver for a donor, and her husband lived.
"From that time on," Mrs. Oliver recollected, "the word seemed to go
round in hospital circles that there was a band of lunatics somewhere
down Camberwell way willing to give their blood to any necessitous
patient in hospital."[72]

Free blood? From prescreened donors who were healthy and who
would come when asked? Hospitals should have loved the idea. Yet
overturning the habit of paying for blood caused problems. Some
donors were treated with puzzling disdain. Sometimes, "having hur-
ried from their business or private affairs, [they] were told curtly that
they were not required and sent back with no explanation that they
could give their employers."[73] Oliver had to exercise constant vigilance
to protect his volunteers. This entailed "a watch against possible injury
to donors through faulty technique at the hands of inexpert operators,
as well as insistence upon observation of the ordinary courtesies."[74]
Donors were largely a polite lot. "Fuss is the thing donors like least of
all," wrote one newspaper. "There have even been protests from at least
one prominent member that they were treated too gently at the hospitals and that they dislike being 'wrapped in cotton wool' and stroked
by a lot of pretty nurses. But possibly this last protest would not be
supported by a majority of the association."[75] Even disgruntled donors
rarely snitched on the hospitals, simply marking politely on their donor
cards that they were unwilling to serve at the hospital again. Eventually, it was discovered that hospital staff assumed that donors were paid
and so felt entitled to treat them carelessly.

Logistics was another concern. Finding a donor on an index card was one thing; finding a donor who had a private telephone in 1920s London was much harder. (Even ten years into the service, when there were 2,050 registered donors, only 400 had a phone.[76]) The Olivers dealt with this with gusto. "When hospitals called," wrote Kim Pelis, "they contacted donors by telegraph, constable, taxi-driver, and sometimes by bicycle."[77] Another option was the police force. "Station Officers showed themselves ready to help," wrote Frederick Walter Mills in a 1949 history of the London Service. "But donors generally did not like being called upon by a policeman, since they found neighbors were disinclined to take a charitable view of the cause of the visit. One donor accepting the kind offices of the police on such an occasion, had his family's embarrassment increased by being returned to his home in the early hours of the morning by a Black Maria."[78]

Other matters were worked out over the years. The ideal donation amount was judged to be 400 cubic centimeters for men and 300 for women, and the interval between donations should be not less than three months for men.[79] Women could donate only every four months: their hemoglobin took longer to regenerate, and they were naturally lower in iron than men, a gender difference seen only in humans and only in menstruating women (after menopause, levels align). With these measures, the voluntary donor was sure to be in better health, spiritually and physically. Also, they were a bargain for everybody. When Oliver learned that in one Midland city, donors were being paid 4 guineas ($20) per donation/sale and an annual retainer, he did some sums. That would cost London £25,000 ($121,000) a year to run the service, when actually it cost a tenth of that.[80] By 1930, sixty-eight hospitals were using the voluntary system—now renamed the British Red Cross Transfusion Service—and were charged £1 and 1 shilling ($5) a call for operating expenses.[81] Other funds came from a partnership with the Ancient Order of Druids, which had—obviously—gained a reputation for collecting and recycling tinfoil. All this cash was put toward finding more donors. Percy Oliver had strong feelings about how to do this. In 1932, he wrote a letter to the *Derby Daily Telegraph* to complain about the reporting of a case in which a young Derby lad had given blood for his father. "Mr. Oliver, without belittling the action, suggests that 'undue prominence to an everyday occurrence is likely to give an utterly erroneous idea of this simple operation.'"[82] Blood donation should not be seen as heroic because most people don't think they are heroes and would be intimidated. It must be seen as a simple medical procedure that did no harm. Already, the service was having trouble with frightened family members. A daughter called to give blood readily consented. It would have been her first donation, but "her mother accompanied her to the hospital and absolutely forbade the surgeon to take the blood." Oliver later pointed out that some two hundred donors had agreed to serve only by keeping it a secret from their parents or wives.[83]

Instead, recruitment should remain personal. A lecture in a drafty village hall, where Oliver could show some slides and answer questions, was more effective than press reports of selfless saintly donors whom no one could relate to. This was ironic. Blood transfusion, previously reliant on the donor and the recipient being in physical proximity, was becoming anonymous and impersonal with the storage of blood allowing a distance between donor and receiver. Donors were also forbidden from finding out about where their blood was going. When one man gave blood, then made his way to the ward to see the patient, the Transfusion Service managers were scandalized. "When after fruitless attempts to get an explanation of his conduct, he phoned to say he was leaving for Australia, that saved us the job of getting rid of him."[84]

Oliver traveled all over the country to give talks, despite poor health. In 1934, he gave 104 lectures, mostly funded out of his own pocket. He thought it should be made clear, wrote Frederick Walter Mills, "that the blood given serves a real need and is not just used for experimental purposes." Stories of real people were more powerful than statistics, and prejudice and myth should be battled with science and fact. "The speaker should be prepared to meet suggestions that, in the blood banks particularly, blood has been wasted, and questioners may insist, without being able to give proof, that they have heard of instances where blood has been used to fertilize the tomatoes."[85]

This outreach worked. People across the country learned about blood donation and knew that in London they should call Mr. or Mrs. Oliver at 5 Colyton Road, where the Olivers had moved after their landlord, annoyed at the Druids' tinfoil that was being stored in the house, had raised the rent. The press wrote positive and glowing stories of ordinary donors such as Mr. Brown, summoned by the "tinkle-tinkle" of a telephone bell in his busy city office, on a cheerful, sunny day, to attend to a man with jaw cancer who needed blood. Mr. Brown went, though he had given blood only two weeks before, but hadn't liked to say, in case they couldn't find another donor. The names were pseudonyms. "The men I have written of are free lance," wrote W. Addison in the *Saturday Review*, "and do not wish to be known. The curious fact applying to these odd fellows is that they all come from odd walks in life. Down with the pick, pen, or drill, off to the hospital, and back to work." They were also all registered with the British Red Cross Transfusion Service, belonging "to the organization which has been formed in recent years to supply donors as and when required."[86]

The success of London was noticed and soon imitated. In Edinburgh, a surgeon named Jack Copland, shocked by the death of a relative who died for lack of blood, set up a donor panel in 1930. Six years later, by the time Copland's donors were activated and plentiful, there were 560 transfusions carried out in Edinburgh.[87] Other provinces and regions had organized donors, more or less successfully. Even so, the model of paid blood continued to persist. One man visiting London from Budapest signed up to the Transfusion Service and was scandalized at not being paid. He complained that he charged £20 ($96) a quart for his blood at home, "dirt cheap for the world's most valuable liquid," and sued the recipient of his blood for £10 ($48). (He did not win.)[88] In 1934, the people of Sheffield were paid £1.10 ($5) a donation. At the Second International Blood Transfusion Congress in Paris in 1937, the London system, and the concept of a network of voluntary donor panels, was accused of being "hopelessly Utopian."[89] Only the Netherlands and Denmark had followed the all-volunteer model of blood donation. Oliver's confirmation that operating costs per case were only 8 shillings (about $5) did not deter the skepticism, even as his service was now receiving seven thousand calls a year.[90]

By the final years of interwar peace, there were all-voluntary systems around the country. The British Red Cross Transfusion Service had created a model that was successful and impressive. It worked. But in a war, in a great city, it would not do.

———

London, 1938. Preparations. Across Britain, thirty-eight million gas masks were given to children and adults.[91] Children's masks were first nicknamed Mickey Mouses to make them less terrifying, then shaped like Mickey.[92] (My mother, a war toddler, thought hers creepy.) People were told to start digging garden plots and to grow vegetables, and they were not told that the production of millions of cardboard coffins was under way.[93] The first air-raid shelters were distributed, named for Home Secretary John Anderson, and sold for an affordable £7 ($28) each. Life looked different: red postboxes were painted with special yellow detector paint, supposed to change color if poison gas was in the air. The sky was filled with what *New Yorker* correspondent Mollie Panter-Downes called "the silvery dermatitis" of antiaircraft barrage balloons.[94]

Coffins, children, cabbages: London's authorities had spent years readying for calamity on all fronts. But they had not thought it necessary to prepare any blood. In 1937, the secretary of state for war had been asked what the nation proposed to do about a mass blood supply. He said, "It is more satisfactory to keep our stores of blood on the hoof."[95] He meant that the best way to store blood was how nature intended: inside a human body. By the beginning of 1939, a single emergency blood depot had been set up in a bombproof building in the outer London suburb of Cheam. It was a depot in name only, with a capacity to store 1,000 pints, but empty. Planners expected it could be stocked within seven days of hostilities. Until then, the actual emergency blood supply for a city of several million people consisted of the stock kept by four London county hospitals for urgent maternity use: eight pints.[96]

Janet Vaughan knew the planners were wrong. Her interest in the Spanish Civil War had introduced her to the astonishing work of Frederic Durán-Jordà, a Catalan physician. In wartime Barcelona, he had successfully pioneered the mass collection, storage, and delivery of blood. He made blood mobile, transporting it to the front line in glass bottles in a converted fish van. Durán-Jordà was visionary but precise and practical. The Barcelona donor center had its own glassblower to make ampoules exactly as he needed.[97] He experimented with sodium citrate, now unquestioned as the standard additive to stop stored blood clotting, and found that mixing in glucose made for healthier red cells. When it was Durán-Jordà's turn to drive the van, he did so while singing the theme tunes "I'm Popeye the Sailor Man" or "Who's Afraid of the Big Bad Wolf?"[98]

He is also featured in the Ministry of Information's 1941 film on blood transfusion. The narrator, supposedly a colleague of Durán-Jordà's, says that "we bled some four thousand five hundred civilians into a citrate solution," and I picture thousands of Spaniards collectively dripping blood into swimming pools of lemony liquid. When Sidney Vogel, an American surgeon volunteering with the Republicans, visited the Barcelona operation in 1937, he was astonished to see workers lining up along the stairways ready to give blood. And then, a bare room where men and women lay prone, all being bled by specially trained assistants. "Bottled blood for transfusions in wartime!," he wrote. "I had used it but I had never seen it bottled." He was more surprised to find a young man, an artist in civilian life, whose only job was to apply iodine to the donor's vein in preparation for a needle, and who was summoned with the call of *Pintor! Pintor!* ("Painter!"). (Barcelona donors must have been hardier than London's: iodine burns made for the most compensation claims in the London Service.) It was a streamlined production line of blood that had never been seen before, and it worked. Withdrawn, mixed with citrate solution, then bottled, Barcelona's blood was good for eighteen days.[99]

The Barcelona service fascinated Janet. It uprooted the on-the-hoof model for a new, efficient method of collection and delivery. Durán-Jordà had also shown that blood need not be collected by surgeons and doctors, who would have more pressing matters to deal with in wartime. The bleeders could be nurses but also "women with BSc degrees and some laboratory training."[100] Blood could be separated from its donor and transported more efficiently. This insight, she understood, would be essential in the war to come. When medical staff would be dealing with the injured, when communications might be bombed into uselessness, calling donors to overburdened hospitals would no longer work. With Durán-Jordà's model, she could turn London's blood supply from mom-and-pop to hypermarket. She had also read of the Russians, who had been taking blood from road accident fatalities and suicides and storing it at low temperatures.[101] Janet didn't think cadaveric blood was a good idea, and nor since then has science (it is difficult to maintain the quality of dying blood and the concept is unpalatable to the public). But she took the Russians' storage method, mixed it with the fish vans and glassblowers and ingenuity, and began to plan.

In late 1938, she approached the medical school dean again and asked him to let her explore how best to store blood, for what was coming. She got assent and £100 ($400) and sent off two assistants in a taxi to buy "immense quantities" of rubber tubing, corks, and clips. They made up transfusion sets Vaughan called "crude," found donors, and set about bleeding. But war didn't come: at a conference in Munich, Neville Chamberlain and other leaders dismembered Czechoslovakia to appease Adolf Hitler, and the war footing sat down. "Everyone said," she wrote, "that the only blood shed at Munich was the blood Janet shed at Hammersmith." It was diverted to hospital use, and for Janet to use on patients. For the next few months of prewar, Vaughan and her team researched the health and usefulness of stored blood and found it good. But even now, in a year that felt like war was coming, there were no government plans to store blood. She thought this dangerous. She would not be exaggerating, she said later, to say that blood transfusion in war was as important as bandages.[102]

In early April 1939, she gathered some fellows—doctors and pathologists—in her Bloomsbury flat and began to plot and plan. The minutes of the meetings of the Emergency Blood Transfusion Service exist in the archives of the Wellcome Library, and they are as rich as blood. The meetings were always in the evening, after the day's work was done, and they lasted hours. Here I let my imagination go for a stroll. Janet Vaughan would be wearing a tweed suit. She would be kind but brisk. Later, someone described her as "down to earth but like air on a mountain."[103] The others might have bow ties. They would smoke pipes. They would drink tea or gin or whiskey and eat crumpets. They would do this while deciding on the size of bottles, or what kind of armrests to put on the "bleeding chairs," and they would change modern medicine.

The revolution would begin with logistics. During the Spanish Civil

War, 10 percent of air-raid casualties had required blood transfusions.[104] If this applied to London, it meant sixty-five thousand casualties a day. The practical thing to do would be to set up blood depots throughout the city. They would be located in areas of London outside areas most vulnerable to bombing, but near enough to easily deliver blood to major hospitals. The depots would control the bleeding of donors in-house and in the community, they would store blood and deliver it to where it was needed, and they would also undertake medical research. There would be four depots: two north of the river and two south.

First, the science. At the inaugural meeting, the minutes record a suggestion that 50cc of 3.8 percent citrate containing 0.1 percent glucose should be added to every 450cc of blood. This was fiercely debated. And there were other issues to resolve. Logistics and equipment, obviously: they planned for eight bleeding couches per depot, at £21 ($84) each. Twelve Cheatle forceps; six sphygmomanometers; one gross of rubber teats; three thousand yards of Elastoplast. And the donors: how to deal with those?

At this stage, the Bloomsbury group—this other Bloomsbury group—took care to consult with Percy Oliver. He was present at the first meeting but not at many thereafter, and I wonder at these medical women and men, and whether they shut him out. In 1936, when Oliver wrote a piece in the *British Medical Journal*, an accompanying editorial seemed defensive about the decision to publish the work of someone who was not a scientist or medical man, while acknowledging Oliver's "unique position as an authority on the problems connected with blood transfusion." There was no call, it wrote, "for a lifting of the eyebrows when a layman like Mr. Oliver refers to the controversial matter contained in headings such as 'cross-grouping' and 'universal donors.'"[105] Oliver wrote with both confidence and apology in his article, pleading for a national blood transfusion conference "with all diffidence as a layman."[106] Early on, his expertise was welcome in the Bloomsbury meetings, because the depots would need blood, so they would need donors, and Oliver had those. As for syphilis, Oliver pointed out that promiscuity increases in wartime. But the committee felt it was a risk that must be taken. They should be screened, because syphilis was definitely a worry. In the end, whether to inform a donor of a positive syphilis test was left to each depot director to decide. Vaughan later tried telling donors, but after they became "extremely indignant," she left well alone, afraid to prejudice donor recruitment. A different view might prevail in a peacetime organization. Oliver's advice on donor cards was more accepted and his specimen cards adopted. These asked for contact details, blood group, whether the donor had any national service obligations, and "the character of the arm vein."

There were many calculations to be made. A population of eight million Londoners, therefore a catchment area of two million per depot. In the event of bombing raids, each depot should plan to minister to ten thousand casualties. For this, each would need a panel of twenty thousand donors, enough to be able to supply five hundred bottles a day of group O in an emergency, and to hold five hundred more bottles in cold store. The plan at first was to use only donors with group O blood, which could be transfused into most people without harm. (By 1940, improvements in cross matching allowed the use of all four blood groups.) Trained assistants would do the grouping. "I can always remember," wrote Janet, "George Taylor of the Salton Laboratory, the English authority on blood groups, saying with great solemnity, you must also enroll girls to determine these blood groups and this should be done at once; it is not easy to procure young girls."

It was not easy to procure the right bottle, either. Without the proper storage vessel, all the sodium citrate in the world wouldn't keep blood safe or make it portable. The choices were few but they were tricky. A Beattie waisted type? A whiskey cap on a United Dairies bottle? A McCartney screw cap? Or a modified McCartney bottle of the L.C.C. type? Vaughan's children grumbled that their home was littered with old bottles, and throughout the summer the deliberations continued. By the second week of June, the Bloomsbury set had decided on a modified milk bottle because they were easy to obtain and also easy to deliver in milk crates. For transport, they would convert Wall's ice-cream vans. Sturdy, refrigerated, and they wouldn't smell of fish.

The committee was informal and unsanctioned—"no authority!" remembered Janet, with glee—but then Vaughan sent a memorandum about their plans to Professor William Topley of the London School of Hygiene, who was known to be organizing emergency services. Professor Dibble—Vaughan's superior—heard of the memo and called her "a very naughty little girl."[107] As if that would stop her. After a good while, someone working with Topley responded positively and asked for a budget. Her friend the dean advised her to triple all the costings. She did, and this was accepted. The organization of the blood supply was handed to the Medical Research Council, which in a post-war report wrote that "from the time of the Munich crisis in 1938, the question of blood during wartime had been much in the minds of medical men."[108] And of one particular woman.

At the same time, the British army was setting up an equally pioneering system of blood supply, under the command of Colonel Lionel Whitby. The Army Blood Transfusion Service would set up a depot at Bristol, ask for blood from the surrounding populations of the southwest, and fly it to the front. No other military facing war had such a plan, and no other military in the war had anything as successful as the Army Blood Transfusion Service. For a logo, the army chose a vampire bat.

Janet was to run the northwestern depot, in the town of Slough. She set off alone to find premises. "How fortunate I was," she wrote in her memoir, "to go to Slough where everyone—mad as they thought me at the time—was more than willing to help me." There was an unshackled energy about Slough that appealed to her. She called it "a frontier town," grown up after the First World War around a vast trading estate, full of migrant workers "with no settled traditions and customs to be disturbed." She was directed to Noel Mobbs, chairman of the Slough Trading Estate, which housed dozens of factories including the chocolate maker Mars and a social center. Mr. Mobbs did not believe a war was coming, but he said the depot could move into the social center, that there was space for cold storage rooms to be built there. There was also a bar.

Premises were found for the Luton, Sutton, and Maidstone depots in a disused part of a hospital, in an adult education center, and in two converted houses. Teams of people were sent out to appeal for donors, advised by publicity experts from companies such as Dunlop and Horlicks who volunteered their time. From July, the press began to publish regular appeals for donors. They should find the nearest "empaneling center," which could be found by phoning a switchboard (Central 8691). Volunteers would be pricked in the ear or finger, their blood group tested and then registered. Transfusions would take place only in the event of war. In the first three days, five thousand people signed up. Three days later, there were eleven thousand volunteers. Keep coming, said the men of the press and from Dunlop and Horlicks. "The empaneling centers are prepared to welcome 30,000 a day." By the end of the summer, when war seemed near, the Slough depot alone had fifteen thousand donors on its books.

On September 1, 1939, Janet Vaughan received a telegram from the Medical Research Council that she described as "laconic." It read, "Start bleeding." The ice-cream vans were driven to Slough, the donors were called, and at eleven fifteen two days later the staff of the northwestern depot stood in the social center bar in white coats and listened to the prime minister, Neville Chamberlain, announce on the wireless that the country was now at war with Germany. "And then," wrote Janet, "we went back to our bleeding."

Slough depot. Wartime. The place bustled with nurses, secretaries, telephonists, medical technicians, drivers, scientists. There were one hundred staff. Diverting from Durán-Jordà's model, the bleeding here was done by medical staff supplied by the Medical Research Council. The drivers were anyone willing to drive Wall's ice-cream vans full of blood through bombs and blacked-out streets. Most were women. These drivers: I picture them as forthright young women in heavy coats, full of pluck. But there was also "Mrs. E. O. Franklin's chauffeur Brady, a mad Irishman," who kept the vans ticking over. When Liverpool was bombed and its transfusion service destroyed, Janet went to the bar to recruit volunteers to drive up supplies to the damaged city. She loved this bar for what it meant to her staff and understood that the depot depended not only on equipment and science. "My young drivers, girls, coming in late at night having driven through terrible weather and blackout, to be able to get some whisky in the war was very important." In many interviews, she said the same: "Someone once said, 'Janet was the only person who had the sense to set up an Emergency Service in a bar.'"

One regular driver was Lady Dunstan, "who must have been at least 70. She always wore a string of pearls and a toque [a small hat] rather like Queen Mary, but she was never daunted." Do not underestimate Lady Dunstan: the conditions were usually hard and often terrifying. "Intimate knowledge of the roads under black-out conditions was essential," wrote the Medical Research Council, "but had to be coupled with a willingness to drive while a raid was in actual progress." Yet the drivers were so adept, sometimes they reached the hospital with their blood before the casualties that had lost theirs.

One day, her ladyship returned in great pride from a Canadian military hospital that had requested blood, and announced with wonderful grandeur: "Yes, the surgeon insisted on me coming into the theatre and seeing exactly what he was doing and why the blood was needed. I think I was able to help him." Another volunteer, wrote Janet, was "a remarkable old lady whose only interest in life before the war had been her string of ponies and her bridge. She came and said she wanted a job and we set her down amongst the young technicians to fix a singular nasty wire filter that was being used at that time for stored plasma." During her frequent episodes of illness, she dispatched her chauffeur in her Rolls-Royce to fetch her filters that she could fix at home while abed. "One of her friends said she had never been so happy in her life before. She knew we depended on her work, as we did, and through us casualties all over the country."

It was a time of stepping up, making do, and derring-do. Katie Walker, a nurse employed at the depot for a couple of years, started work having taken only her British Red Cross exams. "I learnt to drive—you didn't have to do a test or anything then—and I drove the ambulances, took blood around London to where it was needed. We also set up stations for taking blood, all over the place."[109]

And who was giving them this blood? Plenty of people. Much of the recruitment had taken place in the months before the war. At Slough, it helped that the depot was housed on the huge trading estate. Factory workers were approached on-site. Bleeding teams went into the countryside to recruit, with success. Around the country, regional depots were doing similar work. And there was also by now social contagion. Giving blood was an accepted social good. In 1939, when the first evacuees were sent from the city to the countryside, Londoners used to the idea of donation spread the notion. "In a Hampshire village," reported the *Times*, "a London woman who has moved there was asked if she could obtain some more donors for the blood transfusion service. There was, first, herself, she realized, and then, having registered herself, she went out, met a willing response, and in an hour or two 21 more people had registered because of her approach to them."

Bleeding of donors was done both at the depot and by mobile teams sent out into the surrounding small towns and villages where they set up temporary bleeding centers in a town hall, factory break room, church hall, or village pub. Janet was very proud of the quality and sterility of "the bleed." The housewife in the country village or small town, she wrote, "was often a most faithful and regular donor." There was no question, now, of heeding the distaste of the medical establishment for women's blood. Many women gave every three months throughout the war, "feeling it the one personal contribution they could make to the war effort."[110] Iron tablets were often handed out to combat the inadequacies of rationed diets.

The South London depot at Sutton—telephone number Vigilant 0068—was bleeding six hundred to seven hundred donors per week by 1941 when the Blitz began. Forty percent came to the depot to be bled, having been called up by postcard; the rest were bled at "outlying bleeding centers." In an emergency, donors were fetched from nearby factories. The depot didn't need to call upon donors more than once every six months, wrote Sutton's director, because so many donors were available.[111] There were quiet periods, even then. During a major offensive or disaster, hordes of donors came once and then not again. To counter the helplessness of mass disaster, we donate a pint of usefulness: this instinct persists today. After 9/11, 570,000 additional units of blood were donated, but 208,000 were discarded and only 260 units were needed to treat 9/11 victims.[112]

The bleeding teams learned that a loudspeaker van transmitting a special appeal was helpful. If the loudspeaker transmitted a message that the blood needed to be flown at once "to some particular fighting front or blitzed city," donors flocked to give. After the Nazis invaded Denmark and Norway in 1940, one depot raised its weekly donor panel from eight hundred to three thousand almost immediately. Vaughan, the scientist, liked to think that this was not the only thing that drew donors to give blood. "Of course the man or woman in the village on the Chilterns likes to feel he is making a vital contribution to the war. [But] donors are really interested to know what happens to the blood they give and in the scientific advances made in blood transfusion. They respond to facts and figures as well as sentiment."[113]

The press thought differently, preferring to present probably fictitious but vivid accounts of a link between donor and recipient. A bloodline. During the First World War, this desire for connection had been noted. Donors and recipients expressed a wish to know who was on the other end of the bloodline in either direction. "On the 13th June," wrote A. C. Tayler in 1914 to a surgeon, "you took my leg off above the knee, and until I received blood from someone else you considered the betting about 3 to 1 on my pegging out. [. . .] Can you find time to let me know the name and address of the man who gave me blood? I should much like to write to him." A donor in 1917, a gunner named Birditt, asked to know if the patient who now had his blood "is recovering alright."[114]

The North East Regional Blood Transfusion Service, via the *Driffield Times*, presented several colorful though improbable instances of blood brotherhood. On Sunday, November 19, 1945, a Miss M. Lee of Ingle Nook, Stork Hill, Beverley, gave her blood. It was flown overseas three days later, and, on November 30, transferred to Sergeant Howells, wounded by a mine, one amputated leg, who then made favorable progress. Mrs. Backhouse of Beverley assisted Private Cook of Meanwood Road, Leeds, who was almost eviscerated by a mortar bomb launched at him near Venlo. Several feet of bowel emerged from his abdominal wound, but he was transfused with blood from Beverley and he lived.[115]

The official account of the transfusion service, *Life Blood*, used a similar device, presenting personal stories of "a large, middle-aged woman, a soldier, a workman, a young red-haired girl of about twenty, an elderly man with an A.R.P. badge and a limp, and a railway goods guard who has brought his lamp along with him, as he proposes to go on duty directly his visit is over." The middle-aged woman, Mrs. Alice Edwards, is a widow with one son in the army and a daughter in the ATS (Auxiliary Territorial Service). She started to give blood to do her bit, then found "a little blood-letting seemed to do her good, to overcome a feeling of heaviness, and she comes now every three months—'for her own health,' so she says."[116] In the UK, personal contact was restricted to this type of propaganda, but Russians allowed for a more real intimacy. And why not, when people were sharing something as intimate as body fluid? Each bottle of donated Russian blood was labeled with the donor's name and contact details. As most Russian donors were women, and the blood went to serving soldiers, this had a predictable outcome. In 1943, the *Dundee Courier* reported that this friendly transfusion "has led to a number of romances between the soldier patient and the blood giver." In some cases, soldiers wounded for the second time asked for blood from the same girl.[117]

The war embedded the idea of blood donation in popular consciousness like nothing else had done. There was nothing more powerful than the message that blood was going, almost directly, into the veins of a wounded soldier, even if it wasn't. Then again, people gave for all sorts of reasons. The Army Blood Transfusion Service encountered one old gentleman with high blood pressure. Before the war, wrote the *Gloucestershire Echo*, "he used to pay ten guineas to his surgeon to be bled. Now he follows the mobile collecting teams of the Army Blood Supply Depot, and surrounded by pretty V.A.D.s [women from the Voluntary Aid Detachment] and ATS, has it done for nothing." The brigadier in charge of the depot saluted "the faithful crowds of high blood pressure victims who follow our teams around, giving their blood not only for the benefit of our wounded men, but also for their own benefit."[118]

By 1942, the donation of blood had become many things to many people. It could be a weapon of moral superiority. That year, Miss Ivy Standing and Miss Grace Standing were posted some white feathers and accused of cowardice "while young girls who used to be your playmates are doing your bit." The accompanying poison pen letter finished with what was meant to be a powerful flourish. "P.S. We doubt whether you have even given blood to help a wounded soldier." The reporter's flourish was better. "Curiously enough, both girls have given blood transfusions."[119] The same year, James Eric Oldham of South Sale near Manchester, who was on trial for stealing carpets, offered the fact of his blood donations as mitigation. He had given six or eight pints of blood, milord, and it had weakened his will. The mitigation failed.[120]

Another man offered, in his memory of war, perhaps my favorite explanation for donating that I've ever encountered. "1941. War. Blood needed. I had some. Why not?"[121]

Slough, Luton, Maidstone, Sutton, and the Army Blood Transfusion Service: they all gathered blood, they all supplied it, and after a few months their staff also began to transfuse it. Among the civilian blood depots, Janet was convinced that Slough was special. Everyone knew to ask Slough for blood, and they would get it. At first, they would call for it by telephone. But "they soon learnt that Slough could hear and see the bombs falling and would arrive." After a year of waiting, during the period known as "the bore war" or the "Sitzkrieg," the bombing began in the autumn of 1940. The ice-cream vans would get near the bomb sites and deliver their blood to the hospital where the casualties were being taken, or they would perform transfusions on casualties in the street.

On the home front, the depot staff learned to wear electric lamps on their foreheads like miners so that if electricity failed "or the lights were off because the windows were broken and blackout curtains were blowing in the wind," they could see where to stick in their needles and hang their bottles. They changed the needle design because of a house fire on the Great West Road. Vaughan arrived at a hospital filled with casualties and found a little girl, horrifically burned. She left the girl to die, because she had to see who could be saved with a transfusion, but after saving who she could, Vaughan returned to the girl and found her alive. Her legs and arms were so burned she had no veins there. And Janet again remembered something she had read, that you could give blood into bones. "That was the great thing about medicine in the war, you could take risks because people died so they were no worse off if they died because of what you did." She took the biggest needle she could find, stuck it in the girl's breastbone, and told a nurse to pump in blood. (When depot staff couldn't find a vein, they called it "Digging for Victory.") Two hours later, Janet came back to the girl to find that the nurse had got two pints in. After that night, they devised needles to transfuse into the bone marrow. They had special flanges to hold them in place, so they could be used on boats and landing craft "when it might be easier to get into a large bone than an invisible vein." The flanges were used at Dunkirk; the girl lived.

Years later, Harriet Higgens, that burned little girl, applied to Oxford. She remembered Vaughan, who had visited her later in the hospital and sucked blood out of her ears with a glass tube (Harriet doesn't explain why). And all the while, wrote Harriet, "she explained what she was doing and spoke to me as if I was as interested and intelligent as she was." She was allowed to select three colleges on her application, but the only college she put on her form was Somerville, where Vaughan was then principal.[122] "So," said Janet Vaughan of this, "nice things happen."[123]

The depots treated thousands of casualties, but the staff also did scientific research. They learned that a trauma victim needed on average two and a half milk bottles' of fluid, with two bottles of blood used for one bottle of plasma.[124] There was flexibility: overseas, war surgeons learned that "the only criterion was the need of the man." The quantity to be given was the quantity lost. "In cases of necessity," wrote Major General W. H. Ogilvie, a noted surgeon, "blood has been given at the rate of half a pint a minute by using two veins simultaneously, and as much as 18 pints has been given in the course of a two-hour operation."[125]

The war created ongoing innovation, in science and practice. Plasma was now routinely separated from blood, dried, and used. It was much easier to transport than perishable blood, and judged useful at treating blood loss. Also, because plasma has no cells, it didn't need to be cross matched. The army's plasma-drying facility had been built with donations from the Silver Thimble Fund of the women of India.[126] Sometimes,

dried serum—plasma without its clotting factors—was adequate treatment. This was essential for the mass evacuation of Dunkirk in the
spring of 1940, when the depot sent all the blood it had to the coast.
But the casualties kept coming, and the system could not cope, even
after the Americans began sending over blood and plasma on ships
under a program called Plasma for Britain.[127] Vaughan and her staff
had also been working on using plasma, but the plasma looked cloudy
and full of clots, so they hadn't dared risk it until Dunkirk. "We knew
men must die if we didn't transfuse them, so we took a risk on our very
odd-looking plasma." It was another risk justified by war, and the
plasma worked "like magic."[128]

By the spring of 1941, when the air raids stopped, they expected
the demand for blood to drop but it didn't. Transfusion worked and
surgeons and doctors kept doing it. "In many cases," wrote Janet, "no
doubt the pendulum swung too far and unnecessary transfusions were
given, but on the whole the educative value of the war time transfusion service was great." So was its organization. "It was gratifying,"
wrote the Medical Research Council after the war, "to see how easily
the organization, planned without any practical experience of a large
scale transfusion service, swung into action, reflecting great credit on
all those who had given so much time and thought to the preparation
of the scheme."[129]

Vaughan learned to say yes to any request, because "what men and
women need in a desperate emergency is reassurance. They can hold
on if help is coming, and—given the lead—other men and women will
always be prepared to give that help." Just before D-day, Janet received
a phone call from the head of Emergency Medical Services. "Janet, we
have made no arrangements for the Ports, will you look after them?"
She said yes, having no idea what looking after the ports would entail.
"As so often we heard no more, but I can only hope that the Ports
received reassuring messages that Slough would come if needed."[130]

At the end of the war, a memorandum on the South West London
Blood Supply Depot, the only one of the four to maintain comprehensive records, showed that in 1940 it had distributed 9,410 bottles of
blood, and in 1945, 22,397.[131] Some was used for bomb casualties, but
it was notable, wrote Janet, that "even when no bombs were falling,

even for long periods, the steady increase in the demand for blood was maintained. So remarkable was the progress made by blood banks and the transfusion services during the war years that there has been a universal demand for their retention *in perpetuo*." Regional depots were working well around the country. The Army Blood Transfusion Service's work had been equally impressive. At the outbreak of war, it was collecting 100 pints a day. By war's end, when the army had set up 850 satellite centers from Reading to Penzance, it collected 1,300 donations a day. The record amount bled in one day was 1,657 pints.

For Major General W. H. Ogilvie, "the greatest surgical advance of this war, more important even than penicillin, is the development of the transfusion service. A transfusion service, with blood banks sufficient to meet any needs, must be available for the resuscitation of the injured and the restoration of the sick in civilian life." And so it came to pass.

In 1945, Janet Vaughan left the depot. She had had five years of death and burns and bombs, like countless others, but when she was asked to go to the Nazi death camp Belsen to research how best to feed starving people, she said yes. The prevailing medical dogma was that the most effective treatment of starvation was protein hydrolysates, strong proteins in liquid form. She was driven over the Rhine on wooden pontoon bridges, and she waved to troops returning from the front. She saw hundreds of forced laborers in striped pajamas, spat out from their camps and wandering over the countryside. When she got home, she burned all of her husband's striped pajamas.

Before they reached it, they could smell Belsen: a stench of shit and dead bodies. The senior officer at the camp expected that these new visitors had come to help. No, they said. They had come to do research. That sounds brutal now, but Vaughan believed in science, and they would soon need to save all the prisoners of war who would emerge from Japanese camps. The science had to be done, even if she found herself having to inject hydrolysates into skeletal men who saw her medical apparatus and screamed "Nicht crematorium!" because the Nazis had sometimes injected the condemned with paraffin before sending them to the gas chamber.[132] Vaughan writes that this was so that they burned better. This was horror, but Vaughan kept on: when she had to pick the living from the dying in piles of bodies; when she was attacked by naked, desperate men screaming for bread in five languages. She did enough research to show that small amounts of food were a more efficient treatment of starvation than hydrolysates. She wrote to George Minot, her hematologist hero, of seven-year-old children who looked twenty and girls of eighteen who looked fifty, of the constant stench of feces and filth, of men weeping at the sight of her "unable to expect kindness or friendliness."[133] She wrote a letter home that said, "I am here—trying to do science in hell."[134]

She did science in hell for a few weeks, then returned to England and to a job as principal of Somerville College for the next twenty-two years. It was academia, but it was no sinecure: she rose at dawn to dictate all the correspondence needed to run an Oxford college, before setting off every day to put a full day of work in at her lab nearby. If callers to the college wondered where the principal was, she responded, "Do they think I sit knitting?"[135] She became an expert on radiation, researching the effects of nuclear fission on the metabolism of humans and rabbits, who were her test animals. A colleague called her "our radioactive principal,"[136] which was more accurate than he expected, because if there was any risk that radiation had leaked, Janet Vaughan would disappear to have a bone biopsy taken from her tibia.[137] She did this work for decades, and once answered the politician Shirley Williams's question as to why on earth she was handling plutonium at her age with "What could be better than for someone in her seventies to do this work? I haven't long to live anyway."[138] She fought to have women's colleges accepted as full Oxford colleges, and she increased the intake of science and medical undergraduates at Somerville. Early on, she made sure to employ women as her research assistants and, going flagrantly against custom but forcefully in the direction of fairness, kept their jobs for them if they had children. She served on great and good councils, but also, when flu broke out, carried trays of food to sickly students. As principal she was radioactive, but also "accessible and exhilarating."[139] Kindness. She always had kindness. "She was a very human scientist," a former student wrote in a eulogy.[140] Vaughan lived long after retiring in 1967 and wrote books and academic papers until her eighties. She was a Dame, a member of so many associations (my favorite: the Bone and Tooth Society), and loaded with honorary degrees. She was establishment, but a socialist to the end.

Percy Oliver died in 1944, exhausted. He had never been robust, and he had watched his beloved service be overtaken by the Emergency Blood Transfusion Service, without him in it. He had not survived to see, as Janet did, the transformation of both into something exceptional and enduring. "To have seen the Blood Transfusion Service," wrote Francis Hanley, a colleague of Oliver's, in a memoir, "grow from a backroom, one-man enterprise here in London to a multimillion nation-wide organization sustained by voluntary donors, is something not granted to many of us."[141]

It was not granted to Percy Oliver, nor was much recognition. Perhaps medical eyebrows continued to be lifted at the truth: that the modern blood supply system was built on the shoulders of a layman and a very naughty little girl. It took thirty years for a handsome portrait and wooden memorial to be installed at the entrance to the Department of Hematology at King's College Hospital, from where came that first-ever call, as the story goes. A ward at King's is now named for Oliver and offers thirty beds for patients with general medical, respiratory, gastroenterology, and sexual health conditions.[142] In 1979, the local council installed a blue memorial plaque on the Olivers' house in Camberwell: PERCY OLIVER, FOUNDER OF THE FIRST VOLUNTARY BLOOD DONOR SERVICE, LIVED AND WORKED HERE.[143] (So did Ethel Grace, whose efforts are relegated to the unsaid and unheralded spaces between the lines, where wives' contributions generally reside.)

Dame Janet Maria Vaughan has no blue plaque outside her home, although her honors were many and prodigious. Her last years were marred by severe arthritis (she would tell her beloved grandchildren to get on with their lives, not waste time visiting their decrepit granny).[144] She died aged ninety-two in January 1993. She had only recently stopped using her Mini, a common sight and sound all over Oxford, as she drove, in the words of a colleague, "idiosyncratically."[145] Another described her driving style as "like a kangaroo."[146] By then, the Blood Transfusion Service, later the National Blood Service, then NHS Blood and Transplant, had existed for forty-seven years. When she was asked in 1984 how she would like to be remembered, she answered with no hesitation. "As a scientist. That I have been able to solve, to throw light onto fascinating problems. But as a scientist who had a family. I don't want to be thought of as a scientist who just sat thinking. It's important you have a human life."[147]

An accomplished, organized, indefatigable, always human woman. In 1941, Vaughan wrote to Katie Walker, the nurse who had worked at the Slough blood depot for fewer than two years. But what a two years. Janet had no need to write the letter, to one of a hundred staff: turnover must have been brisk. But she wanted Katie to know how much she would be missed. "I cannot in the nature of things see and know you Nurses as personally as I should like to but you must realize that I do love and care about you all very much." She could have signed it formally, as Katie's superior, as the woman in charge of a vast organization with vast responsibilities. She could have written, clinically, "With regards from Dr. Vaughan, Director." She could have signed it with the formality due to a woman who was instrumental in setting up our modern system of blood donation and transfusion; who dared to stick a large needle into the breastbone of a small burned girl; who did science in hell; who never stopped encouraging science in all ways, and women to do more of it. She signed it, "Yours always with love, Janet."[148]

Boniswa, a mural on the Isivivana Centre, Khayelitsha, by Breeze Yoko

FOUR

BLOOD BORNE

Our new home. This is what Khayelitsha means, and this name has never been anything except ironic and bitter because Khayelitsha is the ugly backside of beautiful Cape Town. Khayelitsha, formed when apartheid authorities lifted up its black and colored population and dumped them on the Cape Flats, miles away from the beautiful colonial houses of the city center, from the waterfront and ocean, from the looming, lovely Table Mountain. Khayelitsha is the second most famous township in South Africa after Soweto. It is called a township because that is the name for towns when their inhabitants are black or colored and poor. The population estimates for Khayelitsha vary, but it probably holds around half a million people.[1] Some other figures are firmer and frightening: unemployment is 40 percent, and 50 percent among young people; a third of inhabitants live in "informal housing," also known as shacks; common assault and attempted murder rates, when assessed in 2015, were found to have risen by more than 40 percent in five years;[2] the sexual assault center records a hundred rapes a month and estimates that actually there are nine times more than that; and it is not unusual for girl children to be put on contraception at the age of ten by their mothers in case they are raped.[3]

Khayelitsha now is a place where Uber drivers will drop off a visitor but never return to collect them, in daylight or darkness. No one accepts fares from here. All the drivers who transport me there from the city are Zimbabwean, and all are shocked. They make a *click* sound when we approach Khayelitsha, not because they speak Xhosa, a language of clicks, but because this is a sound of disapproval in many African countries, farther back in the mouth than a *tut* but with the same purpose. Look at this, says one, with a *click-tut*: even in Zimbabwe people would not accept corrugated shacks. They would build, even with mud. Another gives me a running tour of township violence, all these places of danger lined along the N2, a highway that begins in Cape Town and ends on the other side of the country, skirting Lesotho and almost reaching Zimbabwe. The colored township is bad, this driver says. Too many drugs. *Click-tut*. But Langa is the worst. There you can die in cross fire, so easily. He nods to Langa but I see only painted houses and colorful washing, drying draped on the mesh fences, the pinks and yellows and blues and the sunlight making danger seem distant and impossible.

To return to Cape Town at the end of my day, I must rely on kindness and lifts from the staff of Médecins Sans Frontières (MSF), the international medical organization that is hosting me. MSF arrived in Khayelitsha without intending to stay, fourteen years ago, to set up a clinic to deal with HIV. It is still here because so is HIV. The virus is believed tamed in the richer northern countries of Europe and North America, but outside those places the epidemic thrives. Only now it is an epidemic of women.

More than half the 37.5 million people living with HIV worldwide are female. Every week 7,500 young women are infected, and globally HIV/AIDS is the leading cause of death of women aged fifteen to forty-four.[4] In sub-Saharan Africa, young women between fifteen and twenty-four are twice as likely to be infected with HIV as young men.[5] In some parts of KwaZulu-Natal, an eastern province of South Africa, a fifteen-year-old girl has an 80 percent chance of being infected with HIV.[6] Eighty percent! To be interested in HIV is to be pulled to the bottom of the African continent as if by a tractor beam, and to want to know why in 2017 being a black young woman is a death sentence.

As well as having permanent clinics around Khayelitsha (at the Site

B Day Hospital, in Ubuntu), MSF has a satellite facility on a dusty road that is lined with small shacks and cut oil drums used to prepare *braai*, South African barbecue. The clinic is known locally as "the container" but it is actually a mobile trailer. It doesn't look like much, but it provides HIV tests and counseling and maybe tea and coffee. Disco music plays from somewhere among the shacks, these structures of cobbled-together corrugated iron patched with anything possible. The container and the shacks are built on sand. In the summer the shacks are intolerably hot. In the rains they flood. It is now August and winter, and the shacks are cold, though the tiny stalls of the trailer are heated. The reception area is a plastic table outside with a few plastic chairs, monitored by a cheerful woman wrapped in a fleece blanket, a fashion I see often (blankets are cheap warmth), and black-and-white striped stockings, a fashion entirely her own. She looks like a cuddly witch of the west from *The Wizard of Oz* and she has equivalent power: she controls the kettle.

The clinic is open five days a week from seven a.m. until six p.m. The long hours are to enable working people to attend, the same working people who flock on the highway bridges and train station of Khayelitsha in the predawn hours, a murmuration of humans waiting for their ride to where the work is. Without the people of Khayelitsha, Cape Town would have no cleaners, waiters, or drivers. As people are tired and may not come, the clinic entices them with free Wi-Fi, a pull in a country where phone data is expensive. Another enticement is the general nature of the health services: it offers family planning and diabetes and TB (tuberculosis) screenings, so people aren't as ashamed to come along as they might be for a dedicated HIV clinic. There are other clinics in Khayelitsha's twenty-two subdivisions, which the apartheid government named with miserly imagination for the alphabet, so there are Sites A to J, but there is also Mandela Park or Harare. In the HIV clinics in Site B or Ubuntu, people are easy to spot, either by the color of their green appointment card or by their movements. If they turn left, they are going to general health. A right turn, and they are definitely heading for the HIV clinic. They are marked by their orienteering.

At the MSF there is no left or right, only three steps up into one of the snug stalls, so intimate that staff can talk through the walls, and where you can get a free fingerprick HIV test with results in fifteen minutes. South Africans have lived for so long with HIV they speak it fluently, talking easily of status and ARVs (antiretrovirals) and CD4 counts (levels of a particular white blood cell) and viral loads and condomizing. You have no need to tell them that AIDS is acquired immunodeficiency syndrome caused by the human immunodeficiency virus. They have had decades of announcements, appropriately messaged soap operas. Yet South Africa has seven million people infected with HIV, the most of any country.

One of the clinic visitors this morning is Themba, who has come for an HIV test. He tried to get tested the day before but he was too drunk and staff turned him away. That he has come back makes him unusual: men are notoriously difficult to get into health care clinics, anywhere. Other things about him are more routine: that he has been "fooling around" and had sex with a stranger because he was drunk. I look at him when he says this and think of the colossal rates of sexual assault in South Africa, and wonder. But he distracts me by saying he also has three long-term girlfriends. All his friends have several women. "It's what we do. We brag about them." He mentions that he has a box of condoms at home. When I ask him whether he will now condomize, his "yes" is the long one of the liar. But he remembers who he is talking to and he pulls himself up. "Yes," he says, more emphatically. "I will. Because I don't want to live in fear. I'm scared."

The trouble with Themba is that South African women should be scared of him.

Here it comes. In this animated video, the virus, a spherical particle with green waving spikes on its surface, is descending to land on its target. This is usually a CD4-positive T-cell, a white blood cell also known as a helper T: "T" because it is made in the thymus gland and "helper" because it is very useful. If the immune system were *Star Wars*, helper Ts would be the Force: they guide other white blood cells to attack invaders and threats. (Some educational videos, attempting to explain the complicated, marvelous workings of our immune system, liken them to air traffic controllers.) These T-cells release chemicals that draw other white blood cells to the site of the threat and different chemicals that stimulate leukocytes to divide, the more to fight and to conquer. Helper Ts are fundamental and intrinsic, and without them the immune system is the Death Star without its core weapon or a sky of planes in collision. Helper Ts are an astute choice of prey.

In that animation of infection, the HIV particle—a virion—descends to the surface of its target T-cell with the slow grace of a moon lander. Its wavy spikes are sugars on the surface of its protein envelope, the casing that surrounds the virion's core. The spikes become lander legs, waggling like hopeful insect antennae, until they join with other blue spiky forms on the T-cell's surface. "Blue spiky" is my layperson's term: they are more usually known as co-receptors CCR5 and CD4. HIV's legs hook into the co-receptors as a child's hand finds its mother's to hold. The virion injects a spike into the T-cell, sucking up the cell's surface until the invader and the prey have fused. It is unsettling. It is a hunt.

In reality, there are no green spikes or blue co-receptors: like other viruses, HIV is tiny, much smaller than bacteria and too small to have color. I find another video that shows real footage of infection. The color scheme is grays with only the T-cell colored to be distinguishable, and the actions are different: no streamlined moon landing but an incessant jostling and tussling, like a cat pushing its face at you again and again, a sinister nuzzle. I expected to be baffled or puzzled by watching viruses at work but not to rediscover the helplessness of a child watching a pantomime villain: T-cell, he's behind you. Look out. And the hunt proceeds, until the T-cell is breached and the virus sidles in, inevitable.

The color of HIV and AIDS campaigns have always been red because the virus travels in blood and red is what we think blood is. But it is the white blood cells that HIV attacks. Leukocytes, named for the Greek for "white" when actually they are transparent, make up only 2 percent of our blood volume but they are, in the words of a biology textbook, "stupefyingly complex and astoundingly flexible."[7] They are the cogs of the immune system and they are very busy.

A splinter or a cancer. Should anything assault us, white blood cells immediately rush to the threat. Some eat viruses, bacteria, cancerous cells, and toxins; some recognize allergens; others trigger inflammation or produce antibodies, markers that can tag an alien presence so the body can recognize it in the future. White cells do these jobs frequently and successfully. It is rare to find a description of the immune system that does not resort to military terminology: leukocytes are our army, fighters, defenders, protectors. In the 1966 film *Fantastic Voyage*, about a trip in a miniaturized submarine through the human body, the antibodies approaching Raquel Welch are the terrifying villains. "Open it!" she screams at her companion standing at the sub hatch. "Before they get here!" Safely inside the sub, the combined might of four grown men can't remove the antibodies glued to her body. They are indeed terrifying, but they are just doing their job: to patrol and destroy. But they can't yet destroy HIV.

A phone call came in to my desk at a magazine in northern Italy, where from the window I could see green fields and behind them the Dolomites. A clean, fresh view. Nothing wrong with the world. It was sometime in 1998 and the caller was an ex-partner, a young man. His voice was tight. He said, I have herpes and you should get tested and you should also get tested for HIV. My stomach dropped, a sign that my body had diverted blood to my legs and arms away from my gut, in case I needed to run. A sign of fear. I did need to run: I booked a flight to London to get an HIV test, although Italy provided testing. But I was a wounded animal who wanted the safety of home and not Catholic judgment. I invented a family emergency for my manager because how dare I tell the truth? HIV was unspeakable, like death. In 1998 it was the same thing. On the subway journey between the airport and my friend's house, I remember looking at everyone around me. Someone must have it. Statistics say so. Is it you? Or you? Or me?

I was terrified because I was meant to be. And because so were the authorities. The UK government launched one of the most powerful and successful messaging campaigns in public health history. Anyone who watched TV in 1986 remembers the tombstone ad. Frightening music with spooky clangs. A color scheme of black and gray and sinister. An exploding volcano, then a disembodied pair of hands chiseling at rock, carving the word *AIDS* on a dark slab. John Hurt narrated, warning us in an eerie baritone of a danger that threatens us all.

Anyone can get it, man or woman. "So far it has been confined to small groups, but it's spreading." A gravestone rises out of dark mist, then falls back with a crash, as we are told not to die of ignorance, before a bouquet of white lilies is flung on the ground, along with a government pamphlet. On YouTube, someone has uploaded the film with the caption, "This is the ad that scared the fuck out of everyone in the 80s." The clip was criticized for excessive scaremongering. Actually the filmmakers had originally wanted to start with nuclear sirens, but Margaret Thatcher considered that too much for Cold War times. "If we'd kept it like that," director Malcolm Gaskin told a newspaper, "I think everyone would have headed for the beaches."[8]

The scaremongering was necessary because scare was necessary. In 1986 HIV was a rampaging virus that had no cure and that killed inevitably. It could mostly be prevented, then as now, by humans not having unprotected sex or sharing needles, and by properly screening donated blood. But that didn't make the virus less terrifying, or the people who carried it. Some hospital staff treated AIDS patients while wearing full protective gear and never touching them. One mother told me that her young son, dying of AIDS after a contaminated transfusion, gazed at the hospital staff and said to them, "Are you wearing a condom too?" Most people working with HIV or AIDS patients then call them dark days but also times filled with love. My aunty Barbara, chatting with me in her Ontario apartment in the slumbering hours of the afternoon, told me of her time working as a nurse on Ward 5B of San Francisco General, the earliest and most famous AIDS ward. It was horrible, heartbreaking, but also a hoot. "In three weeks sixty-three people died. But there was such fun and everyone wanted to work there. At that point everyone had Kaposi's sarcoma. There was one man who opened his shirt and he was covered and he said, look how ugly I am." Barbara gave him a big hug. On Sundays, Barbara says, a female celebrity came to sing at the piano and always brought tubs of ice cream. Ice cream is easy for a painful throat to swallow. But she was telling me this in the dead time of the afternoon, when sleep is pulling, and she couldn't remember the woman's name. Big hair, she says. A singer. Later, I investigate. Elizabeth Taylor.

Campaigns tried to stop AIDS by appealing to people to use

condoms and warning sternly about injecting drugs and sharing needles. Safe sex was a new concept that had to be incised into people's minds and habits like the lettering on the tombstone. In some of the world, it worked. Public behavior changed. Condoms became conversational along with safe sex. The virus in northern Europe and North America was tamed, contained in small groups and niche populations. There were mistakes: an emphasis on the virus being spread by "bodily fluids," while true, was interpreted too broadly. That emphasis meant that some people still think HIV can be caught from toilet seats (it can't), spitting (no), or eye contact (no comment). The state of Texas has laws that have been used to jail people for transmitting HIV by spitting, something that is biologically implausible.[9]

In the mid-nineties, there was a revolution: antiretroviral drugs were developed that successfully interfered with HIV's ability to replicate. When HIV-positive pregnant women in Thailand were given AZT, an early antiretroviral, the rate at which they infected their children was slashed by an astonishing 50 percent.[10] In 1996, a Canadian HIV specialist named Julio Montaner presented research at a Vancouver AIDS conference on his experiments with combining several antiretroviral drugs. Montaner, an Argentinian, had got the idea from his father, an infectious disease specialist. If a single drug wasn't working, try more than one. "We came up with data," he tells me over the phone from Vancouver, "that demonstrated for the first time, to our surprise, and I'll be perfectly candid, unexpectedly, that if three drugs given by themselves were insufficient, if they were given together they could actually shut down the replication of the virus and keep it at that level." Another trial, using three different drugs, reported the same effect. This was remarkable, stunning. It was one of the most transformative breakthroughs ever. Mortality rates dropped. Now, a person living with HIV who takes his or her medication can suppress the virus to the point that it is undetectable. A person with HIV who takes his or her medication is considered uninfectious and can have a normal life span. In many ways, we are now living in the good-news era of HIV and AIDS. In 2000, fewer than one million people with HIV were on antiretroviral therapy: now it's 18.2 million. In 2000, there were 490,000 new HIV infections in children; in 2015, there were 150,000.[11]

I am not supposed to be scared of HIV anymore. In 2017, the US government changed its website AIDS.gov to HIV.gov, to reflect the fact that now hardly any Americans die from AIDS. Formal entry and travel restrictions for people with HIV have relaxed in most countries, though you will have trouble in Bahrain, Iran, Iraq, and half a dozen others, and you are unlikely to be allowed to settle permanently in Canada and some federal states in Germany. The United States removed entry restrictions in 2010, fourteen years after HAART (highly active anti-retroviral therapy) was developed.[12] Not only can people living with HIV enter the United States "like anyone else," according to the Global Database on HIV-specific travel and residence restrictions, but the United States' visa waiver program no longer considers HIV a communicable disease. This is strange: HIV remains a communicable disease, but now a treatable one. The rule change issued by the Centers for Disease Control and Prevention was more accurate in its language. Its guidelines to physicians about what they must test for in immigrants and refugees read that HIV is no longer "a communicable disease of public health significance."[13] In some places.

The good news in HIV is still followed by the bad. In 2015, there were 36.7 million people living with HIV and 2.1 million new infections. There were 1.1 million AIDS-related deaths in 2015, no great advance on 1.5 million in 2000.[14] The word *plateau* is now used in HIV/AIDS circles: infections have plateaued; AIDS deaths have plateaued. A better topographical analogy is that HIV has forked. In richer countries, it is again a disease of the niche: men who have sex with men, prostitutes, drug users. It is a chronic, manageable disease with a normal life span. Outside the richer countries of the world, HIV has a different face. The Philippines has the fastest-growing HIV rate in Asia (tied with Afghanistan), mostly in gay or bisexual young men. HIV rates across the Middle East and North Africa are rising, fueled by taboo and silence. AIDS-related deaths in eastern Europe have risen by 50 percent.[15]

But South Africa is a puzzle. Other countries in eastern and sub-Saharan Africa are HIV success stories, with dropping rates of infection. South Africa accounts for more than a third of new HIV infections in the southern continent:[16] a quarter of a million in 2016.[17]

———

They are telling me about the levels. The levels vary according to whom you ask, but the most common version is this: Level 1 gets you lunch money. Level 2 is airtime. Level 3 may be a fancy set of extensions, a Brazilian or a Peruvian, which are worth 2,500 or 3,500 rand ($209 or $292). Level 3 may also be a new smartphone. Level 4 is a trip to Durban. And hardly anyone gets to Level 5: that's a Benz or a flight to Dubai. Each level requires a woman or girl to pay for it with sex: that is as sure a fact as that they exist. People talk about them as if they are science and fact, even these schoolgirls at COSAT (Center of Science and Technology), a school in the fancier part of Khayelitsha. Fancy in Khayelitsha means some houses are brick and the roads are wider and you won't see the locked "bucket toilets"—which are what they sound like—that you see in poorer areas. This is a very nice school and has clearly many international supporters and friends. The front yard flower beds are planted with succulents; the main gate is heavy and kept locked; and there is a wellness center staffed by a full-time counselor. The children here have earned their places and are the educational elite of the township. They will go on to further education. They will escape. They talk to me with the tones of youth, kindly yet patronizing. They are adjusting for this white woman who does not know how their world works, who has clearly never been a schoolgirl but was always a person with a notebook, being inquisitive, intrusive, in their faces, disrupting their lunch hour, asking about blessers.

Blesser. The word arose from young women on social media saying they were "blessed" by boyfriends who bought them things. *Blesser* is translated differently depending on whom I ask. For a woman who runs an organization helping women to exit prostitution, it is prostitution. It is girls getting compensated by older men for sex. For people working in public health, it is "transactional sex." For Dr. Genine Josias, who runs Khayelitsha's sexual assault center Thuthuzela, it is rape and blessers are rapists. "The girls are underage. They cannot consent. It is rape." For anyone with any sense, it is a spectacularly efficient way for HIV to spread.

For these schoolgirls, sitting around a table during lunch break, chatting, it is part of their life, like airtime and what they call "free Facebook" (a text-only version of Facebook that doesn't consume data). They talk like they are above such things as blessing, but they know girls at school who have a blesser. One tells me about a school friend who is dropped off at school by a man in a luxury car. "First time I thought it was her father. When they leave, they hug each other. You can see when you hug a father and when you hug a boyfriend." It's disgusting, they say, with fantastic contempt. They say, he is old! with emphatic scorn. They disapprove, these girls who are smart and the elite, who will escape. They tell me the rules of blessing: "Don't have a sugar daddy, if you're not ready to give them sugar. Because obviously the blessers or sugar daddies want something in return and they don't want to call it sex so they call it sugar." In return, the girls get wonderful things: a Peruvian or Brazilian weave, that costs 2,500 rand ($210), a price I would gasp at because my hair is short or my budgeting is thrifty. They get Zara clothes or trips to the malls of the waterfront. One girl had the latest iPhone and then had an even more advanced one couriered to her door. "Couriered!" says the schoolgirl telling this story, with wonder. They act outraged but yet they are indulgent. This is also the attitude of much of South African media to blessing. A blesser named Serge becomes nationally famous when he tells a TV program about his arrangements with young women. He is thought glamorous, not dangerous. There is rapt reporting of an online service named Blesserfinder, defined by Facebook as a "lifestyle service," if the lifestyle is one that encourages the spread of HIV.

I don't feel indulgent of blessers who come to fish for poor girls outside schools and colleges. There is nothing glamorous about the ones who refuse to use condoms, because how can a young woman who wants what you can give resist you? Blesser, blessee: these are new names for something that exists anywhere a young woman exchanges her body for something, but that something is never power. An MSF driver, a local man, gives me a Khayelitsha version of how things work. Level 1 blessers live in shacks. They provide sundries. Level 2s live in the government-provided brick houses, the ones people often sell for cash, then go back to the shacks. Level 3s stay in a township away from here, 4s are even closer to town, and the 5s are probably in Durban.

He is laughing when he tells me this. Most people talk about blessing with humor. I doubt Salim Abdool Karim would. At the Durban AIDS conference in 2016, Karim, professor of epidemiology at the Centre for Aids Programme of Research in South Africa, presented research undertaken in KwaZulu-Natal that mapped the movement of the virus through a population of sixteen hundred people by tracking the similarity of certain genetic sequences. They found that HIV was thriving on blesser behavior: young women aged around twenty were being infected by older men aged thirty and above. The young women then passed this on to long-term partners of their own age.[18]

Another pathway that suits HIV: young women with blessers will also have a boyfriend of their own age. To have multiple partners is unquestioned. I read that this behavioral norm has grown from history and apartheid. Men forced to work away from home in hostels and mines developed the practice of having concurrent partners. (I think: they were forced to work but not be wanton.) The practice became the norm, enough that the COSAT schoolgirls think there is nothing wrong with having more than one boyfriend and are baffled that I do. They pretend to be disgusted at the idea they would sleep with older men but they are happy to admit to what they call a "backup boyfriend."

One girl explains this to me with polite pity and with an appropriate analogy for a writer. "Let's say you are dating. Boys like to cheat, so girls need a backup plan. So when you drop a book it goes boom and it's going to hurt. When you have a backup boyfriend, you will have a backup. Catch the book before it falls." Catch the book, protect your heart. I'm disconcerted by both the poetry she has wrapped around the problem and how the girl speaks with the flatness of the obvious, as if she is telling me the sun is in the sky.

Backup boyfriends have other names, such as armpit. The woman who tells me this is Nelly, who runs a new MSF program that provides pre-exposure prophylaxis (PrEP) to young women. PrEP involves giving antiretrovirals before infection, to prevent infection. "Armpit sounds better in the vernacular," she says, and spells the Zulu word, *Ikhwa-pha*. I wonder how a boyfriend called an armpit could sound good in any language, but the reasoning is reasonable. An armpit is hidden; a second boyfriend is hidden. They also call it a clutch bag, says Dr. Genine

Josias. "You carry a handbag but sometimes you need a clutch bag, too." Josias knows these things because her daughters tell her but also because of her job. She has worked in Khayelitsha since 2004, when she joined MSF. "We used to take histories from patients," says Josias. "Some patients became HIV-positive because they were raped and they never accessed post-exposure prophylaxis." (PEP consists of antiretrovirals given after possible exposure.) This gap in prevention, this rape-size gap, led to the establishment of a dedicated sexual assault center in 2005. It was called Simelela, meaning "to lean on someone." Simelela was taken over by government in 2009 and renamed Thuthuzela, a Xhosa word that means "to comfort."

Thuthuzela has moved to Khayelitsha's fancy new district hospital (it used to be at the hospital in Site B, where MSF founded its HIV clinic). The new hospital is 325 yards away from the MSF office, but I am not allowed to walk there unaccompanied for safety reasons. Violence here is as common as dust. A staff member told me of acquaintances who have been carjacked, raped, mugged, and robbed at home over the last few months. The robbers also cleared out the fridge. Thuthuzela is meant to offer safety for traumatized people, as well as medical examinations, counseling, statement taking, and forensic investigations. It is open all day, every day, all year, every year. Even so, the number of clients has dropped since the clinic moved: women don't want to deal with the screening by security guards who want to know their medical problems. Rape rates haven't dropped. Josias has worked in her field for decades and seen everything. Children raped. Babies raped. Young women. So many young women. Her doctor colleagues raped by doctor colleagues. Rapes so violent the victim must be examined under anesthesia. Nothing surprises her anymore. But at least people understand better about HIV. "They know they need to do something after someone has been raped. They come here and say, I'm not worried about the police or the court, I'm worried about HIV."

How can young women fight off HIV as well as male violence? Women in Rwanda, Tanzania, and South Africa who have experienced violence were calculated to be three times as vulnerable to getting HIV as women who have not. Another study estimated that a woman experiencing intimate partner violence has a 1.5 higher risk of being infected.[19] Josias is worried. Money is not there for marketing and outreach like it used to be. The children they spent years sensitizing have grown up now. There are new children coming, and they know less: they don't know there is Thuthuzela; they don't know there is post-exposure prophylaxis; they don't know their safety depends on active effort, not luck. They don't know their risk factors are shaped just like them, that—in the words of Dr. Josias—here (and not only here), "if you have breasts and a vagina, you are unsafe."

HIV is old. A retrovirus. The retro does not mean old-fashioned but that its creation process is back-to-front. HIV has no DNA, only two strands of RNA, a single strand of ribonucleic acid that existed before DNA. Some scientists believe an RNA world preceded our DNA one. Now, all life-forms are based on DNA except RNA viruses. A retrovirus needs to steal from a DNA cell to survive and to replicate. Inside the target cell, the RNA strands are transformed into DNA by the enzyme reverse transcriptase. Somehow the DNA breaches the cell nucleus by penetrating pores on its surface and splices its DNA into the DNA of the cell. Now it can replicate. Now the T-cell has become a virus factory. New immature virions are created and head to the outer membrane of the T-cell, where they pass through—this is "budding"—and emerge as new viral particles. Again, again, again.

This is a successful infection. It begins with a cloud or a swarm of virions being transmitted. But it is not a foregone conclusion. HIV can be transmitted many times without taking root in the body. Many viral particles will be "dead-end" and cause no harm. Many will be dealt with by the immune system. The trouble with HIV is that it needs to infect only one cell. From one infected cell, a person can develop AIDS.

HIV is flexible. Once inside the body, it can proceed in several ways to infection. Some virions may infect T-cells they encounter in the vagina or anus or blood. Some may cross through the mucous membranes that line these areas into the bloodstream and infect cells there. Some may travel on their own to a lymph node. Others may hitch, often on dendritic cells, another type of leukocyte. I search for images of these and find dreamy shapes with floral folds. A search engine delivers an image of a red rose among the cells and it doesn't jar. HIV may infect some dendritic cells, but usually it uses them to get where it wants to be. In scientific language, dendritic cells "present" the virus to the lymph node, a debutante at a ball.

HIV is shifty. The immune system has plenty of ways to defeat pathogens. It can eat them or tag them with antibodies that will recognize them when they return. It can stick them to the surfaces of cells so that they cannot replicate or bud. But for all this to work, immune cells have to recognize the enemy. In one out of two replications, HIV makes random errors. It produces a particle with slightly different proteins or a slightly different genome and becomes unrecognizable. Eventually, in about two weeks, the immune system will learn to recognize the new versions. But by then the virus will have changed again. It's not that our immune system doesn't work. It works too late.

HIV is both quick and slow. Once HIV has reached a lymph node, it is, says a virologist, "game over." In a lymph node, it will find a surfeit of T-cells. For the first couple of weeks of a successful infection—a stage known as primary or acute infection—billions of new virions are released into the bloodstream every day. Viral activity is intense and the viral load climbs rapidly. The infected person notices none of it. After one to two weeks, the body begins to produce antibodies to deal with the virus. This is called seroconversion, and it produces a seroconversion response that often feels like flu. Fever, aching joints, diarrhea, or malaria-like symptoms are common and can last several weeks or months. The virus plateaus and the T-cell count recovers. But HIV is known as a "lentivirus" because it takes its time. It is only in abeyance. After those first symptoms, there will be hardly any for years. A person feels uninfected and unaffected but their viral load—the number of virions in their body—is high enough to be infectious. The biggest driver for viral evolution is to find a new host and the best hunters are the stealthy ones.

My MSF contact sends me a text. I had asked to meet some HIV-positive young people, but it has proved difficult to get access to a youth club at Site C clinic run by local authorities with MSF support. So she has arranged for some club members to visit me at MSF instead. She texts,

"They've asked about incentives."

I text back my indignation.

"What? I don't pay for interviews. I never pay for interviews."

Another text arrives.

"I think they mean biscuits."

I arrive laden with Cokes, muffins, fruit chutney–flavored popcorn, crisps, smiley-faced biscuits. The smiley faces are because I didn't know how old the youth were going to be, only that they were HIV-positive, and that they were doing me a big favor by coming to the MSF offices to meet a random writer who would ask about the intimacies of their status, lives, and feelings in return for junk food. All three attend a youth club for HIV-positive young people that is meant to give them hope. The club meets only once a month but it is a way for them to be together. Nothing heavy: they play pool, they have fun, they swap tips on how to engage with new partners when you're coming into a relationship bringing HIV with you. It sounds like every youth club ever, but it is part of MSF's groundbreaking adherence programs. The factors fueling HIV in South Africa are complex and many, but one of the most worrying is treatment fatigue. Even people who come to the clinic and get ARVs may not keep taking them. Adults are getting tired of taking drugs, especially when they feel well. Children who grew up with HIV who become adolescents do what teenagers do: they rebel. They stop taking their drugs. It is horribly easy, as MSF pediatrician Ann Moore explains to me later, to create resistance. To protect against resistance, anyone on ARVs must take them 95 percent of the time. I ask her to translate this into doses. Does that mean missing a few doses? No, she says. It means you can risk missing only one a month. This is chilling. Who has never cut short a course of antibiotics or skipped some tablets? Missing one dose a week is a reduction of 12 percent, so a definite danger zone. She tries to explain how this works biologically—it has to do with how the virus mutates—but if I find it taxing, I wonder how they convey that to a teenager. Probably this is why the message to the newly HIV-positive can perhaps seem blunt: take your ARVs, every day, always and forever.

The first two young people to arrive are a stylish young man with a buzz-cut Afro and a young woman in a tight leather jacket and gold jewelry. Neither wants to use a real name, so they choose new ones on the spot. Lisa, says the young woman. Eric, says her friend. She is twenty-five, he is twenty-three. She is originally from the eastern Cape, on the far side of the country and continent, facing a different ocean. But her family moved to Khayelitsha and she had no say in it. Eric was born near the Waterfront, a ritzy shopping district in Cape Town, but lives in Site C. He also has family in the eastern Cape and visits them now and again for family occasions such as a ritual circumcision.

They look at the food but don't touch it. Eric has just learned that people who are on ARVs are at a higher risk of diabetes. I begin to apologize. I should have brought fresh fruit, not these crap carbohydrates. But Eric laughs and opens his jacket. Look, he says, pulling a small packet out of his shirt pocket: "I always have Smarties with me." For some reason ARVs make them want spicy food, chocolate, all the bad good things. And they want them all the time. An MSF doctor tries to explain this to me: it's about calories not working as they should and about what antiretrovirals do to the metabolism. What Eric and Lisa know is that ARVs make you really, really hungry.

Both found out they were positive in 2015. I ask them how they were infected, but their answers are foggy. "I don't know," says Lisa, "because I found out that several people in my family had HIV and never told us. They died. So maybe you know, if someone in the family has a cut, and you rush to help, and . . ." Eric doesn't offer a theory of how he got infected, so I conclude it was sex without condomization. They were both profoundly shocked. You have to be, says Eric, who got home after his diagnosis and sat on his bed, stunned. "I was thinking, how did I get this disease? I questioned myself, and what did I do wrong, is God somehow trying to punish me?" He found himself listening to "Hello" by Adele, a song he had downloaded only the day before. He sings the chorus. "Hello, from the other side." "I remember thinking, why am I playing this song? It's not like I'm dying or anything, there are a lot of people that are HIV-positive and living their lives. I knew that."

Of course there was shock. Even if you know you are at risk, there will be shock. And most people are very bad at admitting they are at risk. An MSF counselor tells me that she used to ask her clients to assess their own risk of getting HIV. Each one said "low," a statistical improbability. "So I would say, do you use condoms? And they would say yes. I would say, 'all the time?' and they would say 'no.'"

Lisa says that HIV has made her stronger. Eric says HIV was a blessing in disguise. They often use the phrase "life after HIV," and I hear the echo of counselors in the phrases, of efforts to induce positive thinking. The drugs were difficult at first: there was nausea and fatigue. When Eric took his first dose of medication, he remembers lying in bed and the bed was rotating. They both laugh. Free hallucinogenics. Since then, they have had no trouble with their fixed dose combination: it is one pill taken at roughly the same time every day. This is a massive advance in treatment to anyone with HIV who remembers taking eight pills at all times of day, but at the same time, and some with food and some without, and some with awful side effects. Dosing was much stricter. People tried to find ways to remember, associating their pills with a favorite soap opera. Now the drugs are more robust so there is more leeway: if they are meant to take it at nine and they forget until twelve, that's fine.

It's a small pill, easy to swallow. No big deal. It's just life with HIV. But despite the confidence and clarity of Eric and Lisa, they hide their status from most people in their lives and definitely from strangers. Lisa has told only her boyfriend and her sister, who is also HIV-positive. Eric told his cousin, but only after she came to him to tell him she too was positive. They hide their ARVs from friends and family. If they are away from their homes and need to take the pill, they scrunch it in a tissue and take it in the bathroom, like contraband.

After a while, a third young man joins us. He wants to give his real name, but it's distinctive, so I'll call him Rollo. He is eighteen and the newest infected: he found out only the year before. Yet he is the most open about his status. "I've told my whole family, my friends. Everyone." The first person he told was his mother, who then revealed that she was HIV-positive and had been for years. "She didn't tell us because she thought we were too young." His mother, healthy and strong, was the best way for him to learn something that public health people struggle to convey: you can be healthy with HIV, despite the huge appetite.

Rollo's brother works for MSF in the male services sector. I don't know his status and I don't ask, but when he goes and does outreach, trying to get men to test, he tells them he is HIV-positive. It's what activists used to do in the bad days. It is to shock them. You have it? A healthy man like you? In the entrance hall of the Khayelitsha office is a photograph of Nelson Mandela wearing a white T-shirt. In black letters, it reads HIV-POSITIVE. It had the same effect. You have it? Madiba? You?

HIV is hard to get. You will not hear this in public health messaging. Rates of transmission depend on many factors from geographical location to your preferred sexual position to how much viral load the infectious person has. Judging risk is a complicated science, so numbers vary widely. The chances a woman may pass HIV to a male sexual partner, for example, are 1 in 700 to 3,000 exposures. From a man to a woman, the risk is 1 in 200 to 2,000 exposures. Needle sharing is 1 in 150. This may sound generous, that you can have sex 2,000 times in safety. Many of the 2,000 times, you may be invaded by HIV but not infected. It is, says Sam Wilson of the University of Glasgow's Centre for Virus Research, "extremely counterintuitive to think of such a successful pathogen being transmitted so inefficiently."[20] But these big numbers can lull and trick: 1 in 2,000 exposures can mean you are infected on the first exposure or sexual encounter. It takes only one virion.

HIV is fragile compared to other viruses. It can be destroyed with heat or bleach, though it can survive outside the body in dried blood or air. (Its survival chances depend on temperature, the medium, how much virus is in the medium, and other factors.) HIV will not be transmitted by a doorknob or toilet seat or handshake or tears or a kiss. If you swallow HIV, the chances of infection are remote unless you have an ulcer or tear that lets it into the blood. HIV's strength is not form but function. It adapts. It travels in sexual fluids, blood, and breast milk because they will get it where it needs to be. It is fragile outside the body because it has no need to be otherwise. It has found a lifestyle and no need to change.

Blood is a sure medium for HIV: a needle puncture or a wound or a cut is a fast way into the bloodstream, where HIV can infect the blood cells or in which it can travel in freely to find others elsewhere. Mucous membranes are another good route. These line the anus, vagina, urethra, and the foreskin and are packed with white blood cells, stationed there to guard against infections that can arrive through the body's openings. The cells are good hunting for virions that target them. HIV can slip through membrane cells to the blood. (Cells in the anal membrane are wider spaced than in the vaginal one, which is why anal sex is riskier.) The best way to get to the membranes is to travel in a fluid that encounters them: semen, pre-semen, vaginal and anal fluids. The best way for HIV to infect humans is for humans to carry on having unprotected sex and sharing needles, which they do.

"They were melting. Literally melting." Eric Goemaere is recalling the early AIDS days. He is the HIV referent, or expert, for MSF, and a genial man, his accent that of a Francophone who has spent decades in English-speaking environments but who retains the audible charm of French. I'm a little awed by meeting him because I know his history. I'm also a little awed by the MSF office in Cape Town, with its statement rattan hanging chairs looking out over the city and the port. "I know," says the MSF press officer. "It's a bit Google." It is an unexpected sight in the context of an organization that is known for its ferocious efficacy and fierce neutrality. Tenacity and glamorous humanitarian doctors in the most difficult working environments such as war zones and disaster, still there when everyone else has left: this is MSF. Despite the chairs, this Cape Town office—the rest of it suitably plain—is still confronting disaster, just one that has better transport links.

"MSF," reads one pamphlet, "does not take political sides or intervene according to the demands of governments or warring parties." On the next page the heading is *Témoignage*. Bearing witness. Where MSF teams "witness gross injustice, violations of basic rights to access care or breaches of humanitarian principles, MSF considers it has a responsibility to raise these concerns in the most appropriate way by addressing governments, the United Nations, other international bodies, the general public and the media." If that's the case, then Eric Goemaere has been one of the best witnesses—one who is vocal—in MSF history.

In the 1990s, when he was working in Chad and other African countries, he remembers people arriving in clinics horrifically sick. Their skin had strange black spots. They were so very thin, yet they had neither diarrhea nor TB. Their life was leaching from them. Then HIV was discovered, and people could be tested, and the numbers were stunning. The virus had made a plague. Yet MSF, this notoriously intrepid and brave association of adventurous doctors without borders, stayed away from HIV. "In the 90s," wrote its chief executive Bernard Pécoul, "we just rejected the possibility of doing anything."[21] Perhaps this was understandable. Doctors treat with medicine and they had none to treat with. Ann Moore, the pediatrician, dates her time at MSF by saying she arrived "before we had any treatment." Her daughter was training as a doctor in Durban at the time and nearly gave up medicine. "She said to me, Mum, on a medical ward at night I'll sign four death certificates for women my age and I can't take it. I mean, [it was] just terrible."

Without any available drugs, MSF would only do prevention. The position was not popular. Many staff were frustrated the organization would not try to do more. In 1995, MSF's newsletter ran a headline describing itself with scorn as a WHITE, HETEROSEXUAL SERONEGATIVE ORGANIZATION."[22] The same year, Goemaere arrived in Johannesburg hoping to set up a clinic in a township in Alexandra. By then, South Africa had the highest prevalence of HIV in the world and he assumed that its government would be desperate to find a solution and a treatment. He expected open arms and he got a fist instead. Because this was the time of Comrade Undertaker and Doctor Beetroot. Comrade Undertaker was President Thabo Mbeki. "Comrade" because of his stern demeanor and the Communist leanings of the African National Congress (ANC), the governing party of South Africa, and "Undertaker" because Harvard academics have calculated that Mbeki's refusal to believe that HIV caused AIDS and to provide antiretrovirals has led to the deaths of at least three hundred thousand South Africans. Doctor Beetroot was Dr. Manto Tshabalala-Msimang, Mbeki's health minister, a woman who told South Africans dying in their hundreds of thousands—including the president's spokesperson—that antiretrovirals were toxic and that they would do better to treat HIV with diet. At the huge AIDS conference in Durban in 2000, the government stall displayed beetroot, garlic, and African potatoes because this is what Tshabalala-Msimang proposed to treat HIV.

It was a shock. Mbeki as vice president had shown no signs that he believed HIV was not related to AIDS, or that ARVs were toxic, or that AIDS was nothing more than a disease of poverty. In 1999 he became president, and—so the tale goes—was on the Internet late one night and found a quack denialist site and made it truth. The government began to support the use of multivitamins and potions to treat HIV.

Even now, it is baffling. Mbeki is known to be intelligent. He was known to understand economics. But he backed, for example, a truck driver who began selling two bottles of potion that he called *uBehjane*. One bottle had a blue cap, the other a white cap. One controlled the CD4 count, the other the viral load. The government promoted *uBehjane*, and later the equally implausible and dangerous "treatment" peddled by Dr. Matthias Rath, a German physician selling multivitamins who still sells multivitamins.[23] The Treatment Action Campaign, a group of activists fighting to get the government to provide ARVs, refers to him as "the charlatan Matthias Rath." The vacuum of ignorance has plenty of room for quackery. Any pandemic unleashes the same mixture of terror and ingenuity. In nineteenth-century Ireland, to beat the new, awful cholera, a bishop devised a pyramid scheme, where Irishmen had to burn "blessed turf," then deliver it, by running, to seven other dwellings. This would save them from the cholera, and exhaust many Irishmen. One ran thirty miles to find a house that had not yet received the blessed sod.[24]

This was the environment into which Goemaere tried to set up a project. It failed. He didn't like South Africa, having been arrested during apartheid for refusing to sit in a whites-only waiting room. He was preparing to leave when he was telephoned by Zackie Achmat of TAC. Achmat invited him to Khayelitsha. He accepted and made his way to the hospital clinic at Site B where he was made thoroughly unwelcome. The nurses didn't want to talk to him. Perhaps they were denialists; perhaps they feared the denialists who controlled their jobs.

But Goemaere followed a limpet strategy and simply hung around. He found that in fact the staff were terrified. They knew nothing about HIV or transmission, despite the fact that they were running a limited PCMT—prevention of mother-to-child transmission—program using AZT that had been grudgingly set up by the government. The staff eventually admitted that they thought MSF's presence attracted HIV. To this, Goemaere pointed out something that was obvious to him: of the people in the waiting room, at least a third were HIV-positive. HIV was already in the clinic, in droves, whether MSF was there or not. This was his supposition. The previous year only 450 people in Khayelitsha had been tested. But by now people were dying of AIDS so frequently, the cemeteries had cut funeral services down to half an hour each, unthinkable in a culture where funerals were usually long, loud, important events.

For three months, Goemaere forced himself to go to the clinic. "I was pushing the door, looking around, trying to speak to the different nurses." They said, "No, no Eric don't come here, we don't want to get infected by those people." Eventually he met the senior health official for the Western Cape government, one of only two provincial governments not under the control of the ruling ANC Party. Goemaere and MSF were allowed to set up a small program to treat people for HIV beyond the mother-to-child project. It was to be known as "private research" and Goemaere was given a small room in the Site B hospital from which to run his clinic. The head nurse told him no one would turn up. "She said, 'Eric, in our culture, no one wants to be identified as HIV-positive.' In one month, there were three hundred and fifty people queuing." They had to build a prefab out back to house them. Some of the people in the queue were in stretchers and wheelbarrows. Some died in the waiting room.

For a while, this limited stealth project was the only possibility. AZT cost $10,000 per person per year, an impossible price for most countries. Anyway, says Goemaere, "treatment was not available in developing countries, as it was considered too sophisticated, as it still is more or less for cancer. Too expensive, too sophisticated." Shack dwellers didn't merit medicine and wouldn't know how to take it. In 2001, Andrew Natsios of the US Agency for International Development told a US government committee that rural Africans would not be able to take ARVs because they "do not know what watches and clocks are. They use the sun."[25] (Later studies showed that sub-Saharan Africans were better at taking their medicine than Americans.)[26] Strict patents prevented the use of generics. TAC launched a Defiance Campaign against Patient Abuse and AIDS Profiteering, and activists flew to Thailand to bring back cheap generic antiretrovirals in their luggage, their hand baggage, and their pockets. TAC launched a court case against the pharmaceutical industry's punitive pricing of AIDS drugs, and won.[27]

The power of the ARVs seemed magical. "The Lazarus effect," says Goemaere. "It was amazing. They were resurrecting." A person who was bedridden, in what Goemaere calls "a half-coma," who could not walk or talk, was up and walking and talking in two months. But MSF did not have enough ARVs for everybody. There were strict criteria, and this was difficult for doctors who felt they were dispensing life or death. It took until 2004 for South Africa's government to change its policy on ARVs. Now, they are freely dispensed, though half of the seven million HIV-positive South Africans do not take them. That number worries Goemaere, still here in South Africa, still working on HIV eighteen years after he arrived at Site B and made himself first a pest and then a savior. In that gap are so many factors, and all of them, says Goemaere, make him "very nervous."

HIV is old. It is an ancient enemy, but for much of history it stayed safely in Old World primates. It was an insignificant simian virus. The tale has been told: that there was a man who hunted a chimpanzee, probably in the early 1920s, probably in Cameroon, and possibly he cut himself while preparing the chimp meat. Then, a virus that had infected chimpanzees for millions of years crossed over into the hunter's blood. It was simian immunodeficiency virus, chimp version, shortened to SIV-cpz. This was not the first crossover of SIV-cpz into a human nor the first "spillover" of a zoonotic virus into humankind. SIVs have crossed into humans countless times, along with viruses from other animals and invertebrates. The crossovers happened often enough that the virus adapted to reproduce inside human cells. The extent of its adaptations is debated, but one adaptation was essential to its success. T-cells have a "restriction factor" called tetherin. This makes the surface of the cell sticky so that invaders can't easily land on it. But when the chimp virus became a human one, it developed a protein that neutralized tetherin. When researchers gave the human virus to macaque monkeys, the protein changed again, the better to infect macaque cells. HIV is not even considered alive because it has no DNA, and many scientists loathe anthropomorphizing a virus. But it is difficult not to see intent and cunning in this ancient thing that is not living until it borrows life from us, then takes it.

SIV-cpz became HIV-1 Group M, the source of the deadliest pandemic in history. There are Groups O, N, and P, presumed to be from other crossovers, but they have never traveled outside West Africa. There is HIV-2, a distinct virus formed from an SIV found in sooty mangabey monkeys. Only HIV-1 Group M found the right circumstances in which to thrive. These circumstances were how humans and their bodies behave. Both suit HIV perfectly. We have unprotected sex, we take drugs, and we are mobile. In the eighty years that HIV spread unnoticed—a virologist uses the term percolated—colonialism pushed people from Africa to Haiti and then to North America and Europe. With colonialism, people moved more: into rubber forests as porters, between Africa and North America and Europe as businessmen, medics. HIV went with them.

"Look at me!" says Lisa. "I'm healthy!" She does sit-ups and push-ups every day to fight her increased appetite, but she is well. She looks great. Lisa, Eric, and Rollo talk of the fact that they know there is life after HIV, that they plan to have children, that they know there are very few barriers to what they want to do. Rollo is getting married soon, then wants to train to be either an engineer or an HIV counselor. Yes, says Eric, who is studying management, I'd like to do that. These choices represent gratitude. Lisa is more adrift: she wants to study but can't afford the fees. All of them expect to have long and normal lives. Why not? Then I ask them about friends. "I don't have friends," says Eric.

"I'm a loner." I don't have friends, says Lisa. Just my boyfriend. Rollo has told his friends of his status, but no one at school. "You have to be able to trust that they will keep your secret." Their conversation with me is less guarded than their lives. For months, Eric resisted going to the youth club, in case he met anyone he knew there. This stigma is a huge problem in HIV care and prevention.

Yet these three young people are success stories. All were put on ARVs within two weeks of their diagnosis. At diagnosis, Rollo had 300,000 copies of viral RNA per milliliter of blood. His viral load. Now, he says with a smile, "it's 28. I've worked hard!" Under 50 is undetectable and—as far as we yet know—untransmissible. Eric, Lisa, and Rollo are what MSF wants to happen to anyone with HIV. But they also represent a failure. These three infections. They were not necessary. They represent what Eric Goemaere calls "a very dangerous moment."

Jonathan Bernheimer is another pediatrician working with MSF. He is American, and seems stern, but the sternness is just a coating for passion. He works with children and adolescents with HIV, and there is much to be passionate about. Eric, Lisa, and Rollo had shown me their drugs with pride. One pill, to be taken once a day, so much better and easier to take. But not for children or adolescents. Children have to take solutions that in Bernheimer's words "taste like crap. I mean, really terrible." The solution version of lopinavir/ritonavir, a common ARV, "is one of the worst-tasting things you'll ever taste in your life. It's incredibly bitter." A drug company did create a pellet formulation and was going to test it in South Africa, but production was abandoned. It said the manufacturing process was too flammable. Bernheimer is sympathetic: it is extremely difficult to produce a child-appropriate drug that works how you want to it work, breaking down in the liver not the duodenum, tasting palatable. Taste masking might affect the viability of the drug. "Look, I get it," he says. "It's not easy to make these drugs." But that human beings going through one of the most turbulent periods of their life have the most complicated, disgusting drugs: this fact, he says, is "maddening." He would like to see better drugs, but in five years not one drug rep has come to Khayelitsha with suggestions or queries.

So many factors conspire to push young people off their ARVs. Sometimes if they were born with HIV, they were never told they had it. Grandmothers, who used to be the caregivers of HIV orphans until ARVs began saving the mothers, would tell children they had asthma. Were they preserving them from stigma? Maybe. "Way back," says Xoliswa, a lay counselor who works in MSF's container clinic and who is HIV-positive, "there was no treatment and stuff like that, people got pregnant, they gave birth to their children, and they died without disclosing their statuses in their families. But the children took their medication. So maybe the grandmother has told the child that they have asthma and have to take the medication every day. Now those kids are fifteen, sixteen, it's very difficult now for the grandmothers to disclose to them. Now the questions with these kids are, what do I say to my boyfriend. For me, they start dating, they meet at school, they start sleeping [with each other]. We will have new infections in South Africa. It will never go away."

If a teenager is a teenager, they may stop taking their drugs. "Adolescence," says Bernheimer, "is the perfect storm for this to go badly. If they don't necessarily feel sick, they think they are invincible so they are going to go off their medicine. We can tell them until we are blue in the face that they are going to die and that might work for a little while but oftentimes it doesn't." The way the virus works in younger people makes matters worse. An adult who goes off medication, says Bernheimer, will "kind of get sick, kind of get sick, get sick, get sick, and then die. Children are fine, fine, fine, fine, fine, fine, fine and adolescents fine, fine, fine, fine, fine, fine, fine, and then they fall off a cliff and they get really sick really quickly." There is very little time to pull them back. The mortality rate for all other age groups with HIV has decreased by about 30 percent over the last ten, fifteen years, he says. "But in the same time period, the mortality rate for adolescents has increased by about fifty percent."

Sometimes barriers to staying on medication come in the shape of fear or husbands or fists. In a report titled "Hidden in the Mealie Meal," Human Rights Watch reported on conditions for women in Zambia who dared not disclose their status to their partners and so had strategies to hide their ARVs. Some dug holes in the ground or flowerpots.

They hid them in boxes of headache pills, or in suitcases under the bed. Some buried the pills in the bag that held the family mealie meal (a maize-like staple food). Some women, wrote HRW, said they missed ARV "doses as a result of all this subterfuge."[28]

Hence the adherence clinics. Stern warnings don't work: any parent knows this. Nags to teens: like oil on water. So the clinics try to do it differently, by setting up buddy systems so a teen is encouraged by a friend to take medication. They have career days, to get teens excited about the future, to encourage them to take ARVs so they have a future. There are sports days, activities. The concept sounds nothing more complicated than any youth club on any grim council estate would offer, but it is a serious and worthwhile attempt to "de-medicalize" care with carrots that have a core of sticks. Bernheimer would like clinics to become not clinics but "centers," so teens want to turn up, to see what's happening, to pick up their meds, to see friends who know how to date with HIV, how to disguise foul drugs with peanut butter or ice cream, how to be healthy while infected, how to stay uninfectious. If they get better at adhering, they can pick up their meds every two months. If not, they must come every month.

Luring children with career days and snooker is one thing. But how do you lure adults to stay on their medication? The realities of life in Khayelitsha do not fit what Goemaere calls the "linear model" of 90-90-90. This is a global target that is as ambitious as global targets always are. By 2020, 90 percent of people living with HIV will know their HIV status; 90 percent of people with diagnosed HIV will be on ARVs; 90 percent of people on ARVs will have successful viral suppression.[29]

Life in Khayelitsha is not so neat. "If I can get you to go for a test," says Goemaere, "and you test HIV-positive, I can get you treatment. Deal done. But life is not like that. Life is not a tunnel." Life in Khayelitsha is competing priorities. Goemaere describes a bridge near the N2 highway. Go there in the dawn hours and it is packed with men seeking work. They are day laborers, and each day must be filled with labor. "For them, to go to the clinic, they lose one day of work. It's a competing priority. As soon as they get better, they say, I'll go to the clinic when I'm sick again, but for now I'll make money because that's what I'm supposed to be doing here." A study in the Western Cape found that the majority of people being initiated into ARV treatment were what health professionals call "non-naive."[30] They had been on ARVs before and dropped out. MSF's male services team is now going to shebeens, taxi stands, train stations, anywhere that men gather. It is piloting self-testing, testing after-hours, outreach testing, all sorts, all hands on deck. When "test and treat" programs were tried in KwaZulu-Natal, only half the people who tested positive went to get treatment within the next year.[31] Test and run.

Blessees may be newsworthy, these girls with their sex-purchased Peruvian and Brazilian weaves. But it is the men who slip from care, from testing. They are the ones, says Xoliswa in the container clinic, who may test, who may be diagnosed, but who never disclose their status. The ones who hide, the ones who let the virus hide in them: the infectors.

I expected more hope and optimism in South Africa. I was fooled by the cheerful version of HIV, the one presented in northern Europe and America. In this version, stigma has been banished (this is untrue). Hardly anyone is infectious (mostly true). Monogamous people, men and women, can take PrEP, a magic pill that will prevent infection. Everything is under control.

In her small stall at the container clinic, Xoliswa is neither hopeful nor optimistic. "When I was infected in 2004," she says, "I thought that by 2017 there would be no new infections. Everyone would know that there is HIV and it's not nice to have HIV. Maybe they see us, beautiful as we are, as healthy as we are, and they think, HIV is nothing." She shakes her head. "It's too difficult."

HIV is not over. I ask Vincent Racaniello, a virologist at Columbia University, what HIV specialists will be discussing in ten years. "We would be talking about the fact that we are still stuck with it and we haven't figured out how to prevent infection. The thing is, with HIV, it's completely preventable, right, you just don't use contaminated needles and you practice safe sex. And that's it. The problem is that a lot of people don't do that, they don't have the means to do it, or culturally it's not acceptable."

HIV is not over. Sam Wilson of Glasgow's Centre for Virus Research says, "The most remarkable thing is that when I lecture undergraduates, AIDS has passed them by and they see it as a history lesson." There are more than thirty antiretroviral drugs available. But there are more than a quadrillion HIV genomes on the planet, in Racaniello's estimation. How much is a quadrillion? Too many zeros to understand, so it is shortened to 10^{15}. The math doesn't really matter as much as this: that among that quadrillion are genomes resistant to every drug we currently have, and HIV continues to mutate.

HIV is not over. Even with successful treatment, the virus hides in the body. Its hiding places are called "latent reservoirs," and they have been found in the gut, the brain, in cells that last for a person's lifetime. Every so often, about once a week, silent viral DNA in an infected cell's nucleus starts to make viruses. Then it stops until the following week, when it does it again, for no obvious reason. It is, says Racaniello, "a big mystery that we have to sort out." Among others. Silent HIV is why antiretrovirals must be taken for life and consistently. There is no leeway. In recent years, a technique known as "kick and kill" was pioneered: drugs that kicked these quiet HIV reservoirs into producing virus, which were then killed. It worked in a two-year-old girl for two years, but her virus resurged.[32] Broadly neutralizing antibodies are the other excitement. BNAbs (really) are produced by a minority of people, though no one knows why. Unlike regular antibodies, which are highly specialized, bNAbs can defeat a wide variety of pathogens. Flame-throwers, not matches. They could defeat HIV. If they could be isolated and understood and reproduced, perhaps they could be a sort of vaccine. A monthly injection. It's a more fruitful avenue than a live attenuated vaccine, where people are injected with a tiny dose of virus. No ethics committee would let people be infected with a virus that is only contained, not cured, and which never leaves us.

For anyone working in Africa, the Middle East, the Philippines, with the gay black population of the United States, with older people (who make up 45 percent of the United States' HIV-positive population), this is a fearful moment. Money is retreating like a tide. In 2016, reported the UK coalition StopAIDS, international funding fell 7 percent. That is $500 million less.[33] Since 2012, funding has fallen by 5.4 percent every year.[34] StopAIDS warns of reprioritization, of competing concerns such as Ebola and Zika. When the Millennium Development Goals, a set of global targets known as MDGs, was launched, HIV merited its own goal. In the Sustainable Development Goals, the updated MDGs, it is one health target among others. AIDS exceptionalism—the sense that HIV and AIDS was an unprecedentedly dangerous epidemic—was why the money poured in. It has been neutered by the medical success of treatment. For Goemaere, AIDS and HIV are no longer seen as international threats. He remembers in the early days of the Khayelitsha clinic, when it received a delegation from the Pentagon. Goemaere couldn't understand why the military was interested, but its representatives were welcome as long as they didn't wear uniforms. And then he understood: they saw HIV as a global risk. They came, they saw, and a year later President George W. Bush launched his greatest presidential achievement: PEPFAR, the President's Emergency Plan for AIDS Relief. PEPFAR continues, and in 2014 it launched DREAMS, a program focused particularly on preventing HIV in young women. But Goemaere is dismayed nonetheless by what he calls "a sort of benign neglect." When he attended the high-profile AIDS conference in Paris in 2017, he was shocked that the French president "didn't even show up."

Pharmaceutical companies lowered the price of first-line drugs (those that are given initially) for developing countries, but not for second- and third-line. MSF has found resistance to first-line drugs at a rate of 10 percent in southern Africa. A study in Kenya, Malawi, and Mozambique found 30 percent of people on second-line treatment had become resistant. Third-line—what is called "salvage therapy"—costs $1,859 a year.[35] That is eighteen times more than the cost of basic first-line treatment, and six times more than second-line, and unaffordable for most developing and even developed nations.

For Eric Goemaere, there will be no true solution for years. "No pandemic has ever been solved without a vaccine," he says. That is years away. As long as girls have blessers and older boyfriends, and boy-friends have many girlfriends, and men rape, there will be new infec-tions, and prevention and treatment are like pouring water into a sink with an open drain. "The dream is slowly being shattered," says Goe-maere. He is sure we are approaching a second wave of the epidemic, "made out of resistance and treatment fatigue." But also donor fatigue. Media fatigue. A lowering of the guard on all fronts. He would like to see injectable ARVs or PrEP implants (you can remove an implant if it goes wrong, but not an injection). He was pleased to see a session on implants in Paris. Meanwhile, he does what he can and is hopeful of dolutegravir, a new and "very robust" ARV that is widely available in Europe but nowhere in South Africa. He has taken to bringing it back with him, in his pockets, as he did in the days of TAC.

People are trying hard on all fronts. Scientists are working on vagi-nal microbicides for women; on vaccines; on viral vectors, other viruses that can transport defenses against HIV. The virologist Paul Bieniasz thinks there is no lack of ingenuity in the fight to beat HIV. "We're limited more by our technology rather than our brains in trying to figure out how this stuff works." In various African countries, cash payments have been made to schoolgirls and young women, in an attempt to undercut the poverty that drives some transactional sex. In some countries payments reduced rates of HIV infections, but in others there was no difference. MSF's pre-exposure prophylaxis pro-gram is young. Only forty young women were enrolled at the time of my visit, but this is still promising. There are serious barriers to be scaled: young women don't want people to think they have HIV because they have medication that looks like ARVs. They worry about stigma and about being accused of sleeping around. The ones who have signed up are the brave ones. In 2017, the health minister of South Africa launched She Conquers, an initiative to stamp out the practice of blessing and to encourage younger women to stay away from older men. It's an unlikable name, the kind devised in dull boardrooms by dulled brains, and the literature isn't much better. But it's a start.

But what about the men? What about Themba and his at least three women and his whiskey sex? Who will protect the young women of South Africa from men like him? In the 1990s, Uganda launched a highly successful behavior modification program with the slogan Zero Grazing.[36] Grazing was having more than one partner. Uganda wanted people to stick to main meals. It worked; condom use increased dramatically and Uganda's HIV prevalence dropped. Then conservative Christian ideology arrived from the United States along with PEPFAR money and found a receptive home in President Museveni and his wife, and condoms were burned, and HIV came back.

HIV is remarkable. Amazing. Elegant. These are words that microbiologists and virologists use, though they always catch themselves. They are quick to counter these words with others. Pandemic. Devastation. Grief. HIV has brought us all these things. But I understand their wonder at the virus. Despite many millions of dollars and forty years of research and countless remarkable minds working on curing HIV, it is only contained and only for now. What does HIV like best? Presumption and assumption that it is beaten.

On the side of the shopping center opposite the MSF offices, the one with the fish and chips shop that is safe to go to if there are enough people around, there is a mural. It is large and can be seen from a mile away (a fact I can prove, as sat navs, given MSF's address, unfailingly chose to send drivers somewhere else, and the mural became my guide, my crumbs through the forest). This mural shows a young black woman, her hair not Brazilian or Peruvian but natural; her clothes patterned with African designs. She stands against a yellow sky, and a dark green forest rises on a hill behind her. I look at her, not just to find my way around Khayelitsha but because of her face. She is grimacing, the face of someone blinded by the sun, and her hands are clasped above her eyes to give her shade. She squints into the distance, this young woman who looks strong but who statistics say is still too vulnerable to violence and abuse and infection. I want her to triumph because she deserves to. I know that countless people are working ferociously to find cures, solutions, innovations, outreach, to marshal every weapon against HIV. I know all that and try to remember it. But I also know the numbers. So I see this young black woman looking into the distance, and I see her seeing that HIV is coming. It is still coming despite our best efforts and it is coming for her.

Coupon incentive for students to sell plasma

FIVE

THE YELLOW STUFF

A boy opens a fridge and removes two small bottles. He takes a syringe, clean and new. He sits himself somewhere comfortable, clean and well lit: the kitchen table will do. He wears a woolen tank top because he is being filmed at some point in the early 1980s when woolen tank tops were normal, not retro. He inserts the needle into the bottle and draws up the liquid, and he mixes it with the other bottle, and he injects ten thousand people into his right arm, into the crook of his elbow. He looks at the camera, and he grins with joy.

Let blood sit. After a few hours, gravity will partition it: red blood cells below, then a narrow white layer—the buffy coat—that holds white blood cells and platelets. Above, there will be a liquid the color of straw. This is plasma, a liquid that makes up just over half the volume of blood. It carries fat globules, water, salts, and at least seven hundred proteins such as albumin, antibodies, and clotting factors.[1] It looks like nothing much next to the insistent red of the blood below. It's just yellow stuff. But it is worth a fortune.

Separated from a donation of whole blood, plasma becomes fresh frozen plasma (FFP) and is used in transfusions to bulk up volume in bleeding or trauma patients. This is usually collected by blood banks and called recovered plasma. But plasma can have an alternate future where it is not a human body part but a product. Plasma companies collect it from paid donors using aphaeresis machines, which spin whole blood, retain the plasma, and return the rest to the recipient. This is source plasma, and it is purified for its components such as immunoglobulins, which boost immune deficiencies, or albumin, the most common plasma protein, which can maintain blood volume and pressure. These components have come to be known as plasma protein therapeutics. A unit of FFP is cheap: the NHS sells one for under £30 ($42).[2] A fraction of source plasma is not. The most popular, intravenous immunoglobulin (IVIG), often costs more than gold (the NHS buys it for about £35 [$49] a gram),[3] and one plasma donation can reap medicines worth ten times what a donor is paid (usually about $30). Human and animal blood is the thirteenth most traded commodity in the world, worth $252 billion. Most of that is products derived from plasma, and most of it is coming from the United States, the largest exporter of plasma. In 2016, the category of "human and animal blood"—actually mostly blood fractions such as plasma products—earned the United States $19 billion, close to what it got from selling medium-size cars or soybeans.[4] The chief of America's Blood Centers, an association of blood banks, has called it the OPEC of plasma.[5] Half of Europe's plasma for medicinal products comes from American veins.[6]

The small boy is grinning because of what was in the bottle. It was plasma-derived clotting factor. Specifically, it was Factor VIII, the name of both a protein found in normal human plasma and a commercial product developed in the 1980s that was usually shortened to "factor." He is in England but the boy has almost certainly just injected a product derived from American plasma sellers. It has revolutionized the lives of millions of hemophiliacs like him. It would also kill them.

When blood escapes the blood vessels, either inside or outside the body, it performs what is called with scientific poetry a clotting cascade. About twelve things need to happen for blood to clot, that seemingly easy procedure that is triggered when you nick your ankle shaving your leg, or cut your finger with a knife, or pick a scab. That thickening of the blood, quickly done, that seems so easy but is not. If any component of the cascade fails, blood will continue to flow. If the blood continues to flow, you are probably a hemophiliac. If a knock to your knee or elbow or head causes you to bleed internally, without stopping, then you are probably a hemophiliac.

One in five thousand boys is born with hemophilia A, says the US National Hemophilia Foundation. (A is the more common variety and means that blood lacks Factor VIII; there is also hemophilia B, due to a lack of Factor IX. The results are the same: you don't clot properly.)[7] The World Federation of Hemophilia suggests a rate of one in ten thousand.[8] Take either figure: hemophiliacs are still rare enough that most people are unlikely to encounter one. A hundred years ago, they would have been rarer because they died young. Hemophiliacs are born with a genetic abnormality inherited from their mothers that prevents blood from clotting. It used to be accepted that women only carried and passed on the faulty gene responsible. But now we know that women can be bleeders: either mild hemophiliacs, with a chromosomal abnormality that means they lack necessary clotting factors; or with genetic abnormalities such as von Willebrand disease or Christmas disease, which cause clotting derangement. The true number of people with bleeding disorders—who have less than 40 percent of the clotting factor in their blood[9]—is unknown. Anyway, the figure is constantly changing, because for several years in the late 1970s and early 1980s, the factor that hemophiliacs were injecting came with HIV and hepatitis C in it. Hepatitis C, a blood-borne virus ten times as infectious as HIV, can be undetected for years and usually leads to cirrhosis and often cancer. A total of 4,689 hemophiliacs in the UK were infected with HIV and hepatitis C, of whom 2,883 have since died.[10] That figure will be out of date by next week: because of the lag time, new cases are continually being diagnosed. A Westminster committee estimated that 32,718 people in the UK were infected with hepatitis C between 1970 and 1991 from contaminated blood products (including transfusions). Only 6,000 have been identified.[11] Globally, 40,000 hemophiliacs have been infected with HIV from contaminated plasma products. No one yet knows how many got hepatitis C, but it is certainly hundreds of thousands.

Neil Weller, a hemophiliac from Oxfordshire, thinks he got off lightly. I met Neil through Facebook, which has become a hub for the rightly furious. The survivors, relatives, the grieving. In the UK alone there are campaigns named Contaminated Blood, Factor 8 Campaign, and Tainted Blood. Neil is cheery and chatty, his accent holding the rounded vowels of the west country that give a disconcerting warmth to statements such as: "I was eleven or twelve when I was told I didn't have HIV." But many of his friends did. Like most hemophiliacs, he spent much of his teenage years in and out of the hospital, dealing with bleeds. He remembers his fellow hemophiliacs there "and one by one they were told they had HIV and one by one they were going in another ward and they were dying." When they died, the cause of death was given as pneumonia or something else. One friend, dead at thirty-two, left a two-year-old child. Another was a teenager. But Neil was fortunate, in his eyes: the contaminated plasma he had injected only gave him hepatitis C, a nasty chronic condition but not, as HIV was then, a certain killer. He walks with difficulty, but he is mobile, although he has had two knees replaced and his ankles are rotten. He has had twenty-two operations, including eighteen on his knees and ankles, and spent more than five hundred nights in the hospital. But he's right: statistically, he is lucky.

Hemophilia is rotten. Hemophiliacs with no access to treatment usually die of bleeding to the brain or into the gut. This was always the case, no matter how much wealth and privilege they had: from Queen Victoria, the most famous transmitter of the faulty gene, hemophilia spread throughout the European royals, lovers of endemic intermarriage. The best-known royal hemophiliac, Tsarevich Alexei of Russia, would probably have died young had he not been shot first in a cellar in Ekaterinburg. Early death is the fate of hemophiliacs in most of the world, where treatment is expensive and rare. Dr. Mark Winter, a noted hemophilia expert, in 2017 recalled a recent visit to Pakistan where "they have got very nice hospitals, experienced doctors, good nurses, they are a nuclear power, but they have no [factor] concentrate." Of the 250 children with severe hemophilia who attended the hemophilia center in Islamabad, he reported, only one lived beyond eighteen.[12]

Even with treatment, life for a hemophiliac was hard and frequently excruciating, but not in the way that most people think. It's "that disease where if you cut yourself you bleed to death," says Neil Weller, and I dare not tell him that that's what I thought, before I began thinking better. Hemophiliacs bleed internally, frequently, and with violent pain. Justin Levesque is a severe hemophiliac who lives in Maine. He calls himself a "bleeder" and his license plate is "bruiser." I asked him to describe the pain to me. This is an impossible request, as pain eludes description no matter how articulate you are. I try analogies. Is it like a broken leg? A burn? A stab wound? He first says, "The worst pain ever. Ever." He tries again. "Your joint is an enclosed space, and then you're trying to fill up a water balloon with as much water as you can but you can't and it can't pop either. It's pressure and pain so you can't think. The only thing I can relate it to is I had a kidney stone one time which was godawful, and it was pretty much on a par with that." Other hemophiliacs say a bad bleed feels like a broken bone. We're still in analogies and I can't conceive of the pain until I see a severe hemophiliac walking: it is a twisted, unnatural movement, the legs stiffened into knock-kneed stilts. It looks like torment.

If a hemophiliac's joint is knocked, blood vessels break. People with the more severe kind of hemophilia can also have spontaneous bleeds. Whatever the cause, there is then a "rush bleed," because within twenty minutes, blood will have poured into the space surrounding a joint. Knees and ankles are most usually afflicted, but also fingers and elbows. The blood keeps flowing until there is no more space. The bones twist and move to accommodate the liquid. That is painful enough. Then the blood that shouldn't be there presses on nerves, and that is agony. If this happens often, as it does with hemophiliacs, the affected joints deteriorate into arthritic ones. Rupert Miller, whose brother Julian was a severe hemophiliac, remembers his brother waking up all the family, screaming, night after night, in their farmhouse in Wales. All they could do was apply packs of frozen peas to his knees—the cold dulled the agony—and stay with him until the pain abated. The family freezer was filled with peas.

When Julian had a bad bleed, he sometimes went to the hospital. At that time, in the 1970s, the only treatment available was cryoprecipitate, a concentrated version of plasma with clotting factors. Hospitals were the safest place because cryo was difficult to prepare, requiring a freezer (then a rarity in British homes) and a complicated process of thawing and agitation. Julian was sometimes there for many weeks, many times a year. Once he missed two months of school. But his parents wanted him to be as normal as possible. Julian wanted to be as normal as possible and to attend the local school with his friends. He resisted going somewhere like Treloar College, a boarding school in Hampshire for disabled children, which had Gothic buildings, good care, and a dedicated hemophilia center.

In a BBC documentary, a former pupil named Ade Goodyear reminisced about life at Treloar.[13] Being there didn't fix your hemophilia, but if you felt the buzzing or tingling or heat that meant a bleed was coming, you could get treatment quickly and expertly. You never had to miss class again. Your schoolmates, unlike Goodyear's at his previous school, were unlikely to beat you up just to see what that did to a hemophiliac. (When Goodyear's previous headmaster introduced him by saying, "You must not hit this boy," his fate was fixed.) At Treloar there were thirty-five hemophiliac boys who knew what it was like. You weren't judged. Treloar was a haven.

For every hemophiliac in the developed world, everything changed when Factor VIII arrived. Easy to prepare and relatively easy to infuse, it could even be used prophylactically, to prevent a bleed coming on. It was, says Rupert Miller, "wow." Factor VIII was a revolution and a liberation. Factor, in the words of a public service video of the time, "is something that will let a hemophiliac live and bleed like a normal person." For the first time, says Justin Levesque, hemophiliacs could have privacy. "They were going from being public, limping, being in the hospital, being very seen with their disorder, into the private."

Cryoprecipitate was expensive. But so was this new Factor. The trouble was that one unit of plasma contains only a tiny amount of clotting factors. To be powerful enough, the concentrate had to come from a pool of thousands of donors. The higher the pool, the cheaper the processing costs and the greater the profit margin. This was a consideration, because the process of fractionation to create the new wonder drug was not cheap. Already, as soon as the new factors arrived—there was also Factor IX, for hemophilia B—some doctors were dubious. They knew that the larger the pool of plasma donors in each dose of concentrate, the higher the risk of infection. It was mathematics. In 1983, the vice president of Armour, a large US fractionator, testified to government that people who supplied the plasma for concentrate had it collected forty to sixty times a year. At that rate, "and given the pool sizes in the United States, four infected persons could contaminate the entire world supply of Factor VIII concentrate."[14]

But home therapy, as it became known, was powerfully attractive. Hemophiliacs understandably wanted it. Doctors began prescribing it for all sorts of bleeding problems, not just to diagnosed hemophiliacs. Ann Hume was a young expectant mother in Shetland in 1974. When we talk on the phone, I hear Norse in her *r*'s rolled like surf, and I picture the wild North Sea from my desk in a landlocked city. How silly and romantic. Ann's story is neither. From a young age, she bled copiously and in alarming amounts, causing dentists difficulty: after a tooth extraction, she could bleed for weeks. Her condition has never been diagnosed, but she was definitely a bleeder. After an abortion—she was young and unmarried, and told it was wise—and then the birth of her first child, she was given cryoprecipitate, preventatively, so that she would not hemorrhage to death. That was fine. For the second child, in 1982, her hematologist said, "'I'm going to use this new stuff, which is better, and it's called Factor,' and I thought Okay, that's fine. I didn't know anything about Factor VIII."

Her daughter was born, she did not bleed too much, and mother and baby returned to their island. That's when it began: the fatigue, and the joint pain, and the illness that would not get better. After about seven weeks, she felt so unwell, it was difficult to rise out of bed, and she was pushed out of it only by the fact that she had children to care for. (Her husband and her mother told her to snap out of it.) One morning she got up and blood clots "like liver" fell out of her body. Her husband was at work; she had a four-year-old son and a baby upstairs. "I asked my son if he would go and fetch the next-door neighbor and he said no, he was in his pajamas." Ann stepped through her blood clots to reach the phone, the neighbor came, and she joined in the panic. The kitchen looked like a crime scene. (She later understood that it was a crime scene, given what she discovered.) Ann got to the island's hospital in Lerwick eventually, and the hematologist in Aberdeen was telephoned. "She said, somebody give her a bottle of Factor VIII and fly her down here."

Julian Miller sits on a chair in a TV studio. Its color scheme is 1980s beige but he is not. He wears a loud checked jacket, a pink shirt, and glasses that are the size that glasses were in the 1980s, so half his face is lenses. He has blond hair and eyes so blue, as serene and charming as his upper-class accent. He is probably the first hemophiliac to go on television and say he is HIV-positive.[15]

He thought he had been infected in the early 1980s. He was told after a routine blood test in 1984. His manner on TV is astonishingly calm. He smiles now and then, when he calls himself a "spontaneous mutation," because he was the first hemophiliac in his family. He speaks of horrific things with the modulated serenity of priest or counselor. "In 1979," he says, "it began to be fairly plain that there might be a problem with AIDS infection of American blood products." He took himself to the venereal diseases clinic at St. Mary's Hospital in London, as that was the only place then to get information on this new disease called AIDS. He sat down in the corridor "with all the other people who were waiting to see him for a different reason"—another grin— then saw a consultant. "He told me in a very grave, serious way about AIDS [. . .] and I left rather shocked." His own hemophilia doctor, when Julian expressed his concern, gave the advice that most other worried hemophiliacs were given at the time. Carry on as normal. He said, "'You're in more danger of a life-threatening bleeding episode than you are of getting AIDS.' So I carried on and in October 1984, having been tested, they told me I was HIV-positive."

Ade Goodyear was fifteen when he was told he was HIV-positive. At Treloar, Goodyear told the BBC, "we went into the office in the medical center in groups of five." The atmosphere was relaxed but weird, because they had heard rumors. All schools have rumors, but not usually the kind that might kill you. "The doctors cautiously informed us that 'you may have heard that Factor VIII is not as clean as it should be.'" Then, with artless horror, "we were told who had HIV by the words 'you haven't, you have, you have, you haven't,' and so on." The boys asked how long they had got. Two years. Maybe. Of those five boys in that room, Goodyear is the only survivor.

The trouble for Julian, and for thousands of other hemophiliacs who injected hepatitis and HIV with their Factor, is who their donors were, what they were donating, and how Factor was manufactured.

By the mid-1970s, the UK's plasma supply had problems. Concentrates were too popular, and the country could not find enough plasma to process. Most countries aim for self-sufficiency in their blood and plasma supply, but plasma concentrates require a level of donations that hardly any country can maintain. In 1973, the UK began to import Factor from the United States. Many other countries did the same. The United States, in the 1970s, was supplying half of Europe's plasma. In his book *Blood*, Douglas Starr quotes Tom Drees, then president of the Alpha Therapeutic Corporation. "'As the US feeds the world,' Drees told a conference of fractionators, 'so does the US bleed for the world. Or, more correctly, the US plasmaphareses itself for the rest of the world.'"[16] How? Because they paid people for it. Sometimes, the kind of people you want nowhere near a safe blood supply.

How do you make blood safe? According to the World Health Organization, you take it from volunteer donors who are not paid and who have no reason to lie about their health. Plenty of studies show that when donors are paid, they are more likely to lie about it. Obviously they do: they want to keep getting paid. Paid donors are also more likely to come from sections of society that have poor health to begin with. But the US blood supply has always rested on commerce and transaction. After the Second World War, countries such as the UK, France, and the Netherlands developed a non-remunerated blood supply, often run by a single government entity. The United States did not. It took a book by a British sociologist to change the world. (That is not a sentence you will read often.) In 1970, Richard Titmuss published *The Gift Relationship*, and it threw the blood business up in the air. He compared the UK's voluntary, altruistic system with the fee-paying blood business in the United States, and his conclusion was damning for those who considered blood to be a product to be bought and sold like any other. Paying for blood made it unsafe. He wrote soberly, but he used words like "moral" and "right." Even so, it was only in 1978 that the FDA required all blood to be labeled as paid or voluntary donation, and payment for blood died out.[17]

But payment for plasma did not. Somehow, plasma taken from blood has become something other than blood: tamer, and less biological. Something you can pay for. This parallel reality began during the Second World War, when a doctor, Edwin Cohn, developed plasma fractionation for military use (plasma derivatives were easier to transport). In the 1950s, trials with gamma globulin, a protein isolated from plasma, proved successful at treating polio. Companies realized how successful plasma might be, if transformed into pharmaceutical products. Trade data lump whole blood and plasma together under the same category but in practice the United States treats them differently. Environmental activists talk of greenwashing: perhaps this was yellow washing. Anyway, it worked. Blood and plasma grew apart in concept and regulation. Now donors gave blood and sellers sold plasma.

But the plasma had to come from somewhere, and lots of it. Even paying for plasma did not bring in the huge numbers required to make good business. What about getting it from a captive resource? In 1947, German doctors were put on trial at Nuremberg for perpetrating "ghastly" and hideous experiments on concentration camp prisoners. Part of their defense was that Americans were also guilty of it. Some prisoners had been infected with bubonic plague in the name of science. Before the war, a doctor named Leo Stanley transplanted testicles from executed prisoners to "senile and devitalized men." Later he switched to animals, injecting several hundred San Quentin residents with "animal testicular substance" from goats, rams, and boars. During the war prisoners took part in experiments that exposed them to gonorrhea, malaria, and the induction of gas gangrene. Sixty-two Sing Sing inmates volunteered in 1953 to have themselves injected with syphilis. Nearly half developed the disease, which attacks the eyes, brain, heart, liver, bones, and joints, and which can kill. In return, they got "a carton of cigarettes brought in by the doctors at Christmas time, a brief note on their records, and the good feeling which comes from having done something to help others."[18]

During the war, more than seventy thousand American inmates gave blood to the Prisoners' Blood Bank for Defense.[19] For the next two decades, there were enterprising efforts to rehabilitate or coerce or coax prisoners into better behavior by using their blood. In the 1950s, writes Susan E. Lederer in *Flesh and Blood*, "prisoners in the Virginia State penitentiary received days off their sentence for each pint of blood donated." Prisons in Massachusetts, South Carolina, Mississippi, and Virginia offered prisoners five days off their sentence for each pint of blood given.[20] Prisoners were used to giving up body fluids for profit; they would be a rich pool for plasma.

Read about the activities of the plasma industry and you will perhaps wonder why they were never turned into a James Bond film. Companies such as Baxter, Grifols, and Octapharma, which control the industry and which are vast conglomerates with secretive billionaire bosses (in the case of Octapharma), set up clinics in skid rows and prisons all over the country. They also harvested plasma from poor countries. As Douglas Starr writes in *Blood*, the Nicaraguan dictator Anastasio Somoza had shares in Plasmafaresis, a plasma harvesting company in Nicaragua, which sold plasma to the United States. Locals called the facility "the house of vampires." In Haiti, a company named Hemo Caribbean paid vendors $3 a sale—$5 if they accepted to have a series of tetanus shots—and flew it out on Air Haiti to buyers in the United States and Europe. *New York Times* reporter Richard Severo, who exposed the unpleasant trade in poor people's plasma, wrote in 1972 that the company was shipping 6,000 liters a month, and looking to expand. Haitian sellers had the motivation of being desperately poor. They also had one of the lowest caloric intakes in Latin America, as well as alarming rates of tuberculosis, tetanus, gastrointestinal diseases, and malnutrition. Hemo Caribbean's technical director, a Mr. Thill, said that "few if any diseased persons slip through," and even if they did, even if they had venereal diseases or malaria, he was confident that the freezing process killed such bacteria. In Cummins jail in Grady, Arkansas, seventy miles south of Little Rock, a plasma center had been running since 1963. Prisoners were given $7 in exchange, and the plasma was sold for $100 a donation. They queued up in the center to lie down on cots and sell. They were "like little cows," said an official who worked with Bill Clinton, twice governor of Arkansas.[21] The little cows were milked and business was good. In 1974, the prison began selling to a company called Health Management Associates. HMA sold the plasma to North American Biologics, a subsidiary of Continental Pharma Cryosan, a Canadian company. (Cryosan later pleaded guilty to "mislabeling" blood from Russian cadavers as coming from Swedish donors.[22] We've all done that.) From there, prison plasma was exported worldwide, to Canada, France, Iran, Iraq, Japan, the UK, Hong Kong.

Titmuss's book caused upheaval in the blood business but not the plasma business. Yet in 1975 a California surgeon named J. Garrott Allen wrote to the Blood Products Laboratory in England, which had been buying American plasma products for years. Did they know, he inquired, that American plasma was "extraordinarily hazardous" and taken 100 percent from "Skid Row derelicts"? It also contained worrying levels of a new variety of unknown hepatitis, later called non-A, non-B, and finally C, which was more frequently encountered in "the lower socioeconomic groups of paid and prison donors." The practice of selling blood became so popular with some socioeconomic groups, the press began to call it "ooze for booze." Allen's research showed that rates of hepatitis in prisoners and skid row donors were ten times higher than in the general population.[23] Drug use, needle sharing, general poor health: all of these were risk factors. But nothing changed. Prisoners and the poor were too good a resource.

Reporters embedded themselves on skid row and wrote firsthand accounts. In 1975, reporters for *World in Action*, a British investigative current affairs TV program, traveled to various US cities to meet people who were selling their plasma. A man named Gary in San Francisco was asked whether he always answers questions about his health truthfully. "No." He reconsidered. "You know, yeah, most of the time." He is scornful of the health screening. "I'm healthy, you know?" Another donor says simply, "Pardon me while I puke."[24]

By 1980 in the US, payment for blood had come to be seen as unethical. But 70 percent of the country's plasma came from paid sellers and no one saw anything wrong with it. BPL did not act upon Dr. Allen's warnings. Hemophiliacs had no intention of giving up Factor and going back to endless hospital stays. They wanted it for three reasons: It gave them a life. They didn't know about hepatitis C. And no one had told them about HIV, insidiously making its way to being a global pandemic.

After his TV interview, Julian Miller starred in a documentary.[25] It opens with a shot of him shaving, a sly reference to most people's ignorant belief that hemophiliacs can bleed to death from a nick. There are scenes of Julian walking around his parents' beautiful home in a beautiful Welsh valley, his stiff walk the mark of a hemophiliac, to those who know the sign language, revealing a person in constant pain. He is twenty-five years old in the film but speaks with the wisdom of the old, with such a calmness it seems like defiance, a refusal to capitulate to expected anger. This beautiful house in this beautiful valley, he says, is his quarantine. He cannot work, he can't easily move around. He has abandoned hope, mostly, of a relationship, of marriage or children, although his brother Rupert, when asked what he wishes for his older brother, says, "A smashing girl who will take him on." He didn't get one. In fact, Rupert later wrote, hemophiliacs often didn't dare masturbate in case they triggered a hemorrhage. "Therefore a shag was out of the question."[26]

His mother, with her cut-glass accent and pussy-bow blouse, also seems serene yet tightly wound. Her only demonstration that there is something behind her public face is the occasional biting of her lip. A small tell of pain. She says she is optimistic, that her son is strong and healthy. "Hopefully," she says, with a pause, "statistics show one thing one minute then another thing the next. So you look on the good side and hope that'll sustain him. That they find something." Under the YouTube video of this documentary, Rupert Miller has written, "They had to edit out my father's part as he could not stop crying."

Julian says, with his calm, "Originally we were told that there was estimated to be a twenty percent chance of developing the full AIDS syndrome. One in five hemophiliacs would develop it and if you develop it you die. A recent survey by one of the leading hemophilia doctors has now estimated that that might rise to eighty percent." At this point he laughs. "So it's getting worse."

I'm not sure why I'm so captivated by this thirty-five-year-old film. Julian is long dead. Perhaps it is his composure in the face of criminal behavior, collusion, and cover-up that would justifiably have made him howl. But off-camera, sometimes he did. In 2014, Rupert published a book about his and Julian's lives, called *Life of a Salesman* (Julian had worked in advertising sales for McCann Erickson, for as long as he was able). Here, at last, is something more than calm. Here is Julian, taken by Rupert to a wedding in a wheelchair because he was in considerable pain. "At the wedding reception a woman, who we had never met before, came up to him, put her nose right in his face, and shouted very slowly: 'CAN I GET YOU AN ORANGE DRINK?' 'No, you condescending bitch,' he barked. 'Fuck off and get me some champagne.'" In his last months, in 1991, Julian developed dementia as AIDS attacked his brain. He "talked gibberish," writes Rupert. "We would sit with him, and every now and again, he'd suddenly sit up in bed, tell us all to 'fuck off,' and then fall back onto his side." While he was dying, he got a letter addressed only to "Julian, North Wales." It was from a nurse in Rwanda who had heard him on the BBC World Service. When he died, he weighed five stone (seventy pounds). At his funeral, "he arrived in a very small coffin. He didn't need a big one."

If Julian had survived five more years, he would have been given antiretrovirals for HIV. He would already have been given synthetic "recombinant" factor, developed in the 1980s as an alternative to plasma-derived factor and now standard treatment for hemophilia. His life span would have been normal. I ask Rupert why he has written his book, and why he still wants to talk about his brother and what was done to him. His answer is simple and pure. "He shouldn't have died."

Heat. What could have saved all those dead was heat (as well as honesty). By 1983, the UK government knew that there was HIV in the blood supply. So did the plasma industry: memos discovered by hemophiliacs suing the giant pharmaceutical company Bayer, which owned the plasma fractionator Cutter Biological, show Cutter managers acknowledging in 1983 that "there is strong evidence to suggest that HIV is passed on to other people through . . . plasma products." Later in the year, Cutter staff predicted a "gigantic epidemic" among hemophiliacs.[27]

Although the new variety of hepatitis was poorly understood and there was no test for it, by the early 1980s it was known that both hepatitis and HIV could be dealt with by good screening of donors, and with heat. But this was costly: the clotting factors weren't destroyed but they were reduced so that more plasma was needed. Also, companies had lots of inventory of untreated factor on the shelves. They decided that it had to go somewhere. At a 1984 conference at the Centers for Disease Control in Washington, DC, a minority argued strongly that all non-heat-treated product should be discarded. The majority disagreed: it should be a matter for local medical personnel.[28] A report by Justice Horace Krever into the contamination by US products of the Canadian blood supply titled one section "The Sorry Story of Blood Product Withdrawals."[29] It was sorry because it didn't happen. An editorial in the *Lancet,* while recognizing the benefits of heat treatment, concluded an editorial with the reminder that "by far the commonest cause of haemophiliac death is bleeding."[30]

By the mid-1980s, heat treatment was adopted as standard, but some companies continued to export their old, unheated inventory. Cutter sold it to Taiwan, Malaysia, Hong Kong, Argentina, Japan, and Indonesia, exporting more than one hundred thousand vials of unsafe, unheated concentrate—even though by then it produced heated product, too—and earned more than $4 million. In response, Bayer officials told the *New York Times* that Cutter had behaved "responsibly, ethically and humanely."[31]

Because they use so much factor, it has been notoriously difficult for hemophiliacs to identify which batch or company infected them. But one hundred Hong Kong and Taiwanese hemophiliacs got HIV from the untreated factor.[32] Carol Grayson, whose husband Peter Longstaff died of AIDS contracted from contaminated plasma, traced back a batch of factor he had been given in Newcastle Royal Infirmary to Arkansas State Penitentiary.[33] To those little cows.

The United States was not alone in its monstrous exports: three hundred Iranian hemophiliacs were infected when Iran, then fighting the Iran-Iraq War, bought factor from France. The French took no sides in the war: they also sold blood products to the Iraqis for years after it was understood that the supply was tainted. More than two hundred Iraqi hemophiliacs, aged from six months to eighteen years old, got HIV. A Baghdadi named Khalid al-Jabor lost five sons. Two were forcibly quarantined by Saddam's regime and died in the hospital. Al-Jabor hid the fourth son until he got too sick, and then he died.[34]

In the UK, hemophiliacs were used as guinea pigs. Or chimps, actually. In a 1982 letter, Professor Arthur Bloom, a hematologist at the Oxford hemophilia center where Neil Weller was treated, proposed testing the new heat-treated product on hemophiliacs. It had been tested previously on chimpanzees, but animal testing was expensive. Bloom decided that quality controls would be better—and less costly—if they were carried out on hemophiliacs who had not yet been exposed to large pooled products.[35] The candidates were called PUPs, for previously untreated patients. Most were children. No PUPs were told that they were a human experiment. One was Colin Smith, treated by Bloom. He died of AIDS at the age of seven. In the months before Colin's death, his parents had to pick him up in two sheepskins because he couldn't tolerate the pain of being touched.[36] Later it was discovered that a product called Factorate, produced by Armour, was heat-treated with a process inferior to the other companies. Armour was told in 1985 of cases of HIV infection in people using Factorate. It did not modify its process until two years later.[37]

Tainted blood activists today say that they are "cheaper than chimps." But bleeders are also used to being told how expensive they are, because of how costly factor is. In 1985, a junior minister named Kenneth Clarke wrote that he was skeptical of heat-treating UK-produced factor. After all, "only hemophiliacs have died." It's an extraordinary memo. "Of course," wrote an unnamed civil servant in the Department of Health's Finance Department a few months later, "the maintenance of the life of a hemophiliac is itself expensive, and

I am very much afraid that those who are already doomed will generate savings which more than cover the cost of testing blood donations."[38] Cheaper than chimps, cheaper when dead.

All this has been explored in huge and expensive reports and inquiries. The Archer Report in England consisted of 114 pages and was produced for $104,000. The Krever Report in Canada: 1,197 pages, four years, $11.5 million. Scotland produced the Penrose Report, finally, in 2016: 1,811 pages, six years, $16 million. In the United States, where 4,000 hemophiliacs have died of AIDS, some redress has come in the form of lawsuits and settlements, with many millions of dollars paid out. The French sent some blood officials to jail. Government schemes worldwide have provided some compensation, although the UK government is unique in its repellent refusal to provide compensation or an apology (it insists on calling payments "ex gratia"). These payments cover people infected by Factor, but also by whole blood transfusions. The blood supply and plasma industry have cleaned up. The reports have been done. Everything is better now and tainted blood is history. Isn't it?

Hardly anyone visits Saskatoon in February. It is in the Canadian prairies, and there is nothing among the flatness to stop the wind, the cold, the snow. The week before I arrived, the temperature had been minus 40 Fahrenheit. This was exciting. Maybe I would be able to stick my tongue to a lamppost again, as I remembered doing—but probably didn't—when I last visited Saskatoon at the age of ten. Maybe I would drive a Ski-Doo. I did neither of these things, because for the ten days I was in the province of Saskatchewan, the weather dispensed an unseasonable thaw. This meant that ice melted during the day and froze again at night. It meant black ice every morning. No one was crazy enough to walk on the streets except me, because the roads were salted and the pavements were not. I walked everywhere, for blocks and blocks, bloody-minded and wet-footed. One day I walked for an hour over black ice and slush, while drivers shook their heads in pity and sprayed me with filth, to reach a low, dull-looking beige building. It was clean and neat and nondescript with ample parking. It didn't look like somewhere that could overturn Canada's public health care system, or enrage placid Canadians, or provoke new laws.

Saskatoon is cut in two by the South Saskatchewan River, from which the city takes its name (the Cree word *kisiskâciwanisîpiy*, which I'm retaining for my next Scrabble game). Affluent Saskatoonians moved to the east side; the rest took the west. The west side is the place where earnest strangers on TripAdvisor advise you not to rent accommodation. Stay east. The river now is not the only thing dividing Saskatoon: health and poverty statistics do the same. The city has some of the highest rates of HIV and hepatitis C in Canada, as does the province. On some reserves where Saskatchewan's First Nations people live, HIV rates are eleven times higher than in the rest of the country and equivalent to Nigeria's.[39] An assessment of Saskatoon's health disparities found that residents from six low-income neighborhoods, all in the west of the city, are 3,360 percent more likely to have hepatitis C than higher-income Saskatoon residents. An infant born in these neighborhoods is 448 percent more likely to die in its first year of life. This mortality rate, wrote the authors of the report, a senior epidemiologist and the city's chief medical officer, "is worse than [in] war torn nations like Bosnia."[40]

The beige building is on the west side of the river. It belongs to Canadian Plasma Resources, a clinic that pays Canadians for plasma. It looks like nothing much, but many Canadians think it is unethical, wrong, and dangerous. Canada's health system consists of a nationwide insurance plan called Medicare. This, according to Health Canada, ensures that "all Canadian residents have reasonable access to medically necessary hospital and physician services without paying out of pocket." Most Canadians translate this to a British visitor as "like your NHS." They see it as a public system for the public good, and something to be proud of. And for many Canadians, the presence of Canadian Plasma Resources is the beginning of its end.

Kat Lanteigne is a playwright and actor in Toronto. In 2013, she had written a play called *Tainted*, about Canada's tainted blood scandal, during which eleven hundred were infected with HIV and thirty thousand—that is a movable number—with hepatitis C. Seven hundred were hemophiliacs and four hundred received transfused blood for other reasons (trauma, cancer, childbirth, surgery).[41] So far, eight hundred people have died, including Lanteigne's uncle. She was in the middle of a year of stage preparation when she read about some new "private blood clinics" that were being advertised on streetcars. "I was in a unique position," she says. "I had just finished a massive three-year research project about the tainted blood crisis and had interviewed people across the country: tainted blood survivors, hematologists, former Red Cross workers, hemophiliac nurses, family members. So I was able to reach out to them and say these blood clinics were here and we probably should stop them, because we all knew they shouldn't exist."

That confidence, that they all knew paying for any kind of blood product was wrong and dangerous, was because of Krever. Talk to any Canadian who knows about their tainted blood scandal, and the word *Krever* is spoken with reverence. Justice Horace Krever was asked in 1993 to chair an inquiry into how Canadians had got HIV from the blood supply. The inquiry was meticulous, comprehensive, and devastating. The judge found fault with people and institutions and Canada's system of collecting and distributing blood. The Red Cross, which had done most of the collecting and had continued to distribute tainted factor long after concerns were raised about its safety, was fired from blood services. Canadian Blood Services, a new national organization, was given full responsibility for the blood supply. Health Canada would be the overseer and regulator. The judge's language was lucid and unmistakable: Canada should not pay anyone for blood or plasma. A voluntary system was safer. After what had happened—he didn't write this—it was also more moral. It is estimated that eventually eight thousand Canadians will die from contaminated blood.

For twenty years, Canada abided by Krever's recommendations, mostly. A small company in Manitoba was allowed to buy plasma, for limited use. Canadian Blood Services never paid for blood, and if you wanted to sell your plasma, you had to cross the border and sell it in America. Canada operated like most other industrialized countries: it collected plasma from whole blood donations for transfusions, and had plenty of that. It bought source plasma from the Americans as well as plasma products that used US plasma. Seventy percent of its immuno-globulin products are derived from American blood.[42]

But in Canada individual provinces have control of health care. If they wanted to pass legislation that allowed paid plasma clinics, they could. Despite Krever's forceful recommendation that both whole blood and plasma should not be sold, only Quebec had passed a provincial law that expressly forbade the selling of plasma or any product taken from the body. Ontario had none. National legislation forbade the sale of sperm, eggs, and embryos. Provincial laws banned selling organs or body parts. But not blood. Even so, when a new company that no one had heard of arrived in Toronto and set up two clinics to buy people's plasma, it was shocking.

So were the locations the company chose for its clinics. Location matters in the plasma business, and the biggest plasma business in the world is just over the border. More than forty years after J. Garrott Allen castigated the use of skid row sellers, the plasma industry in the neighboring United States still had a reputation for locating its clinics in poor areas populated by vulnerable people. Grifols, the global leader in plasma protein therapeutics, has nearly 150 clinics in the United States. Thirteen are along the Mexican border. There are four clinics in El Paso.[43] A paper coauthored by someone with one of the best names in epidemiology, Cameron A. Mustard, examined the location of plasma clinics in the United States between 1980 and 1995 and found that the number of source plasma clinics operating in extreme poverty areas grew from 77 to 136. Commercial plasma clinics were five to eight times more likely to be located in areas considered high-risk than would be expected by chance. What was surprising was that "such clinics continued to operate in these areas well after the epidemiologies of HIV and HCV [hepatitis C virus] and the links between drug use, infection and blood product infection were established." This, in dry science language, "is inconsistent with epidemiologic evidence that locating of commercial source plasma clinics—which provide cash compensation for plasma donation—in the midst of active drug markets and poverty represents a risk to blood system safety."[44] The number of people donating plasma had risen as dramatically as the number of Americans living in extreme poverty, according to Kathryn J. Edin and H. Luke Shaefer, two sociologists who published *$2.00 a Day*, a book about modern American poverty. Over ten years from 2006, the number of people selling plasma grew threefold, to 32.6 million.[45] According to figures from the Plasma Protein Therapeutics Association, an industry body, they were donating more frequently. Today, according to the $2.00 a Day blog, "there are over 500 commercial plasma donation centers scattered across the country, most of which are disproportionately positioned in or around poor areas, where the most likely donors, the impoverished, reside. Of these centers, 100 opened their doors during the Great Recession and almost 200 have opened within the last decade."[46] It makes economic sense: the FDA allows Americans to sell their plasma up to twice a week (because plasma contains no cells, the body can replace it within forty-eight hours) and the payment is usually $30 to $50.[47]

Nowhere else in the world allows people to sell their plasma as frequently as in the United States. European regulations limit plasma sales to twenty-four a year, with at least two weeks between donations.[48] "Selling plasma," wrote Edin and Shaefer, "is so common among the $2-a-day poor that it might be thought of as lifeblood."[49] Selling plasma could raise a $2-a-day income to $3 or $4, about the same inflation as occasionally selling scrap metal or sex. How ethical it is for a profit-making industry to get its source material from poor people who have little economic choice, and what the long-term health costs are for them: that is undecided and unknown. When the journalist Darryl Lorenzo Wellington became a "plasser," as regular plasma sellers call themselves, he experienced extreme fatigue, passing out for five hours. Of three dozen regular "plassers" he interviewed, "more than half of them confessed to frequent, bizarre tingling sensations, pains, rubbery legs, and severe dehydration, as well as to being homeless, having lied to pass medical exams, and having used 'tricks' that allowed them to pass protein-level tests. They lived in circumstances that made plassing a hardship but said, 'I can't eat if I don't plass.'"[50] Plasma donors who donate frequently have been found to be at risk of hypocalcemia, caused by the sodium citrate used to stop blood clotting during the plasmapheresis. Several studies that have compared frequent, paid US plasma sellers with European donors have found that the US plasma contained much lower levels of proteins such as albumin and immunoglobulins.[51]

Canadians looked to their American neighbors and worried that this homegrown plasma company was following similar strategies. Canadian Plasma Resources seemed have taken lessons from the US model. In Toronto, one of the proposed plasma clinics was on Spadina Avenue, close to the Centre for Addiction and Mental Health, as well as a homeless shelter. Another was near Cathedral Church of St. James, which also provided services for the homeless. In hindsight, says Barzin Bahardoust, the company's CEO, these decisions were mistakes. He is a talkative Iranian, now a Canadian citizen, who speaks for an hour smoothly and without pause. In hindsight, he says, "it was a very big optics problem."

It was also a very big political problem. Kat Lanteigne and others lobbied and protested. A hearing was held in Ontario's legislative assembly, during which Doris Grinspun of the Registered Nurses' Association of Ontario told a committee that "blood and money simply don't mix—at least not for nurses." A former deputy minister of health said it was corrosive to have private collection. Objections were not about the safety of paid plasma but the damage that privatization could do to a socialized health care system. If people were being offered money for plasma, why should they freely give blood? Nor was the defense of the Plasma Protein Therapeutics Association particularly persuasive. "The concern about paying for blood," said PPTA spokesperson Joshua Penrod, "isn't really accurate, because that's not what we do." He said, "It's as simple as that."[52]

Nothing about the plasma industry is simple. Nothing about blood supply is simple. The head of Canadian Blood Services, Dr. Graham Sher, gave the committee some sense of this. CBS collects enough whole blood to be self-sufficient in fresh frozen plasma for transfusion. But it collects only 200,000 liters of plasma for protein products, when it needs 800,000 to meet the need.[53] This need is growing: for some reason as yet unclear, Canada's use of immunoglobulins—particularly IVIG—is greater than in most other industrialized countries. "Nobody can explain the demand," says Kat Lanteigne. "Last year it was 4 percent, this year it's 7 percent." Lucy Reynolds, a researcher at the London School of Hygiene and Tropical Medicine who wrote an exposé of the "grubby" global plasma industry, thinks it is due to off-label use by physicians.[54] A pharmaceutical trade publication, remarking upon human blood's "unparalleled healing properties," listed neurological conditions for which IVIG is now being used: multiple sclerosis, neuropathic pain, chronic fatigue syndrome, and asthma.[55] I've seen estimates that up to three hundred conditions are being treated with IVIG. Many of the conditions taken singly would send pharmaceutical company finance departments into a merry tailspin of profit calculation. But what if immunoglobulins could treat dementia? Various studies and trials have explored the ability of IVIG to attack the proteins that build up in a demented, damaged brain. A Baxter product, Gammagard, was disappointing and its development as an Alzheimer's treatment discontinued.[56] But a 2015 study by the Sutter Neuroscience Institute in California, using an Octapharma product, claimed more promising results.[57] There is little reason yet for anything other than caution. But if plasma products can be used to treat dementia, this would create a $7.2 billion market in the United States alone.

Graham Sher's testimony to the Ontario legislature was not what I'd expect from a man who runs an organization set up to deliver a blood service based on voluntary donations of blood products. He did not condemn the practice of paying for plasma. It was something for provinces and territories to determine. "Decades of evidence have proven that drugs made from plasma derivatives today are inordinately safe and just as safe as those made from volunteer donors. This is not the 1980s." In this, he was aligned with Bahardoust, who believes the protests were fueled by Canadian Plasma Resources' mistakes. A misunderstanding. Negotiations had been going on with Health Canada for four years before the clinics arrived, but CPR had not done any public relations. "We very much underestimated the sensibility of this issue," Bahardoust says. "We didn't do any public relations or any government relations apart from meeting with bureaucrats in Health Canada. That was, looking back, a very big mistake on our end."

It was also costly. When the Ontario legislature disagreed and passed the Voluntary Blood Donations Act, the clinics were told to shut down. When they carried on operating anyway, the police were sent in.[58]

Canadian Plasma Resources, unbowed but poorer—it's estimated it lost several million dollars in its Ontario ventures—went looking elsewhere. CPR needed a provincial government that was conservative and free market and could stand up to unions scared of privatization. It found one in the province where Canada's socialized medicine system had begun.

My guide around Canadian Plasma Resources' clinic in Saskatoon is a tall, fair man called Jason. He is a project manager, not an academic or medic. "A jack-of-all-trades, really." He grew up in a small town in Calgary and spent most of his career in oil. This is, then, an appropriate switch, as his experience of dealing with a highly lucrative liquid will be useful in the plasma industry. Trapping the volatile prices of oil barrels to give a stable comparison is tricky, but in a 2012 paper for the *William and Mary Business Law Review*, Sophia Chase gave a persuasive example from 1998, when a barrel of crude oil was worth $13 a barrel, but a barrel of blood would have cost $20,000. Blood separated into its derivative products would have been worth $67,000, while the barrel of oil, even including its derivatives, was worth only $42.59.[59] Yes, Jason says. "There aren't going to be many downs in this business."

None of these calculations appear in any of CPR's literature or online. The slogan is "Give plasma, give life," and the language is that of donation and gift and good, not profit and potential. It is no different from the marketing of voluntary blood donation, until you get to the page on Donor Compensation. But that is for later. First, a tour. Jason fetches me a white coat, more for theater than hygiene, as there are no other requirements such as hand gel or hairnets. The first step for what he calls donors and what I call sellers is an ID test: they must prove that they live within sixty miles of the clinic and not in a shelter or dubious hotel. Local donors, Bahardoust tells me, are more likely to be committed donors. A fixed address is a requirement. They keep a database, says Jason, and update the list of dodgy accommodations as best they can. As for the location limitations, they use Google.

With the right ID and the right answers on the questionnaire, the donor-vendors are screened by a nurse, then go through to the giving room. It's an impressive space, as is the clinic: white, shiny, new. It's not new: when the Toronto clinics were closed down, they simply moved the clinic, cots and all, to Saskatoon. The other clinic is in storage, waiting to be transported to Moncton, New Hampshire, where CPR is planning a second clinic.

There are sixteen cots. Currently the clinic sees about thirty-five donors a day, which is far fewer than CPR would like. Bahardoust would like to see an increase to one thousand appointments a week. That is a reach: business is modest but steady. I ask to interview some people attached to the plasmapheresis machine. Sure, says Jason. He is remarkably casual and open. It's nothing like a later visit I make to Saskatoon's blood donation clinic. It had taken me three weeks of asking a Canada Blood Services PR rep to be allowed to visit. Eventual permission was accompanied by strict instructions. "One of the supervisors will give you an operational tour. But you can't talk to anyone. If you want pictures, we will send you pictures. If you want commentary we will send you commentary." The PR rep finishes with a flourish. "The woman you are meeting is Karina with a K, and she will not give you her opinion on anything." By the time I do get to the clinic and take a picture of the clinic sign, I almost hide behind a parked car in case they see me and drag me into the blood services prison for having taken a picture. There's none of that here at CPR. If Jason doesn't answer something, it's because he doesn't know it. I'd asked Bahardoust for a tour only the day before and been given no restrictions. It is in CPR's interest to be open and apparently transparent. They remember the bad press and want better. But when Graham Sher of Canadian Blood Services later says that they are losing donors to CPR, I wonder whether it's because CBS needs a lesson in getting good press.

The first seller I talk to is Gail Wittig. She's watching something on her phone while the machine takes her whole blood, spins it, retains about 800 milliliters of her plasma, and returns her red and white blood cells. She is exactly the kind of person whom Graham Sher is worried about, because she has given whole blood forty times, and now she gives plasma. She used to do plasmapheresis when Canada's blood was run by the Red Cross. "When I found out about this place, I thought, I'd like to do that again, I think. So I've come over to this side I guess. You can't really do both." Studies have shown that once a voluntary donor switches to being paid, either for plasma or whole blood, they rarely return. When a German company began paying for blood donations, then quickly shut down its business, the local Red Cross blood centers could get back only one in six of the donors they had lost.[60]

Gail works in a lab and knows that this plasma is going to make medication. She says, "They have their little thing out there," meaning a display cabinet in the lobby that holds a few medicine bottles and empty IVIG vials. She's not here for the money, she says, but because she thinks that giving plasma, even when it becomes a product with a huge profit margin, is a good thing. It helps people.

The next seller is also a lost blood donor. He's a twenty-two-year-old student who last gave blood a year ago. Now he's going to sell plasma instead. Why? "Well, one, you get paid. Two, I did blood before, and I felt good. Plus I'm a student." I ask him if he knows what happens to his plasma and he is vague. "Kind of. Like, pharmaceutical companies? I don't know. Not much." I feel like turning a screw, a little. You know this place is controversial? He doesn't. I tell him about Krever and tainted blood, waiting for Jason or the clinic nurse to stop me. They don't. I tell him that people are worried that this clinic is the beginning of something rotten. He is pleasingly concerned, though probably just polite. "I'll definitely go and research it." And he will definitely come back for his money. First-time donor sellers don't get paid until their plasma is quarantined and screened. Seven to ten days after the first plasma donation, they get a Visa card in the post, loaded with $70: $45 for the first lot, $25 for the second. If they sell five lots within three months, they get $50 for their fifth, and $100 for their tenth.

Plasma can be stored for up to three years. At the time of my visit, and probably for a long time after, Canadian Plasma Resources had no market for its plasma. It can't sell it to Canadian Blood Services because it has no contract to do so, and there is no fractionating facility in Canada. What is not set out in the leaflets and advertising is that this plasma sold by Canadians is currently going no farther than a fridge in the back room overseen by a man called Innocent. The IVIG in those now-empty bottles on display in the lobby was made from American plasma, like 85 percent of Canada's source plasma, or fractionated outside the country, like all of Canada's plasma.[61]

This is the case in plenty of European countries whose residents would not countenance the selling of blood but who use medicines derived from paid American plasma donors. Britons have been using foreign plasma for decades, since vCJD—mad cow disease—infected the blood supply (its prions travel in plasma). In 2002, the government bought a US plasma company to supply enough plasma for protein therapeutics (it buys plasma from Austria when stocks fall) that is fractionated by Plasma Resources UK.[62] It quietly sold it off in 2013 to Bain Capital, which in turn sold it to a Chinese conglomerate for a large profit.[63] British people who would be horrified at being paid for blood are now receiving plasma products sold by Americans and governed by the Chinese. On its website, BPL claims that it is regularly audited by government inspectors. When I wrote to ask who would inspect it now that it had been liberated into the private sector, no one replied.

Dr. Ryan Meili is a Saskatoon doctor who has worked for many years in an inner-city clinic. During my visit, he was also fighting a campaign for election to the provincial government. When we meet in a downtown café, I recognize him easily as the man whose face I've been seeing, billboard-size, all over town. He speaks quietly and fluently. His worry about the new clinic is on two fronts. The politician in him is disturbed by the language of donation, which "gives the impression that this is a benevolent operation, a nonprofit whose main concern is the public good. Whereas it's clearly a for-profit organization that is selling these products for the development of drugs." The other worry is medical. The literature into long-term health effects of frequent plasma selling is "not rich," he says, but the people in his clinic have high rates of diabetes, hepatitis C, and HIV. They have poor diets and food insecurity. "So adding this literal drain on their bodies is something that I think should not be done without significant further investigation, and really should not be done at all." I ask him whether his medical or moral concerns are uppermost, and he gives an answer that, if I were from Saskatoon, would have got my vote. "The practice of medicine is a political act however you choose to do it."

Don Davies is the health spokesperson for the opposition National Democratic Party. He objects on the grounds of "two S's." Science and safety. "When you introduce a profit margin, you've now introduced a competing value to safety." Anyone who reads the Krever Report understands that, because of its finding that dangerous inventory was given to people only to cut costs. "We don't think anything should compete with the safety of blood."

For Kat Lanteigne, creeping privatization is what concerns her most about Canadian Plasma Resources. It is not the safety of the product that concerns her but the safety of the system. A private clinic is not linked to the public health care system, so if something did go wrong, it would be harder to trace. "When you go to a Canadian Blood Services donor clinic, you are under the umbrella of our health care service, so a nurse taking care of you is a Canadian public health care nurse, and if something shows up in your blood you get a phone call immediately from a Canadian blood services nurse, and you go to your doctor and it's integrated." She tells me of an artist several years ago who used a bag of blood in an artwork. "A journalist asked him where he got the blood, and he made a joke and said Canadian Blood Services. The agency flipped. But they were able to find that it wasn't from them, that nothing was missing." She thinks this kind of scrutiny will be diluted by privatization. Graham Sher of Canadian Blood Services has said differently. "We may have moral objections and philosophical objections to paying," he told the *Toronto Star*. "But let's not make it an issue about safety when it's not about safety."[64]

In 1987, Ann Hume, living in Shetland with a new partner, was pregnant again. By now there was AIDS in the news, and she thought of this and remembered her terrifying hemorrhages all over her kitchen floor. She asked her hematologist to test her for HIV. "And oh, she went off her head. She said I would never get any viruses out of Scottish blood products." She got the test and she was negative, but only for HIV. "I went ahead with the pregnancy not knowing I had the hep C. I had got hep C in 1982." She sends me her medical records, mostly correspondence between the hematologist and Ann's Shetland GP. Typewritten, first, then computerized. Ann has added handwritten Post-it notes here and there. On a letter dated August 18, 1982, the hematologist wrote that Ann's Factor VIII level "was totally normal." Ann's Post-it: "Yet she still gave me Factor 8?"

She carried on. She was busy with three children and attributed her backaches and joint pains to life or aging. One day she was at the hospital for physiotherapy and saw something in a magazine about the Hemophilia Society. "I wrote to them and got back a letter on leaflets and hepatitis C. I read it and said to my partner, this is what I've got. I've got all the symptoms." When she got her positive test result, she was given another leaflet that informed her that 80 percent of people with hepatitis C would get cancer. Her partner said, We can't tell anyone. The Hemophilia Society said, Don't tell anyone. The stigma of having a blood-borne virus, in the days of AIDS, was high. "You've got this illness but you can't tell people what it is in case it goes all around the town, and everybody is pointing their finger at you." And sometimes they don't believe you. In a letter that Ann sends me from her doctor, the hematologist described this young woman as arriving on crutches and having a vivid imagination. The connection between the two is clear.

Ann hasn't had cancer yet, just persistent joint pain, osteoarthritis, and overwhelming fatigue. She struggled to look after her children. She couldn't work. She got a new hematologist, who had to hunt for her medical records because they had disappeared. Disappearance of documents is something tainted blood campaigners get used to. When Lord David Owen tried to access his records from his time as health minister, he was told they had been shredded according to "a ten-year rule." There is no such thing. But nor are there any records. Ann's new hematologist, a Mr. Watson, also hunted for something else: in her dossier, there are letters to Watson from the Aberdeen and North East Scotland Blood Transfusion Service, assuring him that Ann had not received any blood products—Factor VIII—from their stocks. A few weeks later, another letter, from the director of the Scottish National Blood Transfusion Service Protein Fractionation Centre, who remembered that hemophilia doctors had used American Factor in the early 1980s. Other letters, other official forgetfulness. Another Post-it note: "This Factor 8 bottle brand name was never found." She did finally get a diagnosis, of a platelet clotting disorder. And after taking a drug called Arvon, she has been cleared of the hepatitis virus. But she is not symptom-free. Walking is exhausting and painful. Her joints ache. She blames her first hematologist more than the authorities. "I suppose I blame them for taking it into Britain in the first place. If it was American stuff I got, I don't know, I'll never know. But the genotype of the hepatitis C I had was called 1A and it's prevalent in North America. So that's what makes me think mine was American. But nobody will tell me."

Other victims of contaminated blood were more successful in their efforts at tracing the source of their infection. But when Jason Evans, whose father Jonathan was killed by AIDS at the age of thirty-two, tried to seek his dad's medical notes, he was told they had disappeared. In an interview for the BBC, he said, "Had I not met other people in the tainted blood community, I wouldn't have thought anything of it. But they all said the same thing." Papers had vanished. "I don't think you can help but be suspicious." Evans has now launched another lawsuit—there have been many before, but none successful—using new documents he has found in archives. In early 2017, the English MP Andy Burnham, in his last speech in the House of Commons, said that the contaminated blood issue, usually described as a "tragedy," with the implication of happenstance and accident, was worse. It was, he said, "a criminal cover-up on an industrial scale." Medical records had been doctored. Pages were missing. In 2016, a Conservative MP named Peter Bottomley participated in a debate about tainted blood in Parliament. His mother, he said, had had an HIV test after receiving a blood transfusion, and he supported greater transparency and investigation. He said, of people whose blood was tainted, "People should go out of their way to put arms around them, act not just like a two-armed human being, but like an octopus and get right around them and try to meet all their needs in a way that they find acceptable."[65] This was a rare example of warmth toward a community of people who feel mostly discarded and treated with abominable disdain. These people who were "only hemophiliacs." "We weren't hearts, we weren't cancers," said David Watters, formerly head of the UK's Hemophilia Society.[66]

In Canada, Alberta has become the latest province to introduce legislation to ban the sale of blood, even when it's called "plasma." But the Canadian Plasma Resources clinic in Moncton is now open for business after months of opposition. The plasma protein therapeutics industry is predicted to grow 10 percent a year, with no recession in sight.

Plasma products are as safe as biological products can be. And accusations that the industry targets poor people are unfounded, according to the Plasma Protein Therapeutics Association. "Plasma donors," it wrote in a recent press release (not "plasma sellers"), "deserve our gratitude and respect, not sweeping negative characterizations." To say otherwise is "unfair to plasma donors as well as to individuals living with rare, genetic and chronic disease who rely on access to plasma protein therapies."[67] Perhaps it is also unfair to wonder about the safety of an industry run by only four companies, when conglomerates and monopolies dominate business (the US petroleum industry is in the hands of a handful of corporations; so is its bottled water industry). Pathogen inactivation techniques are sophisticated and widespread. But population growth, deforestation, and climate change are all pushing humans and wild animals closer together. Viruses like that; they like to jump. We can inactivate only the pathogens we know about. Scientists such as Jonathan Quick of the Harvard Medical School believe there will be another major outbreak of something—in our air, food, or blood—in the next fifty years. Since 1975, twenty-five new pathogens have been discovered for which there is no vaccine or treatment. In

2012, a Chinese team detected parvovirus B19, a pathogen discovered in 1974, in more than half of Chinese plasma pools. H7N9, a flu usually confined to chickens, has begun to cross into humans and kill them. China has the largest plasma industry in the world after the United States. How safe is safe? The unhysterical *Lancet Haematology* wrote a recent editorial on "the big business of blood plasma." The financial incentive, it read, "can encourage lying during medical screening and could adversely affect the health of some of the donors as well. With some companies pooling hundreds of thousands of donations together for processing, this can be an important safety risk."[68] We can't protect ourselves fully from what we don't know is coming.

In the UK, the prime minister has promised a proper inquiry into the contaminated blood scandal, though at first the government wanted the Department of Health to investigate itself. There are hopes but not high ones. The death toll that will come from hepatitis C is unknown, because it is not routinely tested for. When doctors in a London hospital recently did a pilot study, they found that rates were three times as high as was believed. Only 250 of the 1,500 British hemophiliacs infected with HIV are alive. Sixty-two contaminated blood victims have died since plans for the new inquiry were announced in July 2017. Of 89 hemophiliac boys at Treloar College during the years of contamination, 72 are dead. I understand why hemophiliacs call themselves the "shut up and die" community: wait long enough, as some governments have, and probably there will be no one left to complain. I watch on YouTube a short film by Bruce Norval, a Scottish hemophiliac infected with hepatitis C, and a vocal campaigner. He filmed it outside, with him leaning against a concrete post. From the exhaustion in his voice, I suspect it is a prop in more ways than a staging one. He sounds weary, but there is force in his quietness. "I shall retain an absolute shame," he says, "to be part of a country that would perpetuate such a crime." He is sure that the British authorities who gave tainted factor to hemophiliacs without telling them and when they knew it wasn't safe are stalling. They just have to wait sixteen more years and all the infected will be dead, of hep C or complications from HIV, of tainted plasma protein products or what he calls "manky blood." It's just math now. But they're not gone yet. And they're not done with wanting to know why something that was supposed to be safe, and that was supposed to give them life, gave them death instead. So "for that last little bit," says Bruce, in his unsettlingly quiet tones, "we're going to scream blue murder."

A *chaupadi* shed, western Nepal

SIX

ROTTING PICKLES

For Radha dinner is served at seven. She crouches down behind a shed, a good distance from her house, then waits. She knows what the menu will be: boiled rice, the same as yesterday and the day before. She knows that it will be her little sister who serves it to her, throwing the rice onto the plate from a height, the way you would feed a dog.

In Jamu, this village in western Nepal, Radha's status is already an inferior one. She is ironsmith caste, a low person. When she menstruates, her status drops further. She is only sixteen, yet for the length of her period, Radha can't enter her family house or eat anything but boiled rice. She can't touch other women, not even her grandmother or sister, because contact with her will pollute them. If she touches a man or a boy, he will start shivering and sicken. If she eats butter or buffalo milk, the buffalo will sicken and stop giving milk. If she enters a temple or worships in any way, the gods will be furious and take their revenge by sending snakes or some other calamity. Radha is allowed to go to school; many girls are not.

Where Radha lives, menstruation is dirty, and a menstruating girl is a powerful, polluting thing. A thing to be feared and shunned.

After dinner, Radha prepares for bed. Darkness falls fast in Jamu, and without electricity the villagers follow old rhythms and sleep with the dark. Radha's parents are both absent: this village, like many in Nepal, has spit out its menfolk, mostly, to be migrant workers elsewhere, but also some women, like Radha's mother. Most Indians know that Nepalese make good security guards. Gulf Arabs know them as construction workers, often dead ones who are crushed in stadia and scaffolding. So Radha lives with her grandmother and her sister, in a house of women. Their home has a solar-powered light, as does the one opposite, where I'm staying with my traveling companions: Anita, the communications and gender officer for WaterAid Nepal, and our photographer, Poulomi. Our hostess is the local schoolteacher. She seems nice.

The solar light is no use to Radha this week because her bed is elsewhere. She leads me over the thoroughfare of pebbles and rocks that passes for a road, suitable only for motorcycles, walkers, and snakes. We hike up a steep hill, through long grass, to a small lean-to structure. It looks like an animal shed, but it is smaller and meaner, its planks rough and scrappy, its shelter imperfect. This is where Radha must sleep because she is menstruating.

In the local dialect, Radha is now *chau*. Originally meaning "menstruation" in the Raute dialect of the far western region of Achham, it has come to mean "an untouchable menstruating woman." This linguistic melt has also happened in English: "taboo" derives from either the Polynesian *tapua*, meaning "menstruation," or *tabu*, meaning "apart." The system of keeping girls and women apart is known as *chaupadi* (*padi* means "woman").[1] The shed is a *goth*. Radha hates it, whatever its name. "I'm forced to stay there. My parents don't let me stay at home. I don't like being there, it's dark, there's no light. In the winter it's cold. I feel so scared."

In the winter, Radha sleeps on the tiny enclosed ground floor, no bigger than a crawl space. The summer accommodation is an earthen floor on a platform above, four foot square, which is open to the elements except for a grass roof. There is not room even for one person to lie down, but tonight there will be three. Radha's relative Jamuna is also menstruating, and she'll be sleeping here along with her one-year-old son. Radha appreciates the company, as Jamuna's presence may be some protection against drunken men who conveniently forget about untouchability when it comes to rape. The stigma keeps women silent, but rapes of those confined to these sheds are common enough to appear as occasional items in newspapers in faraway Kathmandu, and common enough for some women to look down or away whenever I ask them about it. Also common are snake attacks and deaths. (During my visit I see three snakes in three days. Large ones.) In early December 2016, fifteen-year-old Roshani Tiruwa lit a fire to keep warm in her *chaupadi* shed and suffocated.[2] The following summer, Tulasi Shahi was fatally bitten by a snake as she slept in her uncle's cowshed in western Nepal.[3]

Sometimes there are four or five women in Radha's family shed, an unthinkable number. There are always other options, though never the safety and warmth of her own home. Farther along the plateau, I watch with disbelief as a fourteen-year-old girl shows me her sleeping arrangements for the night: the bare ground outside her family's house. She has rigged up a mosquito net, tying it to posts just high enough so her body can lie horizontally under it. She will sleep on dirt and discarded corn husks. A bed of rubbish. It's only the third time she has had her period and already she is resigned. What can she do?

Jamu is remote. We take a two-hour flight from Kathmandu to Nepalgunj on Buddha Air, during which I watch and listen with stupefaction as an old woman asks for the window to be opened so she can spit, then—as the window doesn't open, thankfully—spits all the way to our destination into a bag, with gusto. After that, there is a four-hour drive on a road that is mostly potholes linked by afterthoughts of tarmac. Occasionally Anita gets out and builds makeshift bridges out of boulders so our jeep can cross unexpected torrents. Finally we are tipped out with our packs at a river, where we wrap electronics in plastic safety and wade through thigh-high water because there is no other way. We are in the midwestern part of Nepal but not Himalaya country. Hills, not mountains. Lush greenery, not rocks and sheer drops. It is one of the most beautiful places I have ever seen, and I am here to look for one of the ugliest things I have ever heard of.

A 2010 government survey found that up to 58 percent of women in Nepal's far west regions reported having to live in a shed while they were menstruating.[4] But Jamu is in the midwestern hills, where the survey judged rates of severe discrimination (staying in a shed; being given separate food) to be less than 10 percent.[5] I expected to find progress here. It's easier to get things to lowlands, even by foot with no vehicle access. Things like emancipation and equality and the idea that women and girls shouldn't be banished to unheated sheds because of their biology. So I was worried, with that unforgivable concern of someone who wants a story, that the *chaupadi*s would be gone.

Three more river crossings, an hour of walking, then we reach the village of Narci, one of our stops en route to Jamu. There is a *chaupadi* shed outside the first house, then the second, then every other one. Either the 2010 surveyors didn't like river crossings or people lied to them. Some sheds contain possessions: a comb stuck in the thatch or a bottle of red nail polish. Some hold schoolbooks, ready to be studied by girls who manage to spend all day mixing with boys at school without causing scourges or sickening. Many of the *chaupadi* restrictions are rigorous: whether girls can go to school is more flexible. The cowsheds and storage barns are well kept, with corn husks drying for winter. They are swept and clean. Not the *chaupadi* sheds: they are too small to sweep.

A group of women gather to talk. One said, Where would you like to sit? We move to the ladder that leads to the residential part of the house, on the first floor. But she stops us. "I can't come, I'm on my fifth day."

We sit outside on the ground instead. We are all women: us with our nosiness and notebooks, the villagers sitting patiently, ready for another round with the well-intentioned. One woman sits with a scythe in her hand. Her name is Nandakala and she is also menstruating. She says, along with the others, that *chaupadi* is necessary. If menstruating women don't observe the taboos, bad things happen. A buffalo could climb a tree. Men would start trembling and fall ill. Snakes will be brought by the sin. A woman becomes animated at this: "Yes, it's true. A big snake came into my house. We all saw it." Another says, "If I touch something, I'll get ill, so why should I think *chaupadi* is a hardship?" *Chaupadi* keeps them safe. In this group setting, nobody protests. It is our tradition, they said. It's what our parents and grand-parents did, so it's what we do. I ask what they say when people come and tell them *chaupadi* is wrong. Do they admit to being in favor of it? "We won't lie. We'll say what we've said to you."

But as she has her picture taken in her *chaupadi* shed, a hundred yards away, Nandakala is more frank. She isn't worried about rape. The men have all left to work in India or Dubai: who is left to do the raping? She tells Poulomi, "Of course I hate it." In the winter it's cold. In the summer it's hot. The restrictions are stifling and unfair. "Why should the gods punish us? Why should women be punished? But what the hell can we do?"

In the next village, we stop at a house with a view of the rushing Bheri River, so blue and wild. In this cluster of houses, 90 percent have a *chaupadi* shed. In one, there is a cup and bowl belonging to a female guest who had just left. They will stay in the shed until the sixth day, then be cleansed with fire and taken into the house. The guest must have been unmarried: married women have to observe *chaupadi* for only three days, not the full five or six. The woman of the house told us, "I don't believe in this but my mother-in-law does."

This is not a simple story of patriarchal men imposing evil restric-tions on suffering women. *Chaupadi* is driven by women. It is perpet-uated by the grandmothers and the mothers-in-law and the mothers. Nandakala, brave in private, moved to Mumbai with her husband for six years, where they didn't practice *chaupadi*. And so he fell ill. He got eye pain, knee pain, he shook. I made him sick, said Nandakala. The taboo was wrong yet true. So now she does *chaupadi*.

In the next village, a schoolgirl on her period talks to us. We know she is menstruating because she won't come near us. She says, "I can't go into the house, I can't touch water, I'm not allowed to touch men." Yet she must do chores. She must fetch the shopping. "I have to say I'm menstruating and the shopkeeper throws the stuff at me." Some-times she doesn't use words but shows him, somehow, her unwillingness to take something from him and he understands. Yet she has touched boys at school and nothing happened. "Of course menstruation is dirty," she said, sitting in her *chaupadi* with her schoolbooks that should have told her different. "It's a dirty thing."

Twenty-one liters, give or take. That is how much menstrual blood I've discharged over the past thirty-five years.[6] I've never done that calculation before now, because why would I? I'm not supposed to celebrate, calculate, or in any way highlight my menses. Nor am I supposed to use that old-fashioned word, though I like its lyricism (it comes from "monthly"). Some other words: *Uterus*. Yuck. What a horrible word. *Vagina*: even worse. *Menstruation* sounds like a disease. *Menarche*, *endometrium*: what do they even mean? Euphemisms are everywhere. Having written a book on sanitation, I've become expert at them. Languages have always contained them. Not many diaries or records exist that record what women felt about or called their periods, but the historian Sara Read, in a survey of menstruation in early modern England, gathered a few names: the Visit, the Courses, Terms, Those, Monthly Sickness, Time Common to Women, Months, Gift, or Benefit of Nature.[7] Euphemism: "To use a favorable word in place of an inauspicious one." Euphemizing is the opposite of blaspheming. The same magic was supposed to work when the Cape of Storms was renamed the Cape of Good Hope, yet it stayed just as stormy. Perhaps that wishful thinking is why menopause is known as "the change," a bland word that holds none of the distress and despair of endless hot flashes, depression, brain fog, and eradication of libido. Or maybe euphemisms are a way of sticking women's health in the dark and unspoken corner where it's supposed to belong.

Once in India, I was puzzled when my friend Sabrina started talking about "chumming." Chums are Indian periods. There are at least five thousand other terms, according to a recent survey carried out by the makers of a women's health app called Clue, in partnership with the International Women's Health Coalition. Here are a few: on the rag (a term that always made me look at student rag weeks in a different light), the curse, shark week, having the painters in, Aunt Flo, and the infinitely useful "time of the month." Northern Europeans resort to fruit: lingonberry week in Sweden, strawberry week in Germany. The

French supposedly call it ketchup week, which is disappointing: I'd have expected at least a gastronomic *jus de* something or other. You may want to applaud the creativity of these, but they are all just part of a linguistic scaffolding of shame and secrecy.[8]

Some other things I'm not supposed to confess: the time in an Indian restaurant in Paris when I bled all over a silk cushion, and I'm mortified twenty years later. All the occasions when I had no sanitary products and resorted to wads of toilet roll in my pants. The time when a school friend started her period and none of us told her she had bled through her pale-blue summer uniform. (I'm sorry, Sally.) For something so red and vivid as menstrual blood, it is very, very quiet.

Only half of it is blood, anyway. Every month I and two billion other women discharge blood but also epithelial endometrial layer, the underlying lamina propria, and vaginal and cervical mucus. Most is the lining of the womb, the thick and rich endometrium that is meant to host an embryo. In the words of a 1966 puberty education film, the endometrium makes the uterus "a soft, nesty place." The whole process seems mechanically simple. But menstruation makes no sense. Evolutionary principles dictate that things that cost us should also benefit us. Yet we lose 30 to 50 milliliters of blood and tissue per month and get pain, bloating, depression, and attendant symptoms. What is the benefit? Other species don't bother menstruating because they retain their womb lining. We are in a minority among species, and among mammals, to bleed every month. The only other animals known to menstruate are apes, Old World monkeys, the elephant shrew, and four varieties of bat including *Desmodus rotundus*[9] (which I cite because it is a vampire bat, and its name means "two-thirds of the way around").

There have been many theories. Maybe menstruation is the womb ridding itself of nasty toxins from sperm from all the sex we're having. Except, levels of promiscuity don't correlate with the amount of blood we lose: monkeys have sex like rabbits (which produce an endometrium only when they copulate, a practice I find entrancingly optimistic). But monkeys and apes bleed less than women. Or perhaps it is more economical for the body to rid itself monthly than to keep a constant endometrium, as many other species do.

A better and more interesting theory is the conflict hypothesis. Our

endometrium is so thick and nesty because our embryos are so invasive and parasitical. According to this, humans are one of the rare species to have "maternal-fetal conflict."[10] In other animals, the embryo
and its surrounding placenta attach only superficially to the womb
lining. Human embryos are greedier: the embryo and placenta attach
to the endometrium but then burrow through it, tearing open arterial
walls and diverting them to pass blood to the growing embryo. In this
way, the fetus has a direct line to the mother's main blood supply. "It
can manufacture hormones and use them to manipulate her," writes
the biologist Suzanne Sadedin. "It can, for instance, increase her blood
sugar, dilate her arteries, and inflate her blood pressure to provide itself
with more nutrients."[11] It can also, according to sexual health researcher
Dyani Lewis, dampen a mother's response to insulin so that "a greater
slice of the circulating sugar pie is placenta-bound during its nine-
month residence."[12] This is a hemochorial pregnancy, and it is a battle
between two sets of genomes. Fetal genes want to ensure their own
survival so suck up as many resources as possible.

Because the process of pregnancy is so taxing—I've never been pregnant and I'm exhausted reading about it—the uterus will be extremely
choosy about which embryos get to inflict such a drain on the mother's
body. A total of 30 to 60 percent of embryos are discarded. Anything
not up to standard is ditched, along with the endometrium. "You've
probably read," writes Sadedin, "about how the endometrium is this
snuggly, welcoming environment just waiting to enfold the delicate
young embryo in its nurturing embrace. In fact, it's quite the reverse.
Researchers, bless their curious little hearts, have tried to implant
embryos all over the bodies of mice. The single most difficult place for
them to grow was the endometrium." The biologists Deena Emera,
Roberto Romero, and Günter Wagner wrote that this "evolutionary
tug-of-war between maternal and fetal genomes" was similar to virus-
host interactions. Not everyone agrees with this hypothesis, and
research into what placentas have done throughout history is difficult
when they fossilize rather worse—that is, not at all—than bones. But
whenever species have this kind of invasive pregnancy, and the "spontaneous decidualization" of the womb lining, they also menstruate.

It's a persuasive theory, if an unsettling one. Disquiet and distaste are things you get used to when you read about menstruation.

"The menstrual discharge," wrote the male anthropologist M. F. Ashley-Montagu in 1940, "is most generally conceived to be a particularly noxious effluvium which automatically renders everything unclean with which it comes into contact. That being so, the female during her catamenial flow is considered to be herself unclean and as noxious as the effluvium itself."[13] Anyone writing about menstruation or, as I may call it from now on, the noxious catamenial flow, starts with Pliny. Gaius Plinius Secundus was known as Pliny the Elder and for his multivolume *Natural History*. There are many wonders in the thirty-seven volumes, but even Pliny admitted that it is difficult to think of "anything which is more productive of more marvelous effects than the menstrual discharge."[14] Human females, he wrote, are the only "animated beings" to have a monthly discharge. He was wrong about that. He was wrong in abundance.

On the approach of a menstruating woman, he wrote, nature would cringe and submit. "Seeds which are touched by her become sterile, grafts wither away, garden plants are parched up, and the fruit will fall from the tree beneath which she sits." Her look, also, is formidable, because it will "dim the brightness of mirrors, blunt the edge of steel and take away the polish from ivory." She can kill a swarm of bees, turn iron and brass rusty. She can scare away hailstorms and lightning, as long as she is both bleeding and naked. At sea, she doesn't even need to bleed: a storm will flee before the sight of her unclothed body. What a useful creature. Farmers must employ their menstruating wives with great joy, because "if a woman strips herself naked while she is menstruating, and walks round a field of wheat, the caterpillars, worms, beetles and other vermin will fall from off the ears of corn."[15]

I wish some of these were true: it would save time weeding. The editor of one edition of the *Natural History* adds a footnote to say that Pliny's accounts are "entirely without foundation." But they were built upon centuries of belief about the power of the menstrual woman and others built upon them in turn. It's telling how many of the Pliny powers were judged to be witchcraft. Menstruation must have been unsettling. How could women bleed and not die, when men bled and did? Before agriculture and settlement, as Janice Delaney, Mary Jane Lupton, and Emily Toth write in their book *The Curse*, women's blood was judged to be good. It was like other cyclical processes that seemed magical—the sun, the moon, the tides—and deserved appropriate awe. "Worship and appeasement of the Great Mother and her bleeding fertility would ensure [early man's] temporal safety."[16] The blood turned bad when man became a farmer, life became more stable, and he had less need of magical protection. Then, the menstruating woman became taboo, set apart and separated from things she may damage, like crops and harvests.

By the time of the Old Testament, the evil of menstruation was firm enough to be used as an analogy: the book of Isaiah urged the observant to cast away their sinful silver and gold idols as they would a menstruous cloth. Whoever wrote Leviticus was more straightforward. After pronouncing purity rules around leprosy, he moved on to sperm and menstrual blood. "When a woman has a discharge, if her discharge in her body is blood, she shall continue in her menstrual impurity for seven days; and whoever touches her shall be unclean until evening."[17] At least he was fair: men who emit sperm outside intercourse are equally unclean. Both men and women should offer two turtle doves or pigeons to the Temple at their end of their cleansing. (Knowing something of young men, I'd guess their turtle dove expenditure was higher than women's.) Leviticus's egalitarian pollution was not shared by Aristotle, who knew sperm was a much higher class of discharge.

In the thirteenth century, Nahmanides (Rabbi Moses ben Nahman or Rambam) judged the menstrual woman and found her wanting. "The dust on which she walks is impure like the dust defiled by the bones of the dead."[18] Most religions agree that a menstruating woman should stay away from God or holy books and places, and they are emphatic about cleansing. Buddhists are the most relaxed, but Japanese Buddhism requires women to cleanse for eleven days after a period. Women who have given birth have to cleanse for only ten.

The most creative response to the fearsome catamenial flow comes from the islanders of Wogeo in Papua New Guinea. This place was described by the anthropologist Ian Hogbin in a 1970 book as "the island of menstruating men." Women's menstrual blood is both dangerous—she can kill a man by touch when she is bleeding—and cleansing, enough that men simulate it with a creative technique involving crabs and penises. First, writes Hogbin, the man catches a crab and steals a claw. He spends a peaceful day of nil by mouth, then late in the afternoon:

> He goes to a lonely beach, covers his head with a palm spathe, removes his clothing, and wades out until the water is up to his knees. He stands there with his legs apart and induces an erection either by thinking about desirable women or by masturbation.

Then, he takes his stolen claw and hacks at his penis until blood flows. He must wait until "the sea is no longer pink" (this makes me wonder how much blood Wogeo men contained), then returns ashore. At that point, both menstruating women and fake menstruating men observe the same rituals, though the woman has to stay at home and is not allowed to use doors. When answering a call of nature, she "has to leave and enter through a hole in the floor or the wall."[19]

It is unclear whether the island of menstruating men developed its rituals because men were envious of periods or frightened by them. But anthropologists cite other tribes where the bleeding woman is treated with kindness and respect. Among the Yurok Indians of northern California, menstruating women are spared all chores and duties for ten days because they are on their "moontime."[20] The Kalasha women of the Hindu Kush retire to a prestigious structure called the *bashali*, where women hang out, have fun, and sleep entwined. In this reading of menstrual seclusion, the woman is prized for her blood, because it means fertility and power.[21] She enjoys the time off (who wouldn't?). In Nepal, I was told that some girls like to spend time in the *chaupadi* huts with their friends: they play online games on their phones (because poor people have phones, too) and have slumber parties, even if the slumbering is cold and likely to be disturbed by men and beasts.

Clearly women like to be clean after menstruating. They probably like time off from kitchen and marital duties. But I'm suspicious of ritual purity rules. If dirt is matter out of place, then maintaining purity is a matter of putting people in their place. Imaginary dirt is such an effective weapon of limitation. See India's untouchables, imprisoned in filthy jobs—tanning, body removal, latrine emptying—because they are judged filthy. See the most powerful schoolyard taunts of disadvantaged children: they are dirty, they reek, they are inferior. See "you smell," the hardest schoolyard insult to protest. Such a system is an imaginative phenomenon, wrote Virginia Smith in *Clean*, "that rationality finds so strange—that ritual purity and impurity laws do not refer to observable cleanliness or dirtiness, but to a classified purity status."[22] You can touch something and not be dirty, but you are unclean. You can bathe in the shit-filled Ganges and be filthy, but you are clean. Mary Douglas once wrote that to understand purity rules, you have to ask whom they exclude. "The only thing that is universalistic about purity is the temptation to use it as a weapon."[23]

In 2005, the Supreme Court of Nepal made *chaupadi* illegal, without providing any mechanism to prosecute people who continue it.[24] So it thrives regardless and in enough accessible places for Western media to have become enthralled by it in recent years. They seem less enthralled by the fact that Nepal's menstrual taboos are so far from being eradicated, they are celebrated with a national holiday.

Kathmandu, three a.m. Anita from WaterAid has arrived to collect Poulomi and me from our hotel on a hill above the city. I am grumpy from my mutilated sleep and from the nerves that come with hunting. This morning marks the first day of Rishi Panchami, a popular annual festival that lasts for three days. It is, according to one listing of common Hindu festivals, "celebrated with great joy."[25] This is what it celebrates: Once, there was a Brahmin named Uttank. He lived with his wife Sushila and a daughter in a village. One night, the parents were horrified to see their daughter covered by ants. A local priest was consulted. The cause was obvious: she had committed sins in a previous life. Notably, she had entered a kitchen while menstruating. The answer was to cleanse away this past sin and the ants would depart. In another,

even cheerier version of the tale, the daughter was reborn as a prosti-
tute because she didn't observe menstrual restrictions. To celebrate
Rishi Panchami, the government gives all working women a day off.
This is not to recognize their work but to provide them with time to
perform rituals that will atone for any sins they may have committed
while menstruating in the previous year. (The prepubescent and the
menopausal are exempt.) Women especially enjoy it, I read on the
Hindu website, and "strongly believe that this will wash away all their
sins acquired by them knowingly or unknowingly." The writer help-
fully adds that "in the olden days women were not allowed either in
the house or in the kitchen during their menstrual periods," leading me
to wonder whether he—a sure guess—is writing from space, or a cave,
or the olden days.

We head for Pashupatinath Temple, Kathmandu's grandest. In the
morning dark its beauty is dulled but it's not the star attraction anyway. That is the thousands of women, queueing in that intimate way
that Westerners don't: tightly, and hands on the shoulders of the person in front. It looks like a mile-long embrace. The women are waiting
to pray, and they began lining up at eleven p.m. the night before, but
there is no ill temper or frustration. The atmosphere is one of a concert or a festival. The chatter, the excitement, the Sunday best of red
saris and gold jewelry: this feels like fun. I ask Anita to talk to the
women about the legend behind Rishi Panchami, and wait for them to
say they are just here for the merriment, that they know little about
the truth behind this festival. This is my shameful arrogance. None of
the women Anita speaks to is ignorant about the nature of the day.
Their adherence is not empty but firm and fully aware. "We may have
touched a man by mistake," they say. "We have to do this because our
ancestors did. It's tradition." Nearby I find female police officers watching over the crowds, keeping order. "Yes, we are modern women,"
said one, leaning on her motorbike, gripping a mug of hot tea in the
cold morning, her weapon at her hip. But Rishi Panchami must be hon-
ored. "I can't do the rituals this year because I'm on duty, so next year
I'll do double."

Once they have prayed, the cleansing rituals begin. The rules are
strict: they must enter a river, then brush themselves with a holy twig

365 times to signify that they have purified themselves. Then they cleanse their hair with buffalo dung before washing it with cow urine and milk. This is the theory, but the riverbanks near the temple are empty. No one does the rituals there anymore, says Anita, because too many sewers drain into the river. She calls her mother for advice, and we are directed to the opposite bank and farther downstream. The water looks cleaner, though it probably isn't. But we find five women dressed only in red petticoats squatting side by side on a log facing the river, jewels of vivid color against the dawn and dull water. They haven't yet begun the ceremony, and they gesture to us to sit, to watch, to help them fend off the monkeys. The men doing urgent calisthenics on the far side of the river aren't invited, but they also stay for the duration. They have a good view.

The matriarch is Gita Sharma, fifty-five in age but seventy in looks. She snaps at the youngsters, "You are not doing it properly. You must learn." Muna Dhal is one of the learners. She is twenty-two, from eastern Nepal, and she accepts our strange questions with patience while she manages the bags, potions, and powders that make up the menstrual-sin-washing kit. "Because we maybe committed a sin during menstruation. Maybe unconsciously." Why is it a sin? "Because it's said so." Ask her, Anita, if that means women are dirty. "Yes. They are cleaner now but still we do this."

No one does the full 365 brushes of the ritual: that would take too long. Instead, they will do a symbolic number to stand in for the rest: with the appropriate twig, they brush their private parts, feet, knees, belly button, elbows, heart, underarms, hair, and teeth. Along the way, they chat and laugh and ignore the aerobic voyeurs on the far side, while preserving their modesty. It is an acrobatic and graceful endeavor and I am entranced, right up to their rubbing buffalo shit in their hair and pouring urine on their heads. Anita asks Gita whether she believes she had sinned and Gita responds with superior scorn. "Well, if I didn't, I wouldn't have done all this, would I?"

Rishi Panchami enrages many educated Nepali women. It's not so much the superstition but the legitimacy that the government gives it by providing a holiday that declares women to be dirty and polluting. Why can't the festival simply celebrate women instead? Privately, female

Nepali sanitation activists tell me that their male colleagues—even in NGOs that campaign against menstrual taboo—see no need to object to *chaupadi* or Rishi Panchami because it is tradition. Also, Nepal has made great headway in improving its sanitation, launching policies and promises, even while recovering from a dreadful earthquake. Rishi Panchami is a battle for another day, another year. It can wait. Until then the women will come, dressed and delighted, to atone for sins they didn't commit, in water that won't clean them, removing taints only the gods can see.

On our second day in Jamu, Radha leads us on a ninety-minute walk to Tatopani, a village of ninety-five households where she goes to school. *Tatopani* means "hot water." Cold Water Village is down the valley. Along the way, the *chaupadi* sheds, initially visible in every yard, become rarer. This is because Tatopani has launched a *chaupadi* minimization program, and it's working. In the village offices, a group of concerned citizens has gathered. Some sit on the village water, sanitation, and hygiene (WASH) committee. Some are health workers. Two are young men, a rare sight. These beautiful green paddy fields, dramatic forests, and rushing rivers do not pay wages; leaving home does.

The young men are the most passionate. Their families migrated here from Achham. That is where *chaupadi* is most rooted, but it is also where the first *chaupadi*-free villages emerged, and where a government minister's wife in 1998 became the first menstruating woman in her district to spend a night in her own house. In earlier times, the villagers tell me, the menstrual restrictions probably made sense. Women could have a few days' rest while they were weak from blood loss. The men were around to do the chores and there were family members to do the cooking. Things are different now. The men are gone, the women must work, and the deprivation and damage done by *chaupadi* is greater.

"They have to stay outside but still do all the difficult jobs," says Kabi Raj Majhi, a young man who is the most vocal of all the villagers and the chair of the committee. When WaterAid's local NGO partner NEWAH arrived in the village to build a water point, its staff saw an opportunity to change things. "They said women should be allowed to use the main water point," says Kabi, "even when they were menstruating." Menstruating women are supposed to bathe away from others: in Jamu, we found one girl on her period trying to wash in a puddle. "A traditional healer objected and NEWAH said, fine, you use another water point then." The healer soon capitulated.

An old man in the corner begins to speak: "Before, they were kept outside for seven days. Now it's five and I think that's fine, but it should stay at five." He knows that *chaupadi* is necessary because of what happened during Nepal's civil war, when thirteen thousand people were killed and thirteen hundred went missing.[26] These western regions were full of Maoist rebels. "When the Maoists were here," the old man said firmly, "they didn't observe *chaupadi*. They let women in the house. And then the Maoists died in the war."

The others shout him down. But the problem isn't men like him or traditional healers. "We can change them," said Madan Kumar Majhi, Kabi's cousin and a member of the *chaupadi* minimization committee. "But it's the women who are the barrier." The mothers and mothers-in-law are the worst. A female health worker tells the room how she pretends to be menstruating just so her mother-in-law starts shaking and trembling and pretending to be sickened. "She acts as if she is ill, as if a ghost has come in. But when I am actually menstruating, I touch her and nothing happens." She laughs, but she has to observe the taboos.

Change comes slowly and it is limited. "Sometimes," says Kabi, "we have only got the women to be allowed to sleep inside the compound. We are trying to persuade people to set aside a separate room inside the house for *chaupadi*. We know it's not perfect, but we are trying. There's no electricity in the *chaupadi* sheds so it's damaging girls' education." Even the ones who are allowed to attend school can't study when they get home without light and in the cold. Before, it was worse: menstruating girls were never allowed books because the books were considered symbols of the goddess of knowledge, and they could not be dirtied.

A short walk away, I sit down with a group of girls at Radha's school. They have come in especially to talk to me, even though there is a government strike that day and school is closed. Nepal's government is fragile and any political party can call a national strike, which happens frequently. These girls are not fragile. They are feisty and smart. They say that *chaupadi* is embarrassing. "We know that you don't do it," said Pabitra, seventeen. "They don't do it in developed countries." But only four out of a dozen have been freed from sleeping outside in *chaupadi* sheds. "It makes no sense," says Anjana, whose mother is a health worker. Her mother came home two years earlier and said they weren't going to do *chaupadi* anymore. "Women bleed even more during childbirth but they can stay in the home. Goddesses are women, aren't they? They bleed but they're allowed to stay in the temple. Why can't we?" She knows the answer. "It's a lack of education. People think that because it's an old practice, it's authentic and powerful." She says they talk about menstruation in their health lessons at school. "The teacher tells us it's not a good thing."

Chaupadi is powerful. It is also extreme. But in many countries, you don't need a shed to build a menstrual taboo.

Khushi knew it was cancer. Ankita thought she was injured. Everyone believed they had a sickness. None knew why she was suddenly bleeding, why her stomach was "paining," as Indian English has it, what on earth was causing this sudden earth-shattering blood. They cried and were terrified and then they did their best to find out: they asked their mothers. And their mothers would not answer. So they asked their sisters and aunties. And eventually they were told, you are menstruating. You are a woman now.

I meet Ankita and Khushi in a schoolyard in Uttar Pradesh. I am traveling across India with a sanitation carnival called the Great WASH Yatra. Great, because its ambitions were big: five states, 1,243 miles. WASH, because that was what its ambitions consist of: to spread knowledge about water, sanitation, and hygiene (these are usually given the acronym WASH). And *yatra*, a Hindi word for a procession, pilgrimage, journey.

In each state, the Yatra sets up shop: a central stage, and dozens of stalls housing games and entertainment, all promoting better hygiene, hand washing, the use of toilets. One morning, I wake early and emerge from the dorm room to see half a dozen policemen earnestly playing a game of poop chess (where a blindfolded player has to navigate between turds and find the soap). Every day, the stalls all have long queues of boys and men. But in one corner is something different: a tent of golden yellow and red, which bears the sign FOR LADIES ONLY. This is the MHM Lab. A menstruation tent. (MHM is Menstrual Hygiene Management and the standard NGO acronym for anything to do with periods.) In 2012, when the Yatra and I trundled across India, this lab was revolutionary. I was used to shit being an unspeakable topic of development (things are better now). But menstrual hygiene? The scant level of attention given to periods made fecal sludge management look popular.

That's why the organizers of the lab didn't expect anyone to turn up. But every day, even on religious holidays, there are long queues of women and girls outside the golden tent. Cynics may think they had come for the freebies. There are reusable cloth sanitary pads on offer as well as instructions on how to make more. These are a draw and a good compromise: most women in India use bunched-up cloth (old saris are popular) because they can't afford commercially produced pads. Visitors could also make a bracelet from red and yellow beads, to illustrate the menstrual cycle (twenty-two yellow, six red). Mine is made by a man who clearly should have stepped inside to learn more, because he gives me twenty-two days of blood and six days of relief. But he isn't allowed inside. There, behind the curtain and the man-proof sign, the team is dispensing something that is extremely precious and only for women: information. There, the women and girls can come to find out about their periods, their bodies, themselves. It is this that draws the crowds, not the beads.

Over 70 percent of the 747 women and girls surveyed during the Yatra's travels had known nothing about periods before they began them because their mothers and grandmothers had told them nothing.[27] During one of the Yatra stops, I meet Neelam, a fourteen-year-old girl whose mother had died of breast cancer. (She calls it "something rotten in the breast.") When she started menstruating, she thought it was cancer, because who was there to tell her differently? Nearly a quarter of the MHM tent visitors also said that menstrual blood was dirty. This belief is not unusual. A survey by WaterAid in Iran found that nearly half of Iranian girls and women think that menstruation is a disease.[28] In some cultures—Afghanistan, some Jewish traditions— the acceptable reaction to a girl's first period is to slap her across the face, either as a punishment, because the blood is interpreted as a sign that the girl has had sex, or as a discouragement, so that a slapped girl will not immediately go and have sex with a boy, propelled by her powerful puberty like a jet stream toward sin.[29]

In the Yatra survey, 99 percent of respondents said they faced some kind of restriction when they were menstruating. In the schoolyard in Uttar Pradesh where Khushi and Ankita tell me about periods, they also say this: when we are bleeding, we are not allowed to touch pickles, because we will rot them. This is such a powerful belief in India it inspired Whisper, a commercial sanitary pad company owned by Procter & Gamble, to launch a Touch the Pickle campaign, encouraging girls to break boundaries, smash taboos, and buy Whisper sanitary pads. I am trying to understand how menstruation could damage something suspended in acid when Khushi launches a follow-up. "I don't paint my nails during my period because the varnish will go rotten."

Don't think her dumb or ignorant. She later marched me around her school complaining that her teachers weren't good enough and that she wanted to learn. On the street, walking among fearful dust clouds, she gestured to the air and said, "Ma'am, this is Uttar Pradesh, you can find any pollution you like: noise pollution, water pollution." And women pollution: this charming, memorable young woman thought she was as polluted as anything else in Uttar Pradesh, because she had also been taught that.

Research in the developing world, writes Dr. Catherine Dolan, "paints a picture of menarche as a fraught process, characterized by uncertainty, fear and distress."[30] But the importance of secrecy and hiding is not embedded in growing girls only in developing countries. Shame and embarrassment have nothing to do with poverty or education. Recently I met someone whose mother hadn't told her about periods because she was convinced that if the girl knew, she would start her period sooner. But this girl's mother was a middle-class academic with a PhD.

Stigma and silence spread a long way and in all directions. In the early 1960s, NASA was wondering whether women would make good astronauts, being smaller and lighter, both qualities ideal for cramped space vehicles. A 1964 report, though, found two problems: wombs and hormones. It would be unfeasible to match "a temperamental psychophysiologic human and the complicated machine."[31] By 1983, NASA's understanding about women's biology had advanced far less than space flight. When Sally Ride, the first female astronaut, was preparing for a seven-day space mission, she was asked how many tampons she would need. By scientists. Was one hundred the right number? She said no, that was not the right number.[32] Today, NASA employs Dr. Varsha Jain, a woman with the best business card in science, as it reads SPACE GYNECOLOGIST.[33] Hopefully the agency now knows how many tampons women need, on earth and above it. (Anyway, female astronauts usually opt to suppress their periods in space. You would, wouldn't you?)

"Menstrual taboo" sounds so NGO, doesn't it? Along with "stigma" and "menstrual hygiene" and "menstruation" itself. We, the privileged women who have toilets and privacy and a massive feminine hygiene industry: we are protected from taboos and stigma by our culture, our education, our progress. We say "periods" or "the curse." We send our boyfriends to buy sanitary products. We are advanced and immune.

Glacier National Park, Montana. Sharp mountains, blue-green water, dark conifer forests and woods, majesty and splendor: this place has the kind of scenery that inner-city children don't know how to dream of. It was high summer, 1967, and there were campers all over the park. Glacier contains backcountry, and backcountry is bear country, so there are rules and suggestions to follow for humans who want to share the landscape with *Ursus arctos horribilis*. Keep an immaculate camp. Cover and seal all food and raise it off the ground. Leave nothing of interest. Sleep inside your tent. If you see a bear, climb a tree. Try not to menstruate.

One night that August, twenty miles apart, grizzly bears attacked and killed two young women. These were the first two recorded grizzly attacks since the park had opened in 1910. They were horrific. Michele Koons and Julie Helgeson, in black-and-white photographs, are fresh-faced. They have big hair and excellent American teeth. They look similar, though they weren't related and probably didn't know each other. But they have been paired in history not only because of how they were killed but because of a wrong and poisonous belief as to why.

Koons was camping near Trout Lake with four other young people; Helgeson had headed to the Granite Park Chalet area with Roy Ducat. They knew about grizzlies, because that was what many people came to Glacier Park to see. In Jack Olsen's book *Night of the Grizzlies*, an account of the attacks, a park employee admitted that there was tacit encouragement of grizzly tourism and that garbage dumps were more accessible to bears than they should be.[34] So bears associated garbage with food and with people. Either humans had food or they could become food.

When Michele's companions heard a bear, four managed to escape their sleeping bags and climb to a safe height in time. Michele couldn't unzip hers and died because of it. Over near Granite Park Chalet, the bear was too quick and aggressive. Roy Ducat was badly mauled but escaped. The bear caught Julie and dragged her downhill, where she lay in the open for nearly two hours before an armed ranger and a search party could reach her. Although there were three doctors staying at the cabin, she died of her injuries, which were grotesque and severe. Her legs had been partly eaten.

Michele and Julie are famous not because of their awful deaths but because they both became associated with a belief that endures: that bears and other wild creatures are attracted to menstrual odor, and that having a period in backcountry may be fatal. In its initial report, the National Park Service commented that "the Trout Lake girl was in her monthly menstrual period while the Granite Park Chalet victim evidently expected her period to begin at any time."[35] (Presumably they bothered to name the two young women properly elsewhere.) Michele Koons was wearing a sanitary pad when she died. The other young woman in her party had her period but "was using the internal tampon-type device which supposedly leaves no odor because the menstrual fluid is not exposed to the air." Because the bear came to this girl but then left her alone, and because the Parks Service had received "a number of letters" from women who had been attacked by wild beasts while having a period, the conclusion seemed obvious. Menstruation, and particularly wearing a sanitary pad, "was a plausible reason for the attack."[36] As for Julie Helgeson, two tampons were found in her backpack, so she was obviously expecting her period and must have smelled that way to the bear. The presence of rubbish and food waste from humans attracted the bears, not an externally worn sanitary pad. The bear that killed Michele had hassled several other hikers over the summer. The 1981 pamphlet "Grizzly Grizzly Grizzly Grizzly," published by the US Forest Service and the National Parks Service, advised visitors to abstain from "human sexual activity," to be clean and tidy, not to wear perfume, and that "women should stay out of bear country during their menstrual period."[37]

Scholars took this notion seriously enough to study it seriously. In 1977, Bruce Cushing of the University of Montana's Wildlife Biology Program exposed four polar bears to menstrual odors, using the Churchill Bear Laboratory in Manitoba and a fan.[38] He also had menstruating and non-menstruating women sit "passively" in front of caged bears. Outdoor bears were exposed to used tampons filled with menstrual and venous blood left on stakes, along with seal odor, horse manure, seafood, and chicken. Bears, Cushing found, love seal odor, scorn regular human blood, but are intrigued by the menstrual kind. They chewed the tampons. Cushing ended with prudence. "This study supports the theory that menstrual odors act as an attractant to bears, at least polar bears. However, this should not be taken to extremes as that is not the same thing as saying menstrual odors lead to attacks."

Were the polar bears attracted to the odor? Pheromones? Some peculiar chemical? A later study was more conclusive, and its authors deserve applause for the most entertaining scientific method I've read about in a while. To establish whether black bears were attracted to period blood, they hooked used tampons onto fishing lines and cast them past ursine noses, then dragged them back again. The working hypothesis was that bears would be as interested by the tampons as by garbage and other control substances. So would I be, if a fishing line holding a used tampon had come sailing past my nose while I was minding my own business. They also exposed human-socialized bears to menstruating women by having them hang out together. After six experiments in various conditions over several years with different tampons, different women, and different bears, the scientists came to a conclusion. "No bear showed appreciable interest in menstrual odors regardless of the bear's age, sex, reproductive status or time of year."

In 1988, Caroline Byrd, a forestry specialist, was prevented from working in the backcountry because a bear had ransacked a hunter's camp, and the US Forest Service decided it was the fault of menstruating women. "A few days later," wrote Byrd in her master's thesis "Of Bears and Women," her fury barely contained by the calm Courier typeface of her manuscript, "my crew (three women and one man) was informed that due to the recent bear trouble, women would no longer be allowed to work in the backcountry during their menstrual periods." The policy was eventually rescinded, but not the conviction that period blood was dangerous, which swirls and percolates far and wide, relentless.

Béla Schick was a Hungarian pediatrician who in 1910–11 devised the Schick test, still used to detect immunity to diphtheria. For that, we are grateful. He was also convinced that menstruating women made flowers wilt. This revelation came to him when he asked his maid to put some red roses in water and was shocked that by the next morning they had withered. She told him she was menstruating. Schick experimented further, giving flowers to menstruating women. The flowers died, quickly. He expanded into dough, getting several women to prepare some and noting that the dough prepared by the sole menstruating woman in the group rose 22 percent less than the others.[39] Obviously, he concluded, a woman on her period was expelling not just blood and tissue but some potent chemical that had an abominable effect on botany and bacteria. It was, he declared, "menotoxin." This also conveniently aligned with old-fashioned superstition and taboos. Pliny was right along with every other vividly absurd and usually male commentator on the corrupting superpowers of the bleeding woman. Women really could slay rodents with what a recent TV writer called our "menses badness."[40]

Menotoxin was an attractive idea and an instance of ingenious branding, and it bore much academic fruit. In 1940, the anthropologist M. F. Ashley-Montagu wrote "Physiology and the Origins of Menstrual Prohibitions," which explored recent research on menotoxin. His list of references was unsettlingly long and predominantly German. Male scientists were clearly spending much time, effort, and money on the pernicious question of why dough didn't rise properly when handled by a woman on her period. Some diversified from bakery, injecting menstrual serum, whatever that was, into guinea pigs. Their theories were diverse. Women were emitting choline, or choline transformed into trimethylamine. Or oxycholesterol or mitogenic rays. Menotoxin was being debated in the letters pages of the *Lancet* in 1974.[41] Modernity has not prevented menstrual nonsense: recently, a renowned Japanese sushi chef declared that obviously women were discouraged from becoming sushi chefs because periods spoil fish.[42]

In 2002, psychologists in Colorado enrolled sixty-five students (thirty-two female, thirty-three male) in a study. First they were taken to a room with a woman they are told is a fellow student, asked to fill in a questionnaire, and then left alone. The woman then apparently accidentally dropped either a tampon or a hair clip. She did this "with a blank expression on her face." The goal was to measure disgust. In disgust theory, both the hair clip and the tampon should provoke equal distaste. Cut hair is disgusting because it can carry disease, and so are hair accessories by association. A tampon, as the eminent disgustologist Paul Rozin found, sets off all the disgust alarms: when his research team asked men and women to put the tip of an unused tampon—unwrapped in front of them—into their mouths, 69 percent refused. Three percent wouldn't even touch it.[43]

The Colorado study was striking. Students who had watched the woman drop a tampon then judged her to be "less competent, less likeable," and they avoided her "psychologically and physically." The effect was stronger than when the women dropped "a less 'offensive' but highly feminine item—a hair clip." This aversion was the same in both men and women. When Tampax surveyed one thousand Americans in 1981, half the respondents agreed that a woman shouldn't have sex while she was menstruating.[44] A 2017 survey by ActionAid, an NGO that works with women and girls in poverty, found that half of British women don't feel comfortable discussing their period with men (including their dads).[45] Another by WaterAid found that 42 percent hid sanitary products from their work colleagues on their way to the bathroom, and 56 percent would not go swimming during their period. Nearly 80 percent amended their lifestyle in some way while menstruating, and the amendments were usually the limiting, hiding, avoiding kind.[46]

Surveys are not robust science. But they have to stand in for it because of an absence of data on women's health and menstruation. Here are some things that science could stand to look at more closely: Premenstrual syndrome. Premenstrual disorder. Pain. Hormones. (I wrote this book while fighting menopausal depression, brain fog—a polite expression for temporary dementia—and other symptoms caused by hormonal fluctuations. When I asked the Society of Endocrinologists for an expert to discuss the effects of estrogen on the brain, it said the society didn't have one.)

I did a short and definitely unscientific test by using two search terms on PubMed, a database containing 27 million citations from journals and books. "Premenstrual" had 5,496 citations. "Erectile dysfunction" had 21,672. Erectile dysfunction must be distressing. But does it debilitate 90 percent of men for at least two days a month? Does it damage their ability to work, think, live? Premenstrual syndrome is so poorly understood, its existence continues to be questioned, though not by me. When the psychologist Kathleen Lustyk made applications for grants to study PMS, they were refused "on the grounds that PMS does not actually exist." The magazine writer Frank Bures wrote a book recently, claiming that PMS was a "culture-bound syndrome" created by the level of stigma around menstruation. He aligned it with other culture-bound syndromes such as one where men imagine their genitals have been sucked up into their bodies by voodoo; or the Indian affliction *gilhari*, "in which patients arrive at the hospital with swelling on the back of their necks, complaining that a *gilhari* (a kind of lizard) crawled under their skin."[47] His reasoning for this is that PMS is most likely to be reported by women in western Europe, Australia, and North America, and that "the more time that women of ethnic minorities spend living in the United States, the more likely they are to report PMDD." Premenstrual dysmorphic disorder is a more severe version of PMS. The *Diagnostic and Statistical Manual of Mental Disorders*, the psychiatrists' bible, now lists four main criteria for PMDD. First, you must have symptoms including marked depression, anxiety, persistent or marked anger, and marked affective lability (e.g., feeling suddenly sad or tearful or experiencing increased sensitivity to rejection). These must be bad enough to interfere with daily life, they must be related to the menstrual cycle (and be an exacerbation of symptoms relating to some other disorder), and occur during at least two consecutive menstrual cycles.[48]

I'd gladly switch places with Frank Bures on the several days a month when I have to avoid a nearby road bridge because I don't have the defenses to stop myself from jumping off it, or when picking up a phone and speaking to a human seems the hardest thing in the world to do, and when I must breathe under the weight of invisible kettlebells sitting on my chest. Similar claims are routinely made about the menopause, such as that hot flashes are all in the mind. This is true, as that is where temperature is regulated, though the drivers are not thoughts but hormone fluctuations. Bures should know about this: men's testosterone levels rise and fall on a monthly cycle. That's probably well researched.

Isn't this just the latest round of the wandering womb? Ancient gynecologists decided that hysteria was caused by a uterus moving around a woman's body.[49] How preposterous and old-fashioned. Except it is not. Katharine Switzer, the first woman to run the Boston Marathon, wrote in her memoir that her high school coach—a woman—told her that if women played basketball, an "excessive number of jump balls could displace the uterus." In 2010, Gian-Franco Kasper, president of the International Ski Federation, said on television (publicly!) that a ski jump could cause a woman's uterus to burst.[50] Women were allowed a competitive ski-jumping event only in 2014.

A post on the Public Library of Science (PLoS) blog was headlined, OLYMPIC SKI JUMPING COMPETITION COMPLETED WITHOUT A SINGLE UTERUS EXPLOSION. Unlike men, wrote Dr. Travis Saunders, women's gynecology is safely contained inside the body.[51]

Perhaps claiming that any kind of premenstrual symptoms are akin to imaginary lizards or disappearing penises would be more difficult if research were better or better funded. Although talking of "symptoms" will get me accused of pathologizing a natural process. It didn't feel natural when I was writhing on the floor in pain or when, once a month for several days, no thought was not a black or dangerous one. Last year, endocrinologists discovered that women who suffer from PMDD, which they described as "clinically distressing changes in mood and behavior," may have genes that make them respond differently, and painfully, to hormonal changes or fluctuations.[52]

It wasn't until 2013 that a comprehensive review was undertaken, by a team at the London School of Hygiene and Tropical Medicine, into the state of existing research about menstrual hygiene management.[53] This is astonishing, when it has been understood for a while now that a worrying number of girls drop out of school when they get their period. Often it's when their school lacks a toilet. Ankita and Khushi, the schoolgirls in Uttar Pradesh, used to go in an alleyway behind the school, or behind plants in the dusty yard. Now imagine doing that when you have your period. I'd drop out too.

It's well known that educated girls are better for just about everything: they have fewer, healthier, and better-educated children. The World Bank estimates that getting a higher education is equivalent to a 25 percent increase in wages in later life (compared to a rise of 7 percent for secondary).[54] A UNESCO global report into education in 2014 found that Pakistani women with high literacy skills earned 95 percent more than women with weak or no literacy skills.[55] Among men, the differential was only 33 percent. Overall, if female education rates rise by 1 percent, GDP increases by 0.3 percent. Educated girls are like yeast in the dough of sustainable, successful development (even dough kneaded by a menstruating woman). If girls can be persuaded to return to school because they have a toilet and good menstrual hygiene, then scholars of all ilks should be flocking to demonstrate this.

The London School study found only sixty-five articles to review, a paltry number. It concluded that "menstruation is poorly understood and poorly researched" and that "there is a strong possibility that the best knowledge lies in the hands of those implementing programs."

I wonder. If menstruation were better researched, would my endometriosis have been diagnosed sooner? The average time it takes to diagnose endometriosis is ten years. For twenty years, and by several doctors, I was given prescription-strength painkillers without question. The question should have been: Is there something wrong? Period pain, caused by prostaglandins making contractions in the uterus, is common. But extreme period pain is not. In a paper titled "The Girl Who Cried Pain," the authors Diana Hoffmann and Anita Tarzian explored bias in how pain in men and women is acknowledged and treated.[56] They had plenty of material. Children in postoperative pain: the boys were given codeine, while girls got paracetamol. One study found that male patients who had had a coronary artery bypass graft were given narcotics more often than female patients. The women were more often given sedative agents, "suggesting that female patients were more often perceived as anxious rather than in pain." When researchers reviewed evidence from the American Medical Association's Task Force on Gender Disparities in Clinical Decision Making, "physicians were found to consistently view women's (but not men's) symptom reports as caused by emotional factors, even in the presence of positive clinical tests." Female chronic pain patients, in another study, were "more likely to be diagnosed with histrionic disorder (excessive emotionality and attention-seeking behavior) compared to male chronic pain patients."

The hysterical woman and her wandering womb, her fragile mind. Same as it ever was.

Bihar is the poorest state in India, and despite a decent state government that has improved roads, it is the only place where our Yatra stalls and stage are erected by men who bring their equipment by buffalo. Today's venue is a school that has one computer and only a generator for electricity. The Yatra's MHM team has come to do outreach with teachers. Most are male. Education is the route out of menstrual stigma, but boys and men must also be taught. The state government has insisted that the menstrual hygiene program be carried out even during the festival of Diwali. "I know," says one of the program organizers. "I wouldn't like me for that, either." Perhaps that's why an angry man comes up to me and yells that we are late and how dare we? I am there only as an observer and direct him to the program organizers, because he has confused my skin color for authority. I dislike him for that, and also that during the session he was asked about the menstrual cycle and said, "After twenty-one days, when the egg ruptures, there's a lot of bacteria and it can make them ill and if they make food they can make everyone ill."

In the classroom, men dominate in number and in their bullying. They shout down the women. Vaishalli, who leads the session, hands out a paper that holds three questions.

What are the problems faced by girls when they get periods during school hours?
What problems do female teachers face when they get periods in school?
What problems do male teachers face when they realize female students have periods?

There is confusion. The girls can take two days off school. No, not the girls, the teachers. No, the teachers can't. No, the girls can't. There is no facility in the school to change pads, a man says, so if they need to do that, they definitely need to leave. A male teacher said, "They get very irritable." A female teacher, allowed to speak for once, says, "We help them by giving them permission to leave the school. That's all we can do." Missing education is sad, but period leave sounds better than what happened at my school, when the only concession to periods and pain was a code whereby the games mistress asked, "Who isn't showering?" and I wondered how often you could say "me" and whether Miss Applewhite kept records.

A report by Plan India dating from 2010 claimed that 23 percent of Indian schoolgirls miss school or drop out altogether because they are menstruating.[57] This statistic is cited by almost anyone writing about menstrual hygiene management, even though the report can't be found anywhere. Here is another widely repeated fact: that one in ten schoolgirls in Africa do not attend school during menstruation. This is so striking and large—how do they know?—that I investigate it, finding the original quotation in a UNESCO document quoting the source as a UNICEF study that doesn't actually contain the figure or anything like it. It is one of those shimmering figures that dot authoritative reports (a "zombie statistic," in the words of a WaterAid analyst[58]) that are built only on expectation and belief that it makes sense. I find another zombie figure repeated: that a 2013 study by the University of Nottingham found that 61 percent of girls worldwide had missed some school each month because of periods.[59] The study found no such thing. Believable figures are not global and more guarded: that Nepali schoolgirls missed only 0.8 percent of school days in a year, but half of them missed school at some point while menstruating. In Ghana, 95 percent of schoolgirls reported missing school, and 53 percent in Nairobi, Kenya.[60]A UNICEF survey found that 35 percent of girls in Niger and 21 percent in Burkina Faso "sometimes" missed school because they were menstruating.[61]

I've visited dozens of schools in the developing world, and even when they had toilets, they were filthy and ramshackle and I wouldn't have used them. In Liberia, I once met a young woman who wore two pairs of underwear, a pair of trousers, and two skirts. All at once. She was terrified of staining her clothing, and although her school had just been decently renovated by a Liberian NGO, it had somehow forgotten to build a toilet block. Grace had to go to the bush to change her sanitary cloths, and so she either stayed home or wore her uniform of period-protective clothing.

SHARE, a research initiative at the London School of Hygiene, thinks a link between school absenteeism and periods is plausible but unproven. In sober development language: "Although there was good evidence that educational interventions can improve MHM practices and reduce social restrictions there was no quantitative evidence that improvements in [menstrual] management methods reduce school absenteeism."[62] I'm not sure what's more infuriating: that Grace and other girls are missing or leaving school because they are bleeding or that the reason we can't be sure that's connected to poor menstrual hygiene is because there isn't enough research or science, because there never is, when it comes to women's health.

I read an angry piece objecting to the phrase "menstrual hygiene management," because it perpetuates the belief that period blood is dirty and smelly. But sometimes what pushes girls out of the classroom is because they are terrified that they are dirty and smelly. They don't stand up to answer the teacher for fear their clothes are stained, and many live in hot countries where school uniforms are light-colored. Good for sunshine, bad for period confidence. In Malawi, girls interviewed by WaterAid, who almost all wore sanitary cloths, reported boys taunting them when the cloths fell out of underwear, saying, for example, that they "looked like they had killed a chicken." (This is surely the politer version of what they said.) The cloths they used were so uncomfortable they would make the sanitary pads that my school secretary dispensed—so bulky, we called them bricks—seem like gossamer. The cloths were easily soaked and became visible through uniforms. School toilets had no doors and no water, so they may as well not have been toilets. "We go to the toilet," one girl said, "then we eat with shit or blood on our hands."[63]

Period pain was another reason for absenteeism and dropping out: girls can't afford painkillers and will rarely dare to tell the teacher they are in pain from menstruating. Also, when girls start menstruating, it means that they are sexual beings and can be married. Studies have shown success, though limited, with the provision of sanitary pads and menstrual cups (plastic devices that gather blood and are thought sustainable, including by everyone who writes to me after I publish anything on periods and urges me to try one). When schoolgirls in Uganda were given a combination of puberty education alone, puberty education and sanitary pads, or sanitary pads alone, it was puberty education that kept them in school.[64]

In the Bihar classroom, a male teacher stands up at the end of the session. He is pouring water on the day, in a way, because he said, "It's very easy to talk about menstruation but the social conditions are this: if a male teacher starts to do this, he is going to go through hell. Social conditions don't allow us to talk to girls, and they have problems already talking to female teachers, so forget about us." A female teacher agrees. "If a male teacher tries to talk to girls about periods, he will get beaten up."

Teachers matter when it comes to dissolving stigma. But the girls matter most, the ones like Ankita and Neelam and Radha and her schoolmates, who have the confidence to protest injustice to me, and one day may have more, and spread it further.

I think about Radha sometimes, after I leave Nepal. She wasn't particularly chatty or communicative, not even when we walked together for four hours back to the river where the road arrived, while she carried some of our luggage because she wanted the portering fee. I don't know if she was smart or had hopes. But I knew that her opportunities, as a young woman in rural Nepal, were constricted. Child marriage rates in Nepal are disturbingly high, and with absent parents there was a good chance that her family would think the safest state for Radha to be in was a married one. In 2017, our photographer Poulomi returned to Jamu on assignment and found her. She e-mailed me a picture. Look, she wrote. Here is Radha, nineteen now, and married. She has to endure only three days of *chaupadi* now, not five. That is her pickle, her progress.

Arunachalam Muruganantham at home

SEVEN

NASTY CLOTHS

To change his sanitary napkin and clean himself up, Muruga thought the public well near the burial ground was the best option. People avoided the well as they avoided the dead; he could remove his pad, rinse his stained clothing, and be safe. But Muruga was wrong. People saw, and they talked. This strange man, washing blood from his clothes, wearing a sanitary pad like a woman: What was he doing? The shame of it. The disgrace.

He was doing something revolutionary. It counted as revolutionary in his village, but also in his state, his country, and worldwide. Arunachalam Muruganantham, a poorly educated workshop helper and son of a handloom weaver, was on his way to becoming Menstrual Man.

In the predawn hours, Delhi Airport is packed. Heaving packed. Small children everywhere. The city has been hotter than a heat wave, with temperatures of 50 degrees Celsius (122 degrees Fahrenheit). To stand outside for more than a few seconds is dreadful. One newspaper is running a regular column called Death by Breath, because the temperatures are making the city's air pollution worse than ever. Most people are going north to the cool of the hills, but I'm going south, to a town I'd never heard of a year before. To Coimbatore, Tamil Nadu. When the plane lands, a woman behind me says to her daughter: Look! Red soil and coconut trees.

The air is fresher than I am. I'm tired and traveled and smell like it. But I have a meeting, and I have to wake up: I make a phone call and am told to come for one o'clock, for lunch. (In fact, he tells me later, he never eats before three. He was trying to be considerate of a Westerner's lunchtime and didn't realize this was a Westerner in desperate need of a nap.) His home is only a half-hour drive from Coimbatore center, but the taxi driver gets lost: narrow lanes, no house numbers. I'm sulky-tired and slumping in the taxi when I see him. He stands in the street, looking for his guest, his posture erect, his skin the darker one of South Indians, his clothes white like his house, a smile beneath his mustache. Suddenly I am alert, because I'm about to meet a celebrity. This is Muruga, as he shortens his name, friend of presidents and Bill Gates and Bill Clinton, and a sanitary pad superstar.

He leads me inside, up the stairs, past a child's bicycle, to his one-bedroom first-floor flat. A polite word for it would be "humble": a couple of living rooms, a small kitchen, walls covered with child-made scribbles and doodles. Sit, he says, and offers a chair, while he sits cross-legged on the floor, with the ease of Indians who do not have our uptight hip flexors and rigid hamstrings, our chair-formed inability to sit close to the ground without seizing up. He introduces his family: Shanti, a smiling, round-faced woman in a sari. Preeti, their daughter, nine years old, home on a Sunday in a party dress, energetic in her interruptions, her in-and-outing.

Tea, first, then the story, which began in 1999. I haven't brought a translator, though my e-mails to Muruga were always met with suspiciously short replies. He laughs. "I taught myself English in 2001. From films." In *Menstrual Man*, a documentary about him by Amit Virmani, he says, "If you can't understand it, that's your problem. You educated people, your mind is tuned to correct my English for the right tense and verbs."

We manage.

When the story began, Muruga was not long married to Shanti. He had grown up poor, but he had gone to school until his father died in a road accident when his son was fourteen. The family sold all they could, but he had to work to provide for his mother and two sisters. "I was surrounded by women." (Little did he know how many women would soon surround him.) His marriage to Shanti was an arranged one, as was custom, but "we married and then we loved each other. It is a beautiful system."

One day, he saw Shanti holding something behind her back. He thought it was a tease, a thing that young married couples do, so he tried to see what it was. Dipping, darting, pulling. It was a handful of bloodied rags. Muruga called them "nasty cloths." And Shanti's shame reminded him of his sisters, who used to hide bloody rags in the thatch of the house near the outdoor latrine. He asked Shanti why she was using such things and she admitted that she couldn't spare money to buy sanitary pads. It was milk or Always, and they always needed milk. For Muruga, though he had seen his sisters' rags, as most men had probably seen their womenfolk hiding bloody cloth, this triggered something in him. He wanted to change things. If his wife couldn't afford sanitary pads, then he would make one that she could afford, that would be easy to produce, that would help all women.

Cloth is actually a decent option for menstrual bleeding, and rags or cloth are used by most women in developing countries. The latest National Family Health Survey in India found that 77.5 percent of urban women and 48.2 percent of rural women used "hygienic methods of protection during their menstrual period," namely locally prepared napkins, sanitary napkins (presumably commercial ones), or tampons.[1] As two menstrual experts at UNICEF India told me over lunch one day in the office canteen in Delhi, while other diners worked on their food and blocked their ears, rags are often a luxury. In rural areas, people wear polyester now, so women and girls have little access to absorbent cotton cloth. They accommodate. Some things that women and girls have been known to use: socks, newspaper, ash, sand, sawdust, polyethylene bags, torn-up sacks, grass, or leaves. These may seem horrific, but some make sense: in Rajasthan, women use sand as an absorbent, wrapped in cloth. In 2017, Korea discovered the "insole girls": young women who were making their own sanitary pads by wrapping tissue around shoe insoles, because sanitary pads had increased in price by 42.4 percent in a year (although the price of pulp had fallen).[2] A health worker in India told me the story of a young girl in Madhya Pradesh who was too embarrassed to ask her mother for a clean cloth and used one she found without knowing it had lizard eggs in it. Three months later, the subsequent infection meant her ovaries had to be removed at the age of thirteen. She would be forever tainted as a barren woman, a *banch*, so that people who saw her first in the morning would have to take a bath to wash her stain away. I hear horrible stories from other menstruation activists: a girl who fashioned her sanitary pads from a chocolate box carton stuffed with cotton; another who used a wad of toilet paper, but internally. Women in Kenya who fill a sieve with sand and then sit on it, unmoving, all day.

The problem is not that women use cloth, but the combination of taboo and laundry: Muruga's sisters were secreting their cloths into the thatch because they couldn't dry them openly. The most hygienic way to dry menstrual cloths is openly in sunlight, but millions of girls and women dry their cloths under cots, under other laundry, in roof thatch, because they are ashamed. It is a recipe for infection and bacteria growth. One girl was horrified to see her carefully hidden menstrual cloth being used by her brother to clean his motorbike.

The day after Shanti showed him her bloody rags, Muruga bought her a pack of sanitary pads from the market. I have done this and watched in some stupefaction as the pads were fetched from behind the counter, where they were hidden—although condoms are openly displayed—and handed to me wrapped in a black plastic bag kept specially for sanitary products (the regular bags were transparent). Muruga noted this subterfuge and shame and wondered why it existed. He also noted the price: 20 rupees ($0.50) in 1998, which was expensive. With 20 rupees you could buy food for a family for days. Because he had always been a curious tinkerer, he took one of the pads and ripped it open. It seemed to be nothing more than cotton, which should cost far less than it did. Why were these pads so expensive that most women could not afford them?

Muruga was hardly educated. He calls himself a ninth-grade drop-out but with pride because he knows he is smart. After his father's death, he had worked delivering lunches, then in an ironsmith's workshop as a helper and lackey. He learned welding, tinkering, and curiosity. He had no training but he began to think like an engineer, a solver of problems. He fixed things and made other things. If his wife had a problem, he should seek a solution. First, he bought some cotton from a local mill and shaped it into a rudimentary sanitary pad. He offered it to Shanti and his sisters to test, but he knew so little about the menstrual cycle at first that he didn't realize he had to wait for a month for feedback. "Many boys are becoming men, then husbands and fathers and grandfathers and great-grandfathers, without knowing anything about what is happening to women." At least he managed to do it before grandfatherhood. So he kept pester-ing his sisters and wife for feedback and was shooed away. He per-sisted. He was shooed away again but persisted. Finally, his sisters took to yelling when he approached that Shanti should control her husband and get him to stop bothering them with this sanitary pad nonsense.

Muruga was not deterred. He was too intrigued. But he thought himself an engineer, and an engineer must test things. He had a work-shop for machines and tinkering, but where could he test a sanitary pad if his womenfolk wouldn't help? He thought of another solution. Medical students. Surely as scientists they would be interested in his experiment? He was right: they took his rudimentary napkins along with a carry bag he provided. He gave them a sheet of questions to answer, but one day he saw that three girls were filling in the forms for all, and he was disillusioned. They weren't telling the truth, and the basic truth of the menstrual cycle, that it was monthly, meant his research was taking too long. And he was exhausted, traveling so many miles, handing out napkin samples here and there, having to return for feedback, which sometimes was not provided. He had to discover what a pad felt like for himself.

For his experiment, he needed blood. Cattle in India rarely become beef, so that wasn't an option. Many of his school classmates by now ran chicken stalls, but chickens are small. Goats, though: goat blood might work. If he filled a rubber vessel with blood, linked it by tube to his underwear, and periodically squeezed blood into his sanitary napkin, he might get some idea. He'd know what being a woman felt like. He didn't know how much blood a woman discharged on average, nor that it was half tissue and clots, and goat blood wasn't. But it was good and he thought it viscous enough (he asked "a blood bank guy" what to add to stop it clotting). He fashioned his fake uterus out of a football, and for a week he wore his cotton sanitary napkin under his white dhoti, attached to the football that he wore at his hip, a bloodslinger. If he was walking, if he was cycling, now and then, he squeezed.

I stop him at this point in the story. Are there photographs? And why did he wear white? He laughs at the first question. Cameras cost money. If he had money for a camera, he had money to buy sanitary pads every month for his wife. And white? Because that's what he wore every day. Because he lived in a hot country. Because he didn't know better. He was surprised there was staining, when no woman would be. In *Menstrual Man*, Muruga describes how he was always turning around, checking for leaks. "I'd become a woman." That was when he began to wash near the burial ground, and when he was seen. Rumors began to spread: he must have a sexually transmitted disease. He must be seeing other women. I don't blame the rumormongers, really: they were hardly likely to guess the truth. Muruga describes the place where he lives as a village, but it looks like an urban suburb to me, only with a few more fields and tamarind trees. It has the noise of the city, but the structure and character of a village in the ferocity with which chatter spreads.

The goat-blood uterus taught him about being worried and wet, a combination that every woman knows. It taught him, he says, that "the strongest creature created by god is not the elephant, not the tiger, not the lion, but the woman." But his experience wasn't enough to teach him how to reverse engineer a decent, low-cost sanitary pad that his wife could afford. He had an idea: "Why not ask for used pads from a girl? That napkin will speak to me." By now, village talk about his experiment had pushed Shanti out of the house to stay with her parents for a few days. She was mortified, enough to send him divorce papers two months later and to stay away for five years. He couldn't ask her for her pads. "If I went to my in-laws' house and asked for a sanitary pad, they would definitely think, the son-in-law is going to cast black magic on their daughter. He is going to mesmerize her or something." Who could he ask?

He tried to fish out a used sanitary napkin from a waste bin on the street. The napkin belonged to a woman whose husband was nearby as he had just deposited the napkin and other rubbish there. Muruga had to pretend he was looking for something else in the bin, but it was a close encounter with serious trouble. He thought it was safer to approach the medical students again. He gave them a selection of his napkins and some commercial sanitary pads for comparison. He says, in a rare instance of his pride sliding into something else, "See the cleverness?" The young women accepted. They knew him by now. They called him brother. They said, yes, you can have our used sanitary napkins.

I think about this. I wonder what I would have done had a man asked me to give him my used sanitary pads and conclude that my reaction would have been frank and very Yorkshire. The girls were more obliging, and Muruga headed home with his first load of evidence. "I took the first batch of used napkins like a treasure. I tied my hanky over my nose, then I spread them all out in the backyard at my home." You can imagine the foul smell of dried menstrual blood in Indian heat, so he left them overnight to allow the smell to weaken. He awoke the next morning, excited to pursue his research, and there he must have been, bending over his collection of varied endometrial deposits, when his mother arrived. "She thought I was preparing chicken for Sunday. The moment she realized it wasn't a country chicken she cried like anything. She said, my son is doing black magic, he is mad."

Shanti had already gone. Now his mother was going to leave him as well. The rumors went as mad as her son: he was a vampire, drinking the blood of girls at night. He was dangerous and possessed. "My friends came to tell me that on Friday, after the religious ceremony, they would pronounce a verdict on me." He would be chained to a holy tree, upside down, and the bad spirits would be beaten from him. Then they might chase him out of the village. So Muruga chased himself out instead, moving to the small city of Coimbatore, then to the larger one of Chennai. "I ran away like a thief in the midnight." He was without a wife, without a family, without friends, but undeterred. From the used sanitary pads he suspected that the big companies—Johnson & Johnson and Procter & Gamble are the ones he quotes—were not using simple cotton. Their products were far more absorbent. He knew it was something other than cotton and spent many rupees phoning American manufacturers asking for samples. He talks of the public phone booths he used (because this was before mobiles) with nostalgia. They asked him what kind of vessel he had and finally he understood they were asking what machinery he was using. None. They said the minimum order was a ton. This was no use. Finally he got the help of a teacher friend and sent off letters pretending to be a millionaire mill owner asking for samples. In fact, he was living in a hostel in Chennai with itinerant peddlers, with a dog he loved, but who roamed the streets and came and went. "Not like in England. Not tied with a belt."

The companies sent a FedEx package. (He still has the envelope.) Inside were ten stiff brown sheets. They didn't look like Whisper or Always or Stayfree or like anything you would put between your legs for comfort. They looked like cardboard. Here the stories he tells are different. In the film, one day he simply ripped up the card. To me, he tells a long tale of his dog being left in the room all day, and getting angry, and tearing up the sheets. Either way: "A Eureka moment. I understand the whole secret." It is compressed cellulose. "They compress the fluffy material in a press. Now I have to reclaim the material in a fluffy way." If he wanted to make affordable sanitary pads, he needed to devise an affordable machine. A kind of grinder, first, that could break up the cellulose, then a press to shape it back again into a pad. Finally, some way of containing the fluffy. Gauze, perhaps. All of this, he thought, should be as manual as possible so that it could be used by the widest number of people and income brackets. His focus grew: rather than producing sanitary pads, he would create a machine that could make them, and that could make them affordably.

He began his reverse engineering and inventing with a brilliant idea: he realized that hardly any machine plants worked twenty-four hours, so he asked to rent the unused hours. One part was made here, another there, and after eighteen months he had "an easy-to-operate low-cost mini sanitary napkin manufacturing machine," according to his company website, "in which wood fiber is denigrated, core formed and sealed with soft touch sensitive heat control. It requires single phase electricity for 1HP drive, can be accommodated in a space of 3.5 meters by 3.5 meters and will produce two napkins a minute." It is made of 243 parts, can be produced in a day, requires no water, only a little power, relies largely on foot pedals and manual operation, and can be operated by anyone. It is very Muruga: difficulty dressed as simplicity. The compact cellulose is ground up, then it is pressed, then the napkin is sealed. Simple. He set up an enterprise to market the machine and called it Jayaashree Industries, after his niece. From start to finish, the development of his low-cost sanitary pad machine took eight and a half years.

In 2006, it won first prize at the Indian Institute of Technology Awards.[3] He was given his prize by India's president. Other entries, he says, included "how to extract gold from seawater, how to go to the moon." The machine and Muruga became famous. By now, he has been judged one of *Time* magazine's top 100 most influential people, alongside National Security Agency whistle-blower Edward Snowden and Beyoncé.[4] There is a cupboard in his home where he keeps his awards, which is so crammed that when he opens it they spill out like treasure from a chest. Framed photographs, trophies, gifts, stuff. I notice something: in every picture he is wearing glasses, but he's not wearing them now. He says, "Once I was being picked up at an airport, and I went over, and the man said, 'No, no, we are looking for a VIP.' So now I wear glasses. I don't have any eye problem, they are just to give some lift so the [drivers] will take me. My VIP glasses."

Lunch is nearly ready and though Muruga says I can stay on my chair, I shift to the floor because that is the table and banana leaves are plates. Look around, he says, as my legs go numb. Look at my home. He delights that he lives in rented accommodation, that behind his fame he is humble and real, just as his eyes behind his VIP glasses have perfect vision. His lack of college is an accidental advantage. "If I'm educated," he told a documentary maker, "maybe my mind will be crumpled into a fixed concept, nothing but running after money." To another reporter he says that if he had stayed in school, he would be working in a call center by now. Jayaashree is not a charity: he claims he has never accepted a dollar in donations. He is a businessman, but a particular kind. He has socialized his machine, he says. It does good but it also makes money. Profit means replication and sustainability. The machine costs $1,000 to $3,000, depending on how far it must be shipped. This seems a lot, but it should pay for itself. A detailed breakdown of figures on the Jayaashree website—slogan: "new inventions for a better life"—projects a profit of 25,225 rupees ($425) a month, and a profit margin of 60 percent if the machine produces 480,000 napkins a year.[5]

By now, Muruga speaks fluent development, but in the same way he speaks English: as and when he chooses to. One day, he got a call and the caller asked for "Muganatham." He says, "I knew then that someone educated was calling. They asked, 'Are you supplying machines only at the bottom of the pyramid?'" and he was surprised. "I know pyramids exist only in Egypt, that's why I immediately refused. No, no, I've never supplied any machine at the pyramid, I'm supplying them only in the plains of India. Then I came to know it is called economic pyramid, bottom of the pyramid." The interviewer asks him whether it's fair to say he has come from the bottom of the pyramid.

"Not even the bottom. You can say from a scrap of the pyramid."

I like his humor. I recognize it, from other social entrepreneurs I have known: Jack Sim, who set up the World Toilet Organization, Mechai Virivaidya, who helped to revolutionize condom usage in Thailand and to reduce HIV. There is a connection: humor erodes taboos. Muruga knows this: his interviews are funny and charming. His charm and composure, though, seem excessive when it comes to his wife. She returned only after he became famous, when a TV show broadcast his mobile number. "I am starting to use mobile phone, one kilogram

Ericsson." Shanti called and came to live in the slum in Chennai where he was staying. "Every Westerner surprised!" he says. "You wait five years for the same wife?" He had, and he ignored the divorce papers her parents sent him, and he decided to use her absence to work on his research. And now, he had been proved right. With his marriage restored, they returned to the village that had wanted to chain him to a tree. "I'm coming back as famous. They said, 'We knew Muruga would do something because his forehead is very wide.'"

He and his wide forehead are now "living as a celebrity in my own place," sitting on the floor under his daughter's doodles, no airs, no graces. When I take out my camera, Preeti rushes to her room and returns wearing a green Scream mask, then stands next to her dad under a framed photograph of him meeting the president of India, in his VIP glasses. It's one of the oddest family portraits I'll ever take, and one of the best.

The next day, Muruga takes me to Jayaashree Industries. He parks his jeep in a field that holds a neem tree, which is important to him. Early on, he didn't have an office, so he arranged meetings under a neem tree near the burial ground where he tried out his sanitary pads. When he was working on the different machine parts, he hid them among the branches for storage.

He has cupboards now, and a whole factory. People are working: grinding, packing, posting. The machines are sent in wooden crates, and they still contain some wooden parts so that they remain simple to assemble. They must be robust enough to travel a long way: this month, one is going to Nepal and another to Afghanistan. There are forty-four hundred machines now operational in twenty-five countries, according to Muruga.

Now people approach Jayaashree, but in the beginning Muruga was the Menstrual Salesman. He traveled with his story and his machines. His first expedition was to Bihar. There, he stayed for two and a half months, and one day, he saw villagers running in a flurry, here and there, to and fro. A teenage girl had hanged herself in a tree because she thought she was pregnant. Muruga realized that providing sanitary pad machines was not the only thing he could do to assist women's health. A simple pregnancy test would have perhaps saved the girl's life, so now he provides those alongside machines, and what he calls a cervical cancer stick, a simple urine test for abnormal cells. He says, "India is the most affected country for cervical cancer." Actually, globally it's Malawi, and India isn't even in the top 20, but cervical cancer rates in India are high and disturbing.[6] Is poor menstrual hygiene a factor? Muruga thinks so: he says women were "converting the cloths into bacterial form." They were hiding their cloths, not drying them properly, and getting infections as an inevitable result.

The London School of Hygiene and Tropical Medicine has found that "women who used reusable absorbent pads were more likely to report symptoms of urogenital infection or to be diagnosed with at least one urogenital infection (bacterial vaginosis or urinary tract infection) than women who used disposable pads." A literature survey by SHARE, a sanitation research consortium, found no association between poor menstrual hygiene and poor reproductive health. But only fourteen articles were studied and all were judged flawed, with varying methodologies and an overall quality that was low. "It is biologically plausible," wrote the authors of the SHARE report, "that unhygienic MHM practices can affect the reproductive tract but the specific infections, the strength of effect, and the route of transmission remain unclear."[7]

Muruga wanted to fix things, but to get to the women of Bihar, a conservative state, he had to go through the men. He did this—just avoiding being beaten up—and the machine was installed. From then on, the project was on its own. This is his business model: he calls it the butterfly. Not like a mosquito, a parasite, which is how he thinks large corporations work. They suck what they need and they leave. The butterfly takes from the flower but leaves it undamaged. He is emphatic, though: his is not a charity. People must find a means to pay for his machine. It doesn't have to be money: he believes in the barter system and has accepted payment in goats, buffalo, cattle. The same goes for when the machine is operational: women can buy pads with tomatoes or potatoes. Once, Unilever flew Muruga to London to get his wisdom. I ask if he was paid for it (as he should have been). "No, no, I don't take money." It sent him by business class and gave him good accommodations, but in England he didn't trust the food, so he lived on chocolate. Fueled by big bars of Cadbury Dairy Milk, he told Unilever that the problem with its business model was that it couldn't sell pads for potatoes.

He shows me photos on his laptop: look, here, he took a machine up into the mountains on a donkey's back. "Eighteen-hour drive from Dehradun. The Ganges is forming there. These are tribals. Even if you have cancer, go live there, walk for six weeks with tribals, you will have a six-pack and cancer will be gone." The machine is working, high in the Himalayas. "It's rural marketing. Johnson and Johnson would take another twenty-three years to get there."

He invites me to see for myself, but not nearly two thousand miles away in Uttarakhand. Instead, we travel a couple of miles toward Coimbatore, stopping on the way for him to buy me bananas that taste like toffee. He says we will get to our destination by luck, not by driving skill. No one gets anywhere on India's roads with anything but luck. I was once being driven for ten hours from Dehradun to Delhi, when the driver invented his own contraflow for no obvious reason, suddenly swerving into the opposite lane and driving for a mile against oncoming traffic. It still makes me cold.

Our destination is an association with a Tamil name that translates as the Association for Parents of Mentally Retarded Children. It is Muruga's equivalent of a show home, as the building houses the only operational pad machine in Coimbatore. Most of the work here is done by the parents—I see only mothers—because even this simple machinery is complex for a child with developmental challenges. But jobs are found for them nonetheless. A nonverbal autistic boy turns the compacted cellulose into what Muruga calls fluffy (the industry term is "fluff pulp" or "fluffy pulp"). The grinder is one of the few electrical parts, and all the boy has to do is push a button and time the process. His mother says he loves to do it, though sometimes he gets entranced by his watch and the grinder overgrinds. There is such a thing as over-fluffed fluffy.

After Muruga had his dog-related Eureka moment, he sourced the cellulose in India. Now he says, once self-help groups or entrepreneurs buy his machine, they can source the material from anywhere. He has patented the machine, but he put the design online in 2007. He wants it to spread. He wants a revolution. He says, "Man is dominating the world, and we are turning the world into olive green. I beg men, please give the world to women for at least half a century, they will make better what we did."

The feminine hygiene industry has its eyes and margins fixed on India. With so few women using anything but cloth, it sees profit and possibility in pads. But there are obstacles. Commercial sanitary pads in India are not cheap, usually costing 140 rupees ($2.13). And they are usually to be found in pharmacies or shops staffed by men. Here at the association, there is a cupboard full of sanitary pads for sale. They cost 40 rupees (about 60 cents) for a pack of eight, the standard price for Muruga-machined pads, and each producer can choose a brand name. Here, it is Kangaroo. I'm puzzled by this, and Muruga thinks it funny—"wear it and you will jump like a kangaroo"—but one of the mothers gives a beautiful explanation. Kangaroos protect their young with ferocity and intimacy. These mothers do the same, for their children who can't speak or function fully, in a country where social care is difficult and self-help.

At Muruga's flat, he has a cardboard box filled with various sanitary napkin packets produced by dozens of associations. He and Shanti spread them out and I sit among them, using a few packs as cushions. They're as soft and comfortable as you'd expect. There is Pure and Free and Vings (as Indian English pronounces Wings); and Be Cool, and Mitra ("good"), and Sakhi ("friend"). There are 990 brands across India of pads made on Muruga's machines. Unlike the big companies' products, his can be sold as pick-and-mix of different absorbencies. The big companies do macro; he does something different. "What Muruga did," he says, slipping as he does into the third person, which is never not unsettling, "is made it into a micro and decentralized model of doing things, by the woman, for the woman, of the woman." How stupid it is to produce packets of napkins all the same absorbency, when a woman's flow differs from day to day. With his model, "a woman will come to the place, she will say, I'm having a flow only on the second day. She will take two thicker napkins, three medium, six thin." This is not all she will say. The sanitary pad production becomes something more and better. "The moment a woman uses [a sanitary pad], she goes to another woman, they are talking, they are getting educated. We are the first women-personalized sanitary pad–making machine in the world." Muruga is so famous he has become a Bollywood film called *Pad Man*, a dramatized version of his life that begins with a bass voice-over booming that "Hollywood has Batman. India has Pad Man." It is probably only the latest step on an appropriately colorful path for a man whose first office was a neem tree. "I wanted to show the world, when you do a good thing you don't need a fourteen-floor building with a five-degree slant and a lot of glass panels. You can do good under a tree."

Tangy. Kotex wants me to be tangy. Its 1970s ad featuring a moody beautiful woman draped in clinging orange fabric also wants me to be fresh, and an eye-opener. This, if eye-opening means attracting attention, is a mixed message. Like every other manufacturer of what is known as "feminine hygiene," the last thing it wants me to do is attract attention. Or to smell. I definitely mustn't smell of blood.

There are various figures for how much the feminine hygiene industry is worth. One recent market research estimate is $23 billion.[8] The business of trapping blood in various cotton-based devices is lucrative and unstoppable. It is also single-minded: its entire purpose is to chase away blood into secrecy, silence, and discretion and to make its products simultaneously indispensable and unspeakable. It has been extremely successful.

Since the beginning of humans, women have coped with their periods in the same way: on their own, discreetly. They have been so good at this, historians must guess at what they used and how often and how they disposed of it. Women's menstrual technology rarely made it into written history, as it was mostly written by men. Free-bleeding, into clothing or onto the floor, was probably popular, though periods may have been more infrequent. Compared to us, historical women were more frequently pregnant or breastfeeding or soon dead. This is a common view, but it is too broad. There would have been class differences: wealthier women were more likely to have absent husbands who were off battling at war or at court, making it difficult for them to be endlessly pregnant. So it is that the higher-class and religious-class women are the ones who have left us occasional mentions of periods, little glints in the historical record. Women throughout time have had access to cloth, and that is probably what they used if they used anything. The Old Testament book of Isaiah instructs that gold and silver idols should be cast away "like menstruous cloths," although the Hebrew noun—*niddah*—is translated in more modern versions as "disgusting things" or "unclean cloths."[9] The seventeenth-century poet John Bunyan wrote of the fake righteousness of a Pharisee, which should be condemned "as menstruous rags, as an abomination to God, and nothing but loss or dung."[10] Among Queen Elizabeth I's recorded possessions are "vallopes of Holland cloth," held in place by black silk girdles. Other women used "clouts." The latest *Oxford English Dictionary* lists "clout," though its modern meaning is a punch. Its secondary meaning is "a piece of cloth or clothing, especially one put to squalid purposes." An Italian medical book translated as *The Diseases of Tradesmen*, an early example of occupational health writing, mentioned that surgeons avoided using women's old clothing to dress wounds, "notwithstanding they are frequently wash'd; and that by reason of the Virulency of the menstrual Blood."[11]

Between 1854 and 1914, there were at least twenty patents filed for "napkins, catamenial sacks, sanitary supports, menstrual receivers, monthly protectors, menstrual receptacles, sanitary napkins, and catamenial supporters," according to historian Vern Bullough. I'm intrigued by the catamenial sack, a patent filed in 1903 by Lee H. Mallalieu and Mildred Coke of St. Louis, Missouri. Its rigid ring and flexible sack, used internally, look rather like a condom but may be an early instance of the menstrual cup.[12] But patents give no sense of how women felt about all these devices, nor where they were supposed to access them: the catamenial sacks shop? Mass marketing was decades away. In the 1890s, writes Bullough, Johnson & Johnson began selling Lister's Towels, a gauze-covered disposable pad, but it didn't flourish.[13]

The modern era of menstrual technology really began during the First World War. Probably, throughout many wars, nurses had been aware of the useful absorbency of wads of cotton that they used for

wounds, and put them to other purposes. In the Great War, this practice became formally part of the origin myth of the feminine hygiene
industry, when nurses began using a Kimberly-Clark wound product
called "cellucotton," made from a by-product of sugarcane processing,
as a catamenial device. In 1920, the company launched its Kotex
feminine napkin.[14] Then came the tampon.

There are two ways to deal with catamenial flow: absorb it, internally or externally, or stop it, with hormones. Without recourse to
hormones, women must choose among various types of what the industry calls fluid management devices. Kotex became the most successful
example of the external variety; and then came the internal plug. Tampons existed already, but not for menstrual blood: plugs of cotton
were frequently used by doctors. The historian Sara Read quotes a
seventeenth-century author of *The Royal Pharmacopœa, Galenical
and Chymical*, who described a pessary as "a sort of solid Medicine,
about a fingers length, sometimes somewhat bigger, which is put up in
the Secret-parts with a Riband fasten'd to one end."[15] (Leeches with
strings; plugs with ribbons. These early medicine men were impressively
thrifty with their devices.)

This solid medicine was intended to "provoke the menstruum or
to stop them; to hinder the falling down of the Matrix," a marvelous
word for the uterus that casts a twentieth-century film in a new light.
Wartime medics were probably familiar with plugs of cotton used to
stopper bleeding wounds (they still have recourse to the occasional
tampon for trauma: it still works). In 1879, the *British Medical Journal*
ran a short article on a product devised by Dr. James Hobson Aveling,
an obstetrician, in a section on innovations. This Vaginal Tampon-Tube
was aimed at poorer women who couldn't afford the expensive devices
currently on sale for "passing pledgets of cotton-wool into the vagina."
This new tampon tube also used pledgets but was priced at only a shilling (7 cents). The pledgets could be soaked with glycerin, tied with
string or stout thread, and inserted into a simple glass tube, propelled
by a wooden rod. You insert the tube, push the rod, and hope the pledgets end up in the vagina. It was supposed to be straightforward
enough for women to manage it all by themselves, but in the case of
uterine displacement (perhaps by ski jumping), "further adjustment by

the medical man will be necessary." There is no record of the tampon tube having caught on, despite being the first such device to appear in a prestigious medical journal.[16] (I am delighted to find the same Dr. Aveling elsewhere in my database, because he once saved a young woman hemorrhaging after giving birth by giving her 60 drams—7.5 fluid ounces—of blood from her coachman. She soon recovered enough "to be able to remark that she was dying," though her mental state, noted Aveling, "was not as marked and rapid as I anticipated [. . .] perhaps due to the quantity of brandy she had taken.")[17]

Over the next couple of decades, companies launched cotton plugs, but their sales were not impressive and their names are forgotten. Then there was Tampax. It doesn't take much investigative work to find its history on the American site: there it is, in the drop-down menu headed We Care About All Women, under Building Girls' Confidence, Educational Tools, and Disaster Relief, alongside ads for boxes of Pearl tampons, with applicators, that women and girls in disaster situations could not afford. The company's history is as "storied and colorful" as that of the tampon itself. Except this version isn't. The origin date of March 7, 1936, is heralded as when "Tampax Incorporated is formerly chartered under the laws of the state of Delaware."

A more storied and colorful history is on the British site, because it included Earle Cleveland Haas, a man who is elsewhere described as an osteopath but in the Tampax story is a more respectable general practitioner. The story varies depending on the source, but each version has satisfying Muruga elements: either Haas's wife was a ballet dancer and had been using a sponge to stop her menstrual flow and her dancing also stopped because the sponge was inadequate; or he had traveled to California and met a woman who was not his wife who told him the same thing. Either way, he wanted the dancer to keep dancing, even on those days. In the official history of Tambrands, a book called *Small Wonder*, Haas is a courtly man in a white shirt, a general practitioner who sees women wearing bulky and cumbersome external pads. "I just got so tired of women wearing those damned old rags," he told a newspaper at the age of ninety-six, "and I got to thinking about it."[18]

By 1932 he was working in his basement on a "catamenial device" that consisted of five pieces of cotton sewn together, then stored in an applicator that can be used to insert the device into the vagina. Other tampons had been tried since Aveling's. But these "objectionable" devices, wrote Haas, were "unsanitary; inconvenient and disagreeable to insert; difficult to withdraw when filled with absorbed fluids; and liable to separate in the vagina so as to leave fibers or a portion of the pad remaining therein." Haas's cotton was longitudinally stitched to give it better tensile strength, then shaped under high pressure into a cylinder and inserted in a cardboard tube that also served as an applicator. The innovations were the details: the nature of the applicator—more hygienic than a wooden rod—and the robust stitching (Haas recommended a chain stitch "such as employed for closing flour sacks"). It would be easier to insert, would be easier to withdraw, and would not leave scraps of itself inside the vagina. It was a hands-free, longitudinally stitched little plug of freedom.[19]

I'm fascinated by this patent and the details of it. It's a rare example of relative transparency in an industry that in its 150 years of existence has become highly competitive and consequently secretive. When a Reddit user started a thread called "I design tampons, ask me anything," he was stuck online for ten hours answering a flood of questions. Among his answers: a tube is not the optimum tampon shape because an unaroused vagina, seeing as you asked, "is more like a flattened balloon, since it's being crushed by other organs"; vibrating tampons have been invented but are not for sale; and he doesn't use the word *menses*—or know how to spell it—because a) he's a "dude" and b) the industry word is "fluid."[20]

Disappointed by a lack of interest from companies already producing sanitary napkins, Haas sold the patent in either 1933 or 1936 to a woman named Gertrude Tenderich, who is usually described as industrious or an immigrant or both.[21] She took the patent and the name—Haas had devised "Tampax" from "tampon" and "vaginal pack"—and ran up tampons on her sewing machine, then began larger-scale manufacture in a Denver loft. But pharmacists wouldn't stock them. Tenderich had come up against the same essential problem that has confronted makers of sanitary pads, tampons, condoms, and toilet paper: How do you market the unmentionable? The developing world uses the acronym BPL to mean "below the poverty line." This industry was below the panty line, and it was a tricky place in which to do marketing.

Tenderich tried her best. She was innovative. She recruited nurses to give talks on menstruation, linking commerce with education, sort of, in the same way that massive global corporations today sell themselves as a resource for puberty, biology, and enlightenment. She sent around persistent salesmen (not women). She had some success but not enough, and so sold Tampax to Ellery Mann, a rotund man of terrific charm. "He could talk you off your chair," said Gertrude's daughter, Mary Kretschmar, "and then sit on it."[22] Ellery Mann made Tambrands a huge corporation, and Tampax—and tampons—ubiquitous, at least in some countries.

Today, the average woman in an industrialized country with access to a wide variety of sanitary choices for her menstrual period might use 11,000 to 16,000 sanitary products in her lifetime. Persuasive data on how many women use what are scanty, but a Euromonitor survey of American women aged twelve to fifty-four found they bought on average 116 pads a year, but only 66 tampons.[23] Worldwide the absorbent device of choice is definitely external. Insertion taboos stop many women from using tampons: the only thing you insert in there should belong to your husband, and it's not a tampon. The choices, then, are cloth or a sanitary pad, or towel, or napkin, or whatever we are supposed to call them. Feminine protection (from what?). Feminine hygiene (are we dirty?). Feminine Needs. Feminine Care. Feminine Accessories, as if tampons were earrings. I'd love to see a shop or supermarket that dared have an aisle called Menstrual Products. Blood Absorbers. Plugs and Wads. But plain speaking and feminine hygiene have never been comfortable companions. The advertising of sanitary products on TV and radio in the United States was banned until 1972. Despite one letter writer asking the agony aunt Dear Abby in 1984, "Please tell me what can be done to stop the advertising of personal feminine products on television," the word *period* wasn't spoken on American television until the following year, when a young Courtney Cox starred in an ad for Tampax. The company used the scandalous word, its advertising agency executive said, "because it happened to fit the campaign. We asked the networks to let us prove that it wouldn't be offensive. [. . .] It just doesn't make any sense any longer to show a woman in a long white dress, drifting through a field of wildflowers, saying something like, 'It makes me feel fresh.'"[24] He was wrong about that. This is an industry that has developed like another industry shut out of sight and mind: slowly. Eighteenth-century plumbers could easily fix a modern toilet. In a whole century, there have been only three big inventions for something that affects three billion people: the tampon, the adhesive strip on a sanitary pad, and a menstrual cup.

Today, and then, ads appealing to women overwhelmingly used two things—fitness and fashion—to sell convenience, untainted clothing, and cleanliness (a code for not smelling of period blood). Sporty women abounded, as did glamorous evening gowns. "The filmy frocks that women used to fear now bring security," wrote Ellen J. Buckland, the registered nurse hired by Kotex to write its advertising copy in the 1920s.[25] Back then, there was more emphasis on medical approval. Tampax consistently used the tagline "approved for advertising in the *American Medical Journal*" as if it meant something other than that someone picked up the phone to the ad sales department. It boasted that Tampax was "perfected by a physician" (not an osteopath). I don't find any advertising mentioning a letter to the *British Medical Journal* in 1942 by a female family doctor from Oldham. Shopkeepers had informed her, wrote Mary G. Cardwell, that young girls of eleven, twelve, and thirteen, and young women in general, were asking for these "internal sanitary towels." She had medical misgivings. Some were using them as a preventive "in conditions of promiscuous intercourse." But also, these internal towels entailed "undesirable handling of the vulval area." Many young girls, she wrote, "insert the tampons by the aid of a looking glass. The psychological ill-effects are obvious."[26]

If only a woman could have something that allowed her to entirely hide the fact that she was menstruating, she could not only live her life without being troubled by what one doctor dismissed as "the so-called difficult days," but also not trouble society with it. Kotex, Tampax, Meds, Modess: all the early brands had the same object, to deal with "women's greatest hygienic problem." Kotex offered its products along-side a special deodorizer "thus ending ALL fear of offending." Fear of offending is why girls have grown up wondering why they don't excrete blue liquid, like on TV. In fact, there are no rules governing the use of blood in sanitary pad advertising in the UK, only that sanitary products should not be advertised too close to children's programming or cause general offense.[27] It's tricky to know what this might mean. In 2014 a Tampax ad showing "a woman wearing a red top attending a rock concert and an animation showing how the Tampax worked" attracted twenty-two complaints, mostly about the timing and the imagery. The complaints were not pursued.

If red tops can raise red flags, perhaps advertisers are right to be timid. Internalized shame is the hardest to reach and redress. Once, a female airport security employee, emptying my bag, took my packet of sanitary pads and carefully hid it under my books. I asked her why and she looked surprised. "Most women ask me to." In the UK, where I grew up, I have had access to sanitary hygiene and—though I can't remember learning about it—information. But I still hide a tampon when I go to the toilet. I still remember looking at a lipstick in my pocket and worrying it looked like a tampon, and this was just after I'd given a talk about menstrual hygiene stigma at WaterAid. When a taxi driver retrieved a stray tampon from his car trunk and handed it to me, he wasn't embarrassed, but I was. I can't remember being taught about periods in high school, but I've had years of surround-sound indoctrination anyway. Hide. Dissimulate. Be quiet.

It was only in 2011 that an Always ad dared to show a red spot in the middle of a sanitary pad. In the words of one unimpressed advertising journalist, it looked like a "You Are Here dot on an airport map."[28] (The headline in *Adweek* read, PAD AD TAKES THE BOLD STEP OF SHOWING PERIODS ARE ACTUALLY RED.)[29] In 2016, the sanitary pad company Bodyform released a powerful ad that was widely praised and considered groundbreaking. It was expensively and stirringly shot and showed women boxing, cycling, surfing, and playing rugby to a soundtrack of powerful Native American drums. No different so far from Kotex's tennis girls or Tampax golfers. Except this time the women played sport and bled when they did, because we do. A runner tumbled in a forest. A rugby player bled from a head wound. A dancer peeled back bandages on her bloodied foot, the blood sticky and sticking. The theme, clearly, was power and freedom, and the tagline fitted fine, as it was "No blood should hold us back." I like the advert, but it was missing one thing. I wrote to Bodyform several times to ask why it had shown all sorts of blood except the menstrual kind, and was told it was unable to "progress [my] request." (Bodyform released an ad showing menstrual blood the following year.)

Vagina. A word that should be pretty, with its melodious vowels, but isn't. Perhaps the ill-favor of its name is why the vagina has been so ignored as a topic of study. Until 1992, the US National Institutes of Health had no programs for vaginal research.[30] This is an extraordinary fact, because the vagina is an extraordinary thing. For a start, it is highly absorbent, more than skin. Its lining is covered with layers of defunct cells that form a barrier of mucous membrane. It protects the vagina against infection—despite multimillion-dollar sales of douches and deodorants, the vagina is self-cleaning—but not as forcefully as skin does. This makes the vagina walls highly porous so that they absorb chemicals without metabolizing them. Doctors know this: mine told me that I needed only half the dose of progesterone if I was taking it vaginally, as if it were obvious. Estrogen taken vaginally can produce blood serum levels ten times higher than with an oral dose.[31] The vaginal walls are also loaded with tiny blood vessels, and the vagina is almost anaerobic. All this adds up to a perfect set of conditions for bacteria to thrive and this is desirable: of the thirty-nine trillion bacteria we're now thought to carry (or perhaps they carry us), most are harmless.[32] Even *Staphylococcus aureus* lives in the vaginas of 8–14 percent of women without causing harm.[33] Our vaginal microbiome is useful and protective. But when certain conditions prevail, toxic strains of staph can flourish, particularly the toxic strain TSST-1, which in 1978 was first linked to what the pediatrician James Todd named "toxic shock syndrome" (TSS) after encountering a disturbingly virulent chain of infection in seven young children.[34]

I remember TSS, along with the punishing drought of 1976 and Duran Duran, as having a starring role during my teen years, because I knew it meant I should be scared of tampons. By the early 1980s, most women knew they should be afraid of them. In May 1980, reads the Centers for Disease Control website, "investigators reported to CDC 55 cases of toxic-shock syndrome (TSS) (1), a newly recognized illness characterized by high fever, sunburn-like rash, desquamation, hypotension, and abnormalities in multiple organ systems."[35] By the end of the year, there had been 890 cases, 91 percent of them in menstruating women. By June 1983, writes Philip Tierno in *The Secret Life of Germs*, "of the more than 2,200 cases reported to the CDC . . . ninety percent were associated with women who were menstruating at the time they became ill. Most of these women were young, and ninety-nine percent of them were using tampons."

Tierno is a microbiologist who goes by "Dr. Germ" and who was instrumental in linking TSS to tampon use. The progression of Pat Kehm's illness was typical. Kehm was twenty-five, the mother of two young daughters, and healthy until she wasn't. "She woke up with a burning fever and chills. She soon began vomiting and having diarrhea. Meanwhile her temperature reached 103 degrees Fahrenheit." Later that day, at the emergency department, the admitting doctor noted "sunburn-like rash, red flushing, severe dizziness, low blood pressure, and sore throat. Her chest and other parts of her body showed differences in color, and her legs and arms were starting to turn bluish."[36] Her heart stopped and she died.

Kehm had been using Rely, a superabsorbent tampon made by Procter & Gamble. Rely was great: you could insert it and leave it for hours. It swelled up into a mushroom shape, expanding sideways as well as vertically (and so much more suited to the vagina's actual shape). It contained things like carboxymethyl cellulose, a hyperabsorbent chemical also used in laxatives, toothpaste, and artificial tears. The advertising archive at Duke University holds several TV adverts for the tampon. They're not innovative—here is the swimming woman, the golfing woman, the tennis-playing woman—and the message isn't revolutionary, either, as it consists of "remember they named it Rely," one of the flattest flourishes in advertising.

Women with heavy periods liked it. Women who slept for more than eight hours a night liked it. By 1980, a quarter of American women liked it. But the toxic strain of *Staphylococcus aureus* really liked it, too. Science doesn't yet understand why or how. Possibly, Rely was so absorbent that it dried out the vagina, making it more susceptible to tiny tears and abrasions (common in any tampon use) and giving it a direct line to the bloodstream, where staph could wreak toxic havoc. Or the materials used in Rely were to blame. In 1980, Tierno wrote a letter to the CDC, FDA, and Procter & Gamble that named four commonly used components of tampons: viscose rayon, carboxymethyl cellulose (CMC), polyacrylate rayon, and polyester. All, he determined, made for a very hospitable environment for TSST-1.

By the end of 1980, thirty-eight women had died. Rely and similar superabsorbents were withdrawn; there were a thousand lawsuits.[37] Tampons were now sold only with strict instructions that they should not be used for more than eight hours at a time, and with stern warnings about the symptoms of toxic shock syndrome. There was enough publicity and fearmongering that I grew up scared when I forgot to remove a tampon. Manufacturers stopped using CMC, polyester, and polyacrylates. Tampons still use rayon, a product derived from wood or sawdust, alongside their cotton content. Sanitary pads are 90 percent plastic.

Details of what goes into tampons and sanitary pads are sparse and cloaked in the opaqueness justified by commercial competition. In the case of sanitary products, this secrecy is also facilitated by regulators. The FDA does not require manufacturers to list ingredients on the packaging, unlike with food items. If we were inserting lollipops in our vaginas, we'd know a lot more about them.

Investigations into sanitary product ingredients have tended to come from NGOs and the very concerned, including a 2013 report from Women's Voices for the Earth, *Chem Fatale*. This found the presence in tampons and pads of "dioxins and furans (from the chlorine bleaching process), pesticide residues and unknown fragrance chemicals."[38] Dioxins and furans are both persistent organic pollutants (POPs), which we mostly ingest through the food chain. Ninety percent of human exposure to dioxin, says the World Health Organization, comes via meat and dairy products, fish and shellfish. Dioxins "can cause reproductive and developmental problems, damage the immune system, interfere with hormones and also cause cancer."[39] *Chem Fatale*'s list of damage sounds worse: cancer, reproductive harm, endocrine disruption, and allergic rash.

I blame my endometriosis on disruption caused by environmental chemicals. I don't know if they penetrated me from the air, the water, the soil, ice cream, or a tampon, and I have no proof that they have (although most humans now have background levels of dioxins). Primates—rats, monkeys, and mice—exposed to dioxins show increased rates of endometriosis. One dioxin, 2,3,7,8-Tetrachlordibenzo-*p*-dioxin (TCDD), is considered a human carcinogen, and when inflicted on rhesus and cynomolgus monkeys through diet, caused "a dose-dependent increase in the incidence and severity of endometriosis."[40] Cynomolgus is the monkey's lab name, named by Aristophanes of Byzantium after a tribe of hunting men with long beards (the monkey has a beard) who apparently milked female dogs. It comes from the Greek words for "dog" and for "to milk." I can find no authoritative backup for this, but I want it to be true. Whatever their name—you'll know them better as crab-eating or bearded macaque—they were surgically implanted with endometriosis tissue that was enhanced and flourished after exposure to TCDD.

But, wrote Michael DeVito and Arnold Schecter in *Environmental Health Perspectives*, "no definitive human data refute or support the association between dioxin exposure and endometriosis or other reproductive tract diseases." The sanitary hygiene industry says its products are safe. Dioxins are only trace, there is no chlorine bleaching anymore. On its website, Tampax is quite clear about how safe the manufacturing process is: very. Leading scientists from Harvard, Dartmouth, Wisconsin, Minnesota, and the CDC "have conducted extensive testing confirming that rayon and cotton are equally safe ingredients to include in tampons." The site includes a handy chart listing tampon components, function, and material, including the "Thin Fabric Around the Absorbent Core," which helps with smooth removal and "to remove the absorbent skirt on certain designs."[41] (I immediately Google "absorbent skirt" and lose several hours of life.)

Periodically, the FDA has written, "concerns are raised about dioxin in feminine hygiene products—especially tampons."[42] Somehow the sober FDA author of this manages to sound like he—I bet it was a he—was sighing. In 2009, FDA funded a Jeffrey C. Archer (he is left unidentified further but actually works for the FDA as a chemist, though he does not appear on the FDA's useful list of popular investigators). He was tasked to examine dioxin content in seven brands of tampons. No brand identification was given, but various absorbencies were used. The testing used gas chromatography with a high-resolution mass spectrometer, and found that, although dioxins were present, they were within the recommended limits. How much dioxin are we meant to ingest safely? *Safe* is a dangerous term, but the Joint Expert Committee on Food Additives, a UN body, suggests 70 picograms (a trillionth of a gram) per kilogram of body weight per month. Jeffrey C. Archer's results found that if a woman absorbed all available dioxin from a tampon, weighed 50 kilograms (110 pounds), and used twenty-four tampons a month, she would be exposed to only 0.2 percent of the "tolerable monthly intake" set by the JECFA. DeVito and Schecter's research found that our exposure to dioxins through our food is 30,000 to 2.2 million times more than infants, for example, are exposed to by wearing diapers.

In 2013, an organization with the unappealing name of Naturally Savvy examined various tampons manufactured by o.b., a German brand, using a "third-party certified" laboratory. The lab was looking for pesticide residue, and it found traces of malaoxon, malathion, dichlofluanid, mecarbam, procymidone, methidathion, fensulfothion, pyrethrum, and piperonyl butoxide.[43] The FDA recommends that tampons be "free of 2,3,7,8-tetrachlorodibenzo-*p*-dioxin (TCDD)/2,3,7,8-tetrachlorofuran dioxin (TCDF) and any pesticide and herbicide residues," but does not require manufacturers to monitor pesticide residue.[44] In 2016, a report from Argentina's National University of La Plata found glyphosate, a chemical used in 750 common pesticides, and judged by the World Health Organization to be a carcinogen, in 85 percent of tampons, cotton wool, and gauze.[45] But glyphosate is everywhere: most of us carry traces of it by now. It is in soil and cotton clothing. Dr. Jen Gunter, a Canadian gynecologist who likes to spear pseudoscience, in a post entitled "No, your tampon still isn't a GMO-impregnated cancer stick," wrote that she would "worry a lot more about what you eat and what you wear and where you walk barefoot than a tampon you use four days a month."[46] When the Swiss government's chemical research unit commissioned research into tampon contents, it found nothing of concern.[47]

There are two facts to trouble this moderation. The first is that highly absorbent vagina. The second: that in an average lifetime, a woman might use many thousands of tampons. As for fragranced sanitary pads and tampons, they are a blank yet scented slate: because tampons and pads are classified as medical devices, there are no regulations requiring manufacturers to disclose ingredients.[48] Nor is the FDA currently required to do anything, should toxic shock or a similar health crisis happen again, other than recommend a recall of products. For an industry named for femininity, this seems like a very gentlemanly arrangement. What is a menstruating, tampon-wearing woman supposed to believe?

I side with Carolyn Maloney, a congresswoman in New York State who in 1997 first put forward the Tampon Safety and Research Act.[49] Her trigger was a student who asked her what was in tampons. "I was shocked to learn that the science just wasn't out there. I remember there was more research at the time about coffee filters than tampons." In 1999, she changed the bill's name to the Robin Danielson Feminine Hygiene Product Safety Act, in honor of a forty-four-year-old woman who died from toxic shock syndrome. In 2017, Maloney introduced her legislation for the tenth time. The GovTrack website currently gives the bill a 1 percent chance of success.[50] Back in the late 1990s, toxic shock was more resonant and Maloney easily got more co-sponsors. Hardly anyone is dying from toxic shock now. And things have slightly improved, she tells me. "The feminine hygiene industry has done some self-correction. We've seen changing ingredients and changes in the bleaching process. Some companies have voluntarily labeled at least some ingredients on boxes." Consumers are being more demanding (and the presence of organic-only competitors might be a factor in persuading manufacturers out of their secrecy habit). But, as Maloney wrote in an opinion piece, there is "almost no data on the health effects of the cumulative use of tampons over a woman's lifetime." Imagine, she continued, if we examined the health effects of smoking only a single cigarette. "My bill makes no prejudgments about the safety of products that are enumerated in the bill. It just says show us the research."[51]

These are first-world problems. But not for long. By 2022, according to some market research firms, the sanitary protection industry will be worth $42.7 billion.[52] Where is all that growth coming from? From "significant unmet potential in developing markets."[53] From the world of cloths and rags and sand and chocolate cartons.

But the question of how all these pads and tampons will be sensibly disposed of is as quiet as the subject of menstrual blood. Indians already use one billion noncompostable sanitary pads each month;[54] where do they go? Incinerators are rare and can have unpleasant environmental impacts if used at scale. Also, burning menstrual cloth is thought to bring on infertility. Burying pads is time consuming and modern plastic-filled sanitary pads can take up to eight hundred years to decompose. In a study in Bihar, nearly 60 percent of women simply chucked their used pads and cloths into fields and roadsides.[55] India's waste disposal infrastructure is already overloaded, and with much garbage disposal done by low-caste waste pickers, the many more millions of pads a month that the sanitary pad revolution will supply is both a huge burden and a biohazard for the humans who deal with them. Even when sewers do work, they can be easily clogged by sanitary napkins, which are designed to absorb liquid and expand, exactly what you don't want in a narrow sewer.

Yet despite extensive and expensive outreach and education programs run by Procter & Gamble, Kimberly-Clark, and others, the feminine hygiene market is slow to flourish. P&G has partnered with NGOs and UN agencies to bring school-based and community-based feminine hygiene and puberty education programs to between seventeen million and twenty million young women annually.[56] As Muruga quickly learned, commercial products are too expensive for poor women. A single sanitary pad is a luxury item, let alone a box of tampons. I read accounts of girls who have sex with older men in order to earn money for essential items like sanitary napkins. It's called "sex for pads," and though it is hidden, it is common. A field officer for one NGO called Freedom for Girls in Nairobi reported that 50 percent of girls she encountered in the slum of Mathare had turned to prostitution to afford sanitary pads.[57] When researchers surveyed 3,418 menstruating girls and women in rural western Kenya, one in ten of the fifteen-year-olds reported selling sex for sanitary pads.[58] In Ghana, the Girls' Education Unit of the Ghana Education Service reported that 414 teenage girls in twenty-six districts in the eastern region "were impregnated in the last two years while exchanging sex for sanitary pads."[59] Out of this 414, 229 got pregnant in the first year; 185 in the second. One young Kenyan girl responding to interviewers said, "Some people exchange sex for money. The money is used to buy pads." Another said, "You pay him your vagina."[60]

Only 120 girls were interviewed in that study, and large-scale academic research is still absent, but there are enough anecdotes and small studies to cause alarm. Other women thankfully resort to knockoffs, not sex. Sanitary pads can be faked like anything else. Lebanese customs officials recently seized a half-ton shipment of sanitary pads that were found to be severely radioactive. The pads were made in China by a company claiming they contained anions, groups of negatively charged ionized atoms that apparently had health benefits if you put them in your pants and bled into them. (Another name for anions, the company said, was "air vitamins.")[61] A writer on one Chinese blog, pleased with himself, wrote that the products were "guaranteed to redefine the concept of having a 'hot girlfriend.'"[62] In China, forty-three suspects were arrested in 2013 in a "counterfeit sanitary-napkin ring," a phrase I was unprepared to read, that spread across six Chinese provinces. The operation had been undertaken when women reported feeling unwell after using sanitary pads, and forty-three manufacturing "dens" and twenty production lines were closed down. The net worth of that counterfeit sanitary-napkin ring was 150 million yuan ($22.7 million). Under one of the reports, a woman left a comment. "We women are already in pain during our period and you people (the suspects) produce fake sanitary towels. You are not human. I suggest they be sentenced to death."[63]

An investigation into India's sanitary pad market found dirt and ants in nineteen products widely on sale. Indian sanitary napkin standards haven't been updated since 1980, apart from an amendment in 1981 that replaced the words *showing up* with *staining* or *leaking through*. Sanitary napkins must absorb "30ml of colored water or oxalated sheep or goat blood or test fluid when flowed on to the center of the napkin (at the rate of 15ml per minute) and it shall not show up at the bottom or sides of the sanitary napkin. The pH value shall be from 6 to 8.5." I go through the document carefully, but this doesn't take long: a product that is to come into contact with a highly sensitive part of a woman's body several thousand times gets five pages, double spaced. There are no requirements for what the cotton or material should contain, though the government does worry sufficiently that the filler of the pad "does not cause lump formation with the effect of sudden pressure."[64] I'm not much more soothed by a rare insight into Procter & Gamble R&D, which reveals the behind-the-knee (BTK) test. Designed to assess possible chemical and mechanical irritation, "materials (pad, panty liner, topsheet, uncompressed tampon, fabric, facial tissues, etc.) are placed horizontally behind the knees and held in place using an elastic athletic band of the appropriate size." After six hours, both band and test material are removed, and "the area is illuminated and visually scored for erythema or dryness by an expert grader."[65] Although I'm not convinced that my popliteal fossa looks and feels anything like my genital area, BTK has been accepted by the American Society for Testing and Materials as a global standard test.[66]

Psychologists and NGOs like "drivers" because they bring change. A driver might be someone like Muruga, self-made and eccentric, or period activists like AFRIpads or Irise, all good NGOs doing good work in developing countries. But a driver doesn't have to look like an NGO or charity. Sometimes it can look like Levi's or Timberland or Coke, despite Muruga's scorn of "mosquito" parasitical big business.

In 2010, a women's health initiative began in Dhaka, Bangladesh. It was named HERproject (in which HER stands for Health Enables Returns) and was a program run by Business for Social Responsibility, a membership organization of 250 companies worldwide that includes

Microsoft, Sony, Pepsi, and Coca-Cola.[67] Dhaka was an appropriate place for a business-backed project because many BSR members have their products made in Bangladesh's five thousand garment factories. There are three million Bangladeshis working in the garment industry, and 80 percent are women. They are difficult to reach: the garment industry is suspicious of the press because the press reports things like the Rana Plaza factory building collapse in 2013, which killed 1,129 workers.[68] Under the wings of HERproject, I was given access to a factory that I wasn't allowed to name, on a street I couldn't identify, run by companies I wasn't permitted to mention who produced clothing for Western brands that had to remain anonymous.

Most garment workers in Bangladesh arrive in the cities with little education. They come aged sixteen or younger, but with a certificate from the village leader saying they are eighteen. They leave their villages too soon to benefit from NGO education programs, and there are barely any NGOs working on hygiene or education in urban areas. So Nazneen Huq, HERproject's Bangladesh director, set up schemes to improve awareness about nutrition, sexual health, and HIV/AIDS. But she knew she also needed to talk about periods. Factory managers admitted that their female workers were not turning up for work for several days every month. When a factory operates on tightly controlled production lines, a missing pieceworker is noticed. A missing day is definitely noticed, by man, machine, and spreadsheet. Huq, who has worked with garment factory workers for years, had two simple implements: frankness and economic loss.

"They knew that women were going absent, and they knew that it had to do with periods, but they didn't dare talk about it." Huq's tactic was to focus on what managers would talk about, which was business. "I would say, if you have a thousand workers who are women, and each woman is absent one to three days . . . They answered, embarrassed, yes, yes. If five hundred are absent for one day you lose five hundred productive days. Then they get very shy, but they say, yes. It makes sense."

All these managers knew about compliance, as Bangladeshis call workplace safety. They just hadn't thought about health, beyond providing the medical center required by laws. They didn't realize that poor health costs them money. They didn't know, for example, that 80 percent of their female workers didn't use sanitary pads because they couldn't afford them. One of the first factories targeted by HERproject is in Ashuria, a suburb only a few miles from downtown Dhaka, but a three-hour trip in Dhaka's appalling traffic. It is a good factory, as factories go: it is properly ventilated and it has decent toilets. Downstairs, there is a crèche. The workers are protected by face masks, obviously produced from material covered in cartoons that will become children's leggings or pajamas, a sight that is jolting among the machinery and industry.

Panna, a twenty-five-year-old worker, comes away from her post on the Finishing Section to talk. The HERproject model works with "peer educators," women who are trained, then responsible for disseminating information to twenty other female workers. Panna came to Dhaka four years ago, and as soon as she started work, she managed her period as everyone else did, by using scraps of cloth from the factory floor. They call it *joot*. "We got it from the cleaners. They would sweep it up and give it to us. It was very itchy." Every month, she stayed at home for one or two days. "I was in pain a lot, and the *joot* only lasted half an hour. There was leaking, discharge. But we didn't know about reproductive health, even though we are women." (She uses "reproductive" to mean "gynecological.") In another smaller factory nearby, a woman named Banani leads me to the cutting room. Here, barefoot men run along long tables pulling billowing lengths of cloth behind them, for fellow workers to cut. It's quite beautiful to watch. Banani, the factory's welfare officer, leads me to a bin near one of the tables. "There. That is *joot*." It used to be her job to gather the *joot* in secret and hand it over discreetly to whoever asked for it. Every woman I spoke to had regularly had discharge and health problems from using *joot*. There were insects in the bins; people threw water into them. Seventy percent of women, says Huq, had white or smelly discharge. "They thought white discharge was a part of life." Up to 10 percent of workers were going absent. Once they were asked, the factory managers—all male—admitted that they knew about *joot*. One was getting requisitions to clean the sewer pipes all the time, because they were clogged with menstruous rags.

Most women here now use sanitary pads. A central part of HER-project is to persuade factories to buy sanitary pads from a local supplier and sell them at a highly subsidized cost. In this factory's medical center, a basic but clean room off the factory floor, there are several boxes of pads, sold to workers for 31 taka, about 30 US cents (compared to a market price of 90 taka). The compliance officer here is an intense young man named Hasan. "We had real problems with absenteeism. And obviously this is a production-based factory. But even so, the general manager was embarrassed to talk about menstruation." Absenteeism is now down to 6 percent, according to the factory's records, and embarrassment is redundant. Sewer pipes are no longer clogged. In another Dhaka factory, absenteeism dropped by half. A factory manager said that although she initially saw the program as "just another project," she had to change her mind.[69] Now the male workers are asking for a health project.

Managerial blockages still occur, as surely as *joot* blocks a sewer. The program model requires an hour a week for the training in the first years, and not every manager is willing to take that time away from the production line. Progress may not have the speed of those barefoot men on the cutting tables, but it is going in the right direction. And hopefully unblocking one stigma unblocks others. "The HERproject has also helped my relationship with the women workers," a factory manager named Mr. Riaz said. "They are not so shy to talk to me anymore. If there are problems, I now hear about them."[70] HER-project has now reached eight hundred thousand women in seven hundred factories and farms. In Bangladesh, its data show that in ten factories, the proportion of women using sanitary pads has risen by 49 percent.[71]

As I sit in a HERproject refresher course for peer educators, surrounded by young women wearing pink HERproject aprons and head scarves made from the school uniform material the factory produces, someone brings me a brown paper bag. All around me, women are talking with frankness and freshness about a subject that would usually be confined to whisper and shame. It's great. As the women discuss sexually transmitted infections, the health and hygiene of the reproductive tract, and the unquestionable nutritional value of pumpkins,

I open the bag and find a sanitary pad taken from the factory's subsidized supply. It is produced in Bangladesh, manufactured by Savlon, and its name is Freedom.

A driver might be a blocked sewer. It might be a large dose of bloody-mindedness. It might be an end of a tether. Whatever the driver, the last few years have seen something tip and topple. A male gynecologist told me a few years ago with conviction that menstruation was not taboo in the UK. I looked at him in disbelief and said, "The day that menstrual blood doesn't have to be represented by blue mouthwash, because it is thought dirty and distressing, you will be right."

I'm still waiting for the blood, but there is plenty to keep me hopeful that the borders of the taboo, so fixed until now, are wavering and shifting. In 2015, *Time* magazine showed a tampon (unused) on its cover, which held the headline YEAR OF THE PERIOD. Period blood has appeared all over the place, from artists using it to make gorgeous swirly images; to Rupi Kaur, a poet, faking it in an Instagram post, showing her bleeding through gray sweatpants; to the musician Kiran Gandhi running the London marathon as a "free-bleeder," bleeding freely into her conveniently red sweatpants. (She claimed that she did it because she wouldn't have any privacy to change a tampon, an odd statement when the London marathon is probably the only time of the year that Londoners can find sufficient numbers of public toilets.)

Suddenly there was tennis player Heather Watson saying she lost a match because of "girl things," or, more straightforwardly, the Chinese swimmer Fu Yuanhui telling an interviewer, as she stood dripping water after an Olympic event, that she had performed poorly "because my period came yesterday, so I felt particularly tired."[72] She hadn't won her race, but this statement got her more coverage than any medal would have. Behind the color and noise, there is real money. Venture capitalists, those male-dominated businesses, are funding women-run period businesses like Thinx, which makes underwear you can bleed into, and advertising—using graphic and cunning pictures of vagina-resembling grapefruit, a category unknown to fruit growers—you can't avoid; or Flex, a "menstrual disc" that captures blood and in 2016 secured $4 million in funding. There are organic cotton tampon companies such as Lola and Maxim building on environmental worries. Lola raised $10 million in investments just in 2017. Clue, an app that allows women to monitor and track their menstrual cycles, has raised $20 million.[73]

Menstrual Hygiene Day, founded only in 2014, saw 349 events in fifty-three countries in its third year of existence. The government of Kenya has now promised to provide free sanitary pads to "every girl child registered and enrolled in a public basic education institution."[74] The government of Uganda, in the form of First Lady and Minister of Education and Sports Mrs. Janet Museveni, promised the same thing in the months before an election.[75] None have been provided, and when Dr. Stella Nyanzi criticized the broken promise, as well as calling the president "a pair of buttocks," she was jailed for four weeks.[76] This is the same Mrs. Museveni who once proposed nationwide virginity tests as part of a national "virgin census."[77]

In India, the commercial-led campaign called Touch the Pickle was supplanted by a ground-up one called Happy to Bleed, which saw young Indian women nailing sanitary pads to trees, the pads' top-woven liners covered with messages such as "We bleed! Get over it!" Founded by the twenty-year-old Nikita Azad, Happy to Bleed began as a protest against the banning of menstruating women from temples. (At one temple in southern India, a temple elder proposed installing a "menstrual-scanning" machine, though he hadn't yet invented such a thing.)[78] The temple ban was small-minded and stupid. But, as Azad wrote, any menstrual taboo is about much more than blood. "It is not a question of pure versus impure or men versus women. Our fight begins from our homes and workplaces. Relatives who beat our mothers to abort us, to in-laws who burn us, to those who rape us, to temples that denigrate us."[79] Struggling against stupid taboos was also a fight against entrenched misogyny and its side dish of endemic violence.

There are campaigns now to provide pads to homeless women, also users of socks and whatever works. When I spent a night driving around Saskatoon with a youth outreach charity, which served mostly the poor and indigent, the most commonly requested items were tampons and diapers, and the outreach worker referred to herself with no irony as

That Tampon Lady. The American Civil Liberties Union has filed lawsuits on behalf of several incarcerated women because of assaults on
their access to menstrual hygiene. In women's prisons, access to sanitary pads is a far cheaper weapon of control than a firearm. Female
prisoners report that they are given sanitary pads in insufficient quantities and of dire quality. Londora Kitchens, an inmate at Muskegon
County Jail in Illinois, testified that in 2014 she had been menstruating and had run out of sanitary napkins. "You're shit out of luck," said
a certain Officer Grieves, and told her to bleed on the floor.[80] In a lawsuit filed by Brooklyn Defender Services before the New York City
Council on Women's Issues, an attorney wrote that her client, incarcerated at Rikers, "asked her social worker not to visit her while she
was menstruating because she was worried about leaking through her
uniform and having to walk the halls of the jail with a bloodstain."[81]
Rikers gave its inmates twelve sanitary pads a month, and they were
thin and inadequate. In Arizona, women who use up their twelve allotted pads must ask for more. Tampons are forbidden because they are
a "security risk," according to ACLU. State senator Athena Salman
recently proposed a bill that would allow prisoners unlimited menstrual
hygiene products and not allow prisons to charge for extra. Jay Lawrence, chair of the all-male committee, said, "I'm almost sorry I heard
the bill [. . .] I didn't expect to hear pads and tampons and the problems of periods."[82] He still had to listen to testimony about heavy flow
and leaks. Nurse Molly Nygren, who has worked with women prisoners, told reporters of women who fashion tampons by twisting several
maxi pads. "That can increase bacteria and cause toxic shock syndrome," said Nygren, "and I think they shouldn't have to do that."
The bill passed 5–4.[83]

Other campaigns focus on tampon taxes. In the United States, tampons merit sales tax though condoms, hair-growth products, and lip
balm do not. In the UK, they were taxed at 20 percent until 2000, when
the rate dropped to 5 percent.[84] This remains one of the few tax cuts
that was never publicly announced in the budget, according to Prime
Minister Gordon Brown's former spin doctor Damian McBride, "due
to Gordon's reluctance to refer to tampons at the despatch box."[85]
Despite the European Union relaxing its laws to allow states to exempt

sanitary protection from tax, that 5 percent tax rate means feminine hygiene products count as luxury items, unlike essential items exempt from VAT such as ice cream, houseboats, and incontinence pads. In 2017, New York State voted to provide adequate sanitary provisions to prisons, homeless shelters, and schools for free. It also revoked the tampon sales tax, as did eight other states.[86] Ireland, Spain, and the Netherlands have removed the tampon tax; France has reduced it to 5.5 percent.[87] In Switzerland, women filled thirteen Zurich fountains with red dye to protest the taxing of feminine hygiene at 8 percent (the rate reserved for luxury items; day-to-day products get 2.5 percent).[88] In the UK the Conservative government pledged to remove the tax, did nothing, then used £250,000 (about $355,000) of sanitary product tax revenue to fund an antiabortion charity (despite promising that tampon tax revenues would support women's shelters).[89] Before he left office, President Barack Obama expressed surprise when he was told that forty US states tax feminine hygiene products as luxury items. "I suspect," he told the interviewer, "that it's because men were making the laws when those taxes were passed."[90]

How different things would be if men had periods. This is an old idea, and never better expressed than by Gloria Steinem in her peerless and perfect 1978 essay "If Men Could Menstruate." In this alternate world,

Men would brag about how long and how much.

Young boys would talk about it as the envied beginning of manhood. Gifts, religious ceremonies, family dinners, and stag parties would mark the day.

To prevent monthly work loss among the powerful, Congress would fund a National Institute of Dysmenorrhea.

Sanitary supplies would be federally funded and free. Of course, some men would still pay for the prestige of such commercial brands as Paul Newman Tampons, Muhammad Ali's Rope-a-Dope Pads, John Wayne Maxi Pads, and Joe Namath Jock Shields "For Those Light Bachelor Days."

Street guys would invent slang ("He's a three-pad man") and "give fives" on the corner with some exchange like, "Man you lookin' good!" "Yeah, man, I'm on the rag!"[91]

———

At the Association for Parents of Mentally Retarded Children in Coimbatore, Muruga invites me to try the machine. The cellulose is deconstructed by the young lad at the grinder; then I tamp it into a steel box. I compress that, using a simple lever, to form the liquid-absorbing core of the pad. With another manually operated machine, I enclose the pad in fabric, glue the adhesive strip on it, place it under UV light for thirty seconds, like shellac nails in a salon, and it is done. Photos are taken, and in them I am grinning with abandon, because I've just made one sanitary pad, regular, called Kangaroo.

Maybe it looks like nothing much. But in the hands of a girl or woman who dries her clothes with shame and risk; who stays away from school because she fears smelling and pain; in the hands of women who earn a living from producing low-cost sanitary pads where before they had none, this pad is holding a lot more than cellulose. It may be called Friend, or Peace of Mind, or Purity, but actually Wings—or Vings—is its best name, because it is freedom, prospects, possibility. All in the shape of 12 grams of compacted, affordable fluffy.

London's Air Ambulance helicopter landing near Tower Bridge, London

EIGHT

CODE RED

Code Red. Open chest. These phrases become a noise in the room, repeated by staff. They say it almost with wonder to each other, though they are experienced emergency and trauma professionals and they deal with calamity and catastrophe daily and many times a day and night. Adult trauma call. Adult female. Open chest. Code Red. Eight minutes.

As soon as the nurse hangs up the red phone, the one that delivers news of incoming emergencies, the one a passing nurse calls the evil phone, and says "Code Red," then "open chest," the atmosphere changes. Now there is tension, intensity, focus. I hear "open chest," and my mind thinks of pirates and black caskets, florins tipping out on a beach, because I have no idea what an open chest is going to consist of and suddenly everyone is too intent and busy to be asked. My stupid fancies fill the minutes that pass in heightened anxiety until there she is, our Code Red, being wheeled into the resuscitation room at the Royal London Hospital Major Trauma Centre. Her trolley is surrounded by stern staff, some in orange, some in green. By stern, I mean the kind of unsmiling authority you want in someone who is too involved in saving your life for levity.

Now I see what an open chest is.

She is horizontal, and I am sitting away from the bay so as not to disturb, so my eye line sees only her bare feet on the plastic trolley, then strange pink mounds rising out of her torso rising and falling, rising and falling. Her lungs. I have never seen anything like this. I am frozen with shock while everything around me speeds up and intensifies: the number of people in Bay Eight, where she has been taken (ten, then fifteen, then thirty); the activity and movement (constant); the beeping machines (maddening and constant). There is no panic. They are just busy. They have things to do.

A doctor has his hands in her chest and he is massaging her heart. Resuscitation. From the Latin, "to raise again." The Lazarus ward, this.

And Jesus called in a loud voice, "Lazarus, come out!" The dead man came out, his hands and feet wrapped with strips of linen, and a cloth around his face. Jesus said to them, "Take off the grave clothes and let him go."[1]

The French call it "reanimation," the Germans use "revival." To restart life; to make someone live. Before, the places where the dangerously and seriously injured went were called moribund wards. In wartime they were also called "dying tents."[2] During the First World War and because of the power of blood transfusion, the name changed to something more hopeful, to possibility, not foregone conclusion. In a trauma department, the resus room is for the most dangerous and serious. At the Royal London this is where paramedics bring people who have endured falls from more than two stories, "bull's-eyes" (heads that hit a windshield), burns that cover more than 30 percent of a body, traumatic amputations, mangled extremities, stabbings, gunshots. Here they take in "one-unders," as people who go into and under trains are known, and Code Blacks, people with severe brain injury. They bring the Code Reds. There are very precise ways to assess a Code Red involving blood pressure and hemoglobin levels, but it amounts to this: severe hemorrhage. Bleeding to death.

In a nearby bay there is a man who fell from a height who is a Code Black. The Code Blacks don't speak. The Code Reds might for a while. Later, in the operating room, the Code Black man becomes a Code Red as well. Two Code Reds in one morning: the Royal London gets several Code Reds a month, but this is still a startling way to start a week. An open chest is even more unusual and so spectators arrive. Soon there is a crowd of medics drawn from all floors of this huge hospital. Their curiosity is the educational kind, not vulgar. She is dying: they want to see how she can be saved.

A London morning. Rush hour on a busy road that heads into the city, that is shared by buses, cars, bicycles. The woman was cycling to work. She did this every day, as it was a clear ride from her home in one part of the city to her workplace in another. She did this every day, because she probably lived with the belief that all inhabitants of a major city must hold to themselves like treasure: that nothing bad will happen today. Nothing will hit, or crush, or crash into, or stab, or shoot, or damage them. They will not stumble, or slide, or trip, or slump. City dwellers do not think like trauma medics, who designate cases as Car Versus Pedestrian or Pedestrian Versus Truck. This versus, so adversarial, is the truth of city life but one that we cannot afford to accept if we are to continue to function.

She cycled alongside cars and motorbikes and trucks and buses and mopeds and pedestrians and other bikes, and she made it all the way, nearly, until a bus pulled out from a side street; and then for some reason (the investigation is ongoing) she went under the bus and its wheels went over her body. They ran over her pelvis and it was smashed. Smithereens, from an old Irish word meaning "small pieces." Her bones were smithereens, but also the blood vessels in her pelvis: veins, arteries, capillaries, blood vessels that can be so small, cells pass through them in single file. They are all over the body and they can be broken like bones. All these broken vessels were leaking blood out of her circulation, into places where it should not be. The body needs an adequate volume of blood to feed everything it needs to feed: this is blood pressure. If blood is pouring out of the circulation, then the body starts to starve, sputter, shut down. A vehicle with holes in the fuel tank.

At 8:56 a.m., the police were informed of an accident. The London Ambulance Service arrived, paramedics in dark green uniforms, then the HEMS team in orange. HEMS is the Helicopter Emergency Medical Service, a service that incorporates doctors from the Royal London Hospital and elsewhere; paramedics on attachment from the London Ambulance Service; and flight crew (and helicopters and fast cars) from London's Air Ambulance, a charity. HEMS is those glamorous types in orange flight suits, moving urgently and importantly with large boxes and bags of equipment like soldiers. It is one of the most advanced trauma teams in civilian life. The HEMS team had come by car, a fast Škoda that is used at night or in poor weather, when the helicopter can't fly. (The short distance made a helicopter ride pointless.) Two doctors, one paramedic. Quickly, the medics in orange suits set to work alongside the green ambulance staff. They had already made a transformation because this grim roadside on this busy highway was now a place known as "pre-hospital." Every accident scene everywhere and every time is potentially pre-hospital. But the term has now come to mean a particular thing: that a small team of paramedics and doctors can now perform procedures that they would never have done before in a location like this. In the military, there is a place called "an austere setting." It is where special forces work, and where everything is limited except danger and difficulty. Although this was a busy gritty road in the middle of a great city, it was austere enough for medics about to do procedures usually performed in sterile, safe, and supervised operating rooms.

At first, trapped under the bus, the woman was talking. Emergency staff call such cases "talk-and-die": people who have had horrific injuries, who are crushed and bleeding, can at first talk normally, because at first their body is not yet shutting down. At first it is pausing under the enormity of the insult. This is not my term: medics talk of the insult done to a body. Don't mistake "talk-and-die" for callousness. It is an alert, a warning to medics not to be fooled, for them to plan for what is coming.

Bleeding, says Karim Brohi, a trauma and vascular surgeon at the Royal London, "is the biggest disease you have never heard of."

Nearly six million people a year die of injuries.[3] Trauma accounts for

10 percent of the world's deaths, says the World Health Organization, or "32 percent more than the number of fatalities that result from malaria, tuberculosis and HIV/AIDS combined."[4] In Africa, more people now die from trauma than from AIDS.[5] Of people who receive traumatic injuries in civilian life, up to 40 percent die because they were bleeding. Add up all the common infectious diseases, says Brohi, and they don't even "touch the sides" of bleeding's death toll. Hemorrhage causes up to 80 percent of all potentially survivable deaths in combat.[6]

The World Health Organization classes trauma as a disease, because it has a cause—a profound injury to the body—and a treatment. Bleeding has no category of its own, among the categories of things that kill us, but you can see it, seeping through others. Brohi adds up. Post-partum hemorrhage, more than a thousand a year. "Gastrointestinal hemorrhage, two hundred thousand. The WHO categorizes them as injuries or maternal deaths but there are lots of ways to die if you are injured and there are lots of ways to die if you have maternal deaths. Actually these people are dying because they are bleeding."

Brohi is a tall, gray-haired man with a calm air and quiet words, some including a London glottal stop (his Twitter biography reads "Londoner"). He is the director of the Centre for Trauma Sciences, known to thousands of medical students and conference attendees as Mr. Trauma and by most measures he is eminent. Brohi is a vascular surgeon by training because when he was a student the specialism of trauma surgery was not available to him. It didn't exist.

He was a junior emergency medicine SHO (a senior house officer, a junior doctor who is actually quite senior) when he saw a couple of trauma patients who, even with his limited knowledge, he could tell "were badly treated, clinically, and I was pretty sure we could do better." It wasn't that clinicians knew what to do and did it badly: they didn't know what to do. Since then, his career has consisted of trying to rectify this. It is an ongoing task. He set up a website to aid medical professionals here and abroad and called it trauma.org. It is a place where I can spend many hours, running through its moulages, where you play at being a trauma doctor in simulated situations. I fail consistently. For example: a motorcyclist traveling at around seventy miles per hour who went into the back of a stationary vehicle on the side of the motorway and was thrown some fifty to sixty-five feet from the bike. I'm given three options: assess for intra-abdominal bleeding, check the airway, or chat to the fire crew. I pick the most medical-sounding option and check for bleeding.

> YOU ARE SO UNBELIEVABLY WRONG. One hundred chimpanzees with one hundred computers wouldn't have chosen that option in a hundred years. Go back and try again.

I choose next to check the airway.

> To the shouts and cheers of the fire crew you run toward the patient. A blur of orange, dashing through the puddles. The fire crew are waving and jumping about. OK, you look good, but you've never had this reaction before. Then a thought goes through your head . . . puddles? It hasn't rained in a week. Looking down you realize that you are in fact running through a pool of petrol. The next thing that goes through your head is the back wheel of the motorcycle as the car explodes. You would feel stupid, but you can't feel anything anymore.

I tell Brohi it's funny. He dismisses this by saying he wrote it "back in the day." But he also wrote it so it could be used in developing countries. That's why its layout seems old-fashioned, as analog as digital can get, so it can be useful in places that don't have hi-tech or Wi-Fi. "There's a reason it's shit." There's also a reason it's funny: we remember humor, like smell. It sticks.

In the same simulation, I eventually choose to check for pelvic instability. The patient is bleeding, and I'm given two options: give crystalloid or colloids. Both of these are volume expanders. TV hospital dramas would call them "fluids." I choose crystalloid.

> Despite the number of IV sites you find, and the amount of fluid you keep pouring in, the patient's pulse and blood pressure keep on

falling. Finally you get control of the situation—that is, if a pulse and blood pressure of zero is "control."

I have killed the patient, again. But this time I can't really be blamed for my choice. Until the last decade, modern resuscitation relied on crystalloid. The clear stuff in bags, as one trauma surgeon described it to me, was pumped into trauma victims in the hope of boosting a circulation that was being drained by hemorrhaging. Refill the circulation with fluid, even if it's not blood, and the body's damaged blood cells will have enough strength to get oxygen to the tissues and organs, as it is supposed to do.

But this was wrong. This is why my moulage patient's pulse and blood pressure kept falling. "Now we know," says Brohi, "that if you give lots of fluids when someone is actively bleeding, you dilute down all the good stuff in the blood, which is desperately hard to deliver and restore." The good stuff is what the blood needs to deliver oxygen to organs and tissues, as well as remove toxins and waste. A person who is bleeding is like a bucket with a hole in it. "All you do is pour fluid into the top, and you're putting more pressure in the bucket for it then to bleed more." And hypoperfusion begins. This is the medical term for the body's failure to deliver oxygen to tissues and organs. You'll know it better as shock.

In 1870, the German doctor Hermann Fischer described a patient he had encountered. The young man had been hit by a runaway team of wagon and horses. The wagon's shaft had struck him in the pelvis. There were no outward signs of bleeding; the shaft had not penetrated his flesh. But soon,

he lies perfectly quiet, and pays no attention whatever to events about him. The pupils are dilated and react slowly to light. He stares purposelessly and apathetically straight before him. His skin and such parts of the mucous membranes as are visible are as pale as marble, and his hands and lips have a bluish tinge. Large drops of sweat hang on his forehead and eyebrows, his whole body feels cold to the hand. . . . Sensibility is much blunted over the whole body. . . . If the

limbs are lifted and then let go, they immediately fall as if dead. . . .
The pulse is almost imperceptible and very rapid. . . . The patient is
conscious, but replies slowly and only when repeatedly and impor-
tunely questioned.[7]

His temperature is taken: he is cold. His arteries are contracted and
of "exceedingly low tension." Cold, low pulse, rapid breathing: all these
things are now known to signal a person in shock. Shock, which has its
own society (the Shock Society) and dozens of journals and 130 years
of scientific expertise devoted to its study. But shock is still imperfectly
understood despite our best efforts. And despite medical advances—
clever blood clotting bandages, new tourniquet techniques, better
drugs—also imperfectly understood are why trauma can make people
bleed so catastrophically and why sometimes we can't stop it.

On the road, under the bus, she had stopped talking. It's not always
easy to guess at internal bleeding, but good clinicians can read clues,
on the patient and elsewhere. In the trauma moulage, my first step
should have been to talk to the fire crew, for example, before doing
anything medical: they would have told me about leaking petrol and
when it was safe to minister to the casualty. Good emergency doctors
will ask bystanders and first responders about the mechanism of injury:
if eyewitnesses describe a head hitting a windshield, that is a clue. Head
injury, but probably not bleeding. They will read the body. Someone
who has been screaming in pain, then becomes drowsy? Probably
bleeding. Is the heart rate fast? Probably bleeding. Does the pelvis feel
broken? Probably bleeding. Cold, unresponsive? Bleeding.

Bleeding is caused by what clinicians call "derangement" of the
body's physiology. It happens fast. A quarter of all bleeding deaths
occur within three hours of injury, says Brohi, and most severely bleeding patients are dead within six.[8] Nearly half of patients with severe
truncal injury—a penetrating wound to the chest or abdomen, for
example—die within thirty minutes.[9] There is a concept, now, of the
"golden hour," when trauma intervention must be done. Brohi thinks
this is a marketing concept, and in a way not useful: some people bleed

fast, others bleed, then stop. Bigger wounds, more holes, more blood loss. But definitely the first minutes and hours are a critical window. What happens to the body after massive injury? Brohi truncates or translates an explanation for me. It is "really really really complicated."

There is an insect bite on my leg and I scratch it. It's a bad habit, but I like to see the blood flow. Rich, and red. And then I like to count. Usually within five seconds, depending on what I have scratched and where, the flow stops, the blood clots. It looks like something ordinary, but it is complex and amazing. It probably takes up to one hundred proteins to form a blood clot.[10] All that complexity, and the body does it selectively. Cut yourself shaving, and you will clot only at the cut, not everywhere. Break a bone and a clot will form around the break, but only there. "Clotting happens in the right place," says Brohi, "but it has to be turned off everywhere else. If you were a wildebeest on the plains of Africa and some saber-toothed tiger took a chunk out of you but you survived, you've lost some blood so you've got less blood flowing more slowly around your body." In consequence, blood becomes less coagulated—less likely to clot—so that you don't get clots where they shouldn't be.

A removed chunk: that can usually be dealt with, by either the body or medicine. Those injuries are the survivable kind. But the people who come to the Royal London and other major trauma centers have complex, severe trauma that turns their blood "rampantly anticoagulant." Nothing is clotting; nothing is stopping blood from bleeding. Patients with such severe trauma, by the time medical professionals reach them, already have this severely disordered clotting. For reasons that are still opaque, their platelets—which should help clotting—also stop working properly. It is a complex derangement of biology known as acute traumatic coagulopathy.

The woman's blood vessels, so badly crushed, were leaking blood into the body, out of the circulatory system in which it should have been contained. Her blood volume and her blood pressure were dropping. Her heart, which should have been pumping ten pints of blood a minute, was slowing and emptying. Her organs and tissues were being starved of oxygen that the blood could no longer deliver. She was shutting down. Internal hemorrhage sounds like a quiet thing: the body bleeding inside itself, politely. But the violence caused by it is massive. Blood cells starved of oxygen were producing lactic acid, making the body acidic (the medical term is *acidotic*). They were also releasing potassium. Too much potassium can stop the heart, which is why potassium chloride is a popular component of lethal injections. The reduced blood flow, along with the environmental context (many accident patients are lying on cold ground) lowered her temperature. Blood normally carries heat around; it stopped doing that. The colder she got, the worse platelets got at clotting. The worse the clotting, the more blood she lost, the colder she got. The more blood lost, the greater the acidosis. Everything was working together to make things worse. That is why these three conditions are known in trauma circles as the "lethal triad."

She stopped talking and her heart stopped. It could not pump blood when there was not enough blood to pump. The medical team needed first to let her breathe, so they passed a tube into her windpipe. There was no point trying chest compressions: the pelvic trauma meant she had "bled out." Her heart had nothing to pump and was still. Five years ago, she would have been declared dead at this point. But HEMS had two more options. First, they took her out from under the bus, they cut open her chest, and they manually massaged her heart to get it going. This is an emergency thoracotomy. Royal London has been doing the procedure since 1993, and now does about twenty a year. But it still looks awful and dramatic. The other option was a procedure HEMS began to do only in 2012. Passersby would think nothing of it. The groundbreaking thing was to give her blood, there and then.

The passage of medicine toward accepting blood transfusion was one of bumps and jumps. In the linear, smooth version, James Blundell's nineteenth-century experiments led to blood transfusion being thrown into the darkness because no one could understand how to stop blood clotting. Then Karl Landsteiner shone a light, and all was well. In fact, the debate about transfusing blood continued throughout the nineteenth century after Blundell's efforts, and medical gentlemen had regular spats in medical journals about its merits and evils. They were also debating what to transfuse. Blood clotted, so it was bad. But sometimes blood revived, so it was good. Saline was better, because what a bleeding person needed was volume. This made some sense: blood was not seen to be the highly complex living tissue that it is now, so to replace volume, why would the more straightforward saline not do instead? It did not spoil like blood; it did not clot and clump. It was easier all around. Supporters of blood transfusion scorned the London Hospital as being the headquarters of the "salino-alcoholic disciples," but saline was more popular for a good while.[11] Even Landsteiner's discovery did not immediately trigger a rush toward blood transfusion: he ignored his findings for decades, as did most medical journals.

The usefulness of blood in transfusion was still being debated in the Second World War. The British used it; the Americans preferred plasma. Its proteins and clotting factors are useful. It worked very well healing burned people. But plasma cannot, alone, fix a body in shock. In 1944, Colonel Frank Gillespie, a British liaison officer stationed with American forces, wondered whether American shock victims were made from different stuff than the British. "I have often wondered at the physiological differences between the British and American soldier. The former, when badly shocked, needs plenty of whole blood, but the American soldier, until recently, has got by with plasma." In Normandy he found "American surgical units borrowing 200–300 pints of blood daily from British Transfusion Units, and I'm sure they were temporarily and perhaps permanently benefited by having some good British blood in their veins."[12] This was happening elsewhere thanks to the efforts of Colonel Edward Churchill, who traveled around Europe reporting back that blood was better than plasma, but who was not heeded. After American medics resorted to doing stealth transfusions in stealth transfusion stations, policy changed and American soldiers were given blood from their own compatriots, not borrowed from the Brits. During the four years the United States was at war, the American Red Cross collected over 13 million units of blood, and blood taken from a donor on the Atlantic Seaboard could be in the veins of a wounded soldier the following day.[13] The template seemed set. Blood worked. By the Vietnam War, 38,000 units of blood were transfused each month, the largest ever wartime blood program.[14]

Then came the rats. A notorious study in rat resuscitation in the 1950s seemed to demonstrate that clear fluids were better than blood. (Even though, Brohi says, the rats were first treated with blood.) But a switch was flipped, a new dogma installed, and resuscitation became associated with the clear stuff in bags, not blood. For the past few decades of trauma care, ambulances and paramedics have not given blood, not even to a severely hemorrhaging patient. They couldn't: they didn't carry it. Most ambulances still don't. First they would check the airway, assess for injury, and then deal with bleeding by giving fluids. For decades, this was the standard protocol: liters and liters of clear liquid. It was thought to stabilize a patient long enough, by maintaining the blood volume, for them to get to hospital or to an operating room. But they were giving the wrong things, at the wrong time, in the wrong way.

About ten years ago, a new way of thinking was resurrected (because it was actually an old way of thinking). "We've completely shifted," says Brohi, "from trying to restore volume with whatever to trying to maintain competence of blood to form a clot." Two three-word phrases can describe this approach. In the military, it is "damage control resuscitation." Or as a guiding principle: "stop the bleeding."

In London, at the accident scene, the HEMS team got out its blood bag. They did this because war, again, had changed the minds of medics. In Afghanistan, British MERTs—mobile medical emergency response teams—began trying to deal with all their hemorrhaging soldiers by taking blood to the soldier, not filling the casualty with fluids and transporting him or her to blood. It was no different from the Royal Army Medical Corps man with his transfusion kit, or the transfusion officer on a northern French beach, a blood bag hanging from a rifle stabbed into the sand. In Afghanistan, the Americans, seeing what a difference it made, followed suit. (Transports that carried blood were called "vampire flights.") In 2012, HEMS began to carry "blood on board." This was exactly what it sounded like. For the first time in its history, the service began to carry blood products on board the helicopter, in thermal containers used by the British military and nicknamed "Golden Hour Boxes."[15]

That is why, on this London street, the woman could be transfused with three units of blood. Or near enough. In fact, the red liquid in the bags was something different. It was packed red blood cells (PRBCs), and it was a component of blood, but missing a few things: plasma, platelets, white blood cells. In the UK, all white blood cells must be removed from blood products because they can transmit vCJD. She couldn't be given plasma because it is stored frozen and it would be logistically impossible for HEMS teams to carry and thaw it. Platelets need to be kept warm: same logistical obstacle. So PRBCs were the best option, and for the rest, she needed to be transported quickly to the hospital, where she could be given other products to help her blood clot, where she might have a chance. She was fortunate in only one thing that morning: that she was within the target area of the busiest trauma center in Europe. Within an hour, she was transferred to Bay Eight of the resuscitation room at the Royal London, and it was the best place she could possibly be.

The screen in Bay Eight reads CODE RED, 30S CYCLIST, TRAUMA CARDIAC ARREST, THORACOTOMY. The nurses had been talking while they waited for her to arrive: she was a woman who went under a bus. No, a young woman. Maybe a girl. A truck, not a bus. A cyclist. She was definitely a cyclist: the screen says this. Now she lies with her chest open, her feet yellow, and she has almost disappeared under all the people working on her. Someone says, Can we get some volume? Someone asks for insulin. Will someone fetch some bicarb? Let's get her into bypass. No one shouts. Everyone is politely urgent: will you, would you, would you mind.

I thought she had been in the bay for at least half an hour. Perhaps even more. No, says Brohi, who was one of the fourteen staff involved in the resuscitation. She was in resus for only fifteen minutes. Time dilates in there, yet time is what they have least of. Priority: potassium, calcium, and blood. The anesthetist immediately checked the woman's potassium levels, even though it was obvious they would be high. The dangerously leaking potassium could be addressed with insulin, and this was provided. A regular dose of insulin for a diabetic will be 10 units. By the end she had been given 9,000 units. She needed calcium to help with the derangement of her clotting, and she needed blood.

At the blood infuser, there are two nurses. Their job is to watch for the blood bags; to read the labels out to each other, the reassurance of repetition. O negative at first, as this is the universal and emergency blood and was stocked in the resus room fridge, the one with warnings not to waste blood because it costs £123 ($175) a bag. But then she was cross-matched and blood more suited was brought down from the blood bank. All must be checked. All must be right. Later, I learn that the IT system had gone down, so all blood orders had to be taken by hand to the blood bank, over the road in another building, not ordered by computer. First principles, Brohi calls it, and it worked fine. What was not fine was her heart: when she came in, it was flat, like a tire. It had stopped beating because she had no more blood in it to pump, and even the transfusions could not stop it from stopping. Now that it has blood, another surgeon keeps massaging her heart with his hands, and I watch all this furious calmness, dazed by it.

The composure was deceptive, said Brohi a few days later, when I asked him for a debrief. What seemed calm was actually the most aggressive trauma care you would ever see outside a military context. He calls it "hugely aggressive," and "the extreme of trauma care." By the end of that day, she had had 30 units of packed red blood cells, more than three times her own blood volume. Also, 31 units of plasma, 8 units of platelets, and 8 units of cryoprecipitate. In civilian hospitals, this is the gold standard of care, carefully developed over the last ten years at the Centre for Trauma Sciences. When patients were given a ratio of one to one to one (1 unit of PRBCs, 1 of plasma, 1 of platelets), survival rates increased. A study of 246 patients with massive blood loss at a US combat support hospital in Iraq found that the risk of death dropped from 65 percent to 19 percent when the plasma-to-red-blood-cells ratio was increased from 1:8 to at least 1:14.[16]

Her heart recovers enough to start beating, and she is taken upstairs to an operating room. There, Brohi takes over the cardiac massage. Everything that was done in resus and the operating room had the same goal: to get her heart to work on its own. One way to do this is to block off parts of the body, so that blood has less area to cover. The Royal London helped pioneer a balloon device called a REBOA (resuscitative endovascular balloon occlusion of the aorta)[17] to do this:

inserted into the aortic artery, it expands and blocks off the blood supply to the lower half of the body. You hope that the lower half of the body survives long enough for you to deal with bleeding in the upper half. It is all balance, risk calculation, courage. But REBOA can be used only on patients with a pulse. Here, instead, they had to do things manually. A surgeon dealt with her lacerated left lung by tying off its artery and taking it out of circulation. Someone opened up the abdomen to deal with the bleeding there. The liver was packed with gauze pads, to protect it. The spleen was taken out, as that's what you do with damaged spleens. And, all the while, the blood. Not too much, not too little. A heart can be overfilled with blood and not be able to beat. If it is underfilled, it can't beat. And this heart was so damaged already. Brohi's hands were the best measure of this: he was operating by feel, telling the other operating staff what to do from what his hands were telling him: start to fill, stop filling, let's give more calcium, let's give more insulin. They carried on.

Death from bleeding is preventable. In theory. That is the awful truth of hemorrhage, and a source of extreme frustration. "Brain injury we can't do a lot about," says Brohi. "[But] if you get to people in time, know what to do with them, stop them bleeding, restore their circulatory volume, correct all the bad physiology that goes on, you could potentially save everybody." The trouble is, he says, even now when we give them blood, it's "a bit rubbish." As Landsteiner proved, not all blood is alike. Red blood cells, plasma, and platelets are supposed to perform like whole blood. But some trauma experts believe that they don't.

The problem is that in the 1970s blood met cancer. In the United States, the National Cancer Act of 1971 and President Richard Nixon's declaration of "a war on cancer" started the chemotherapy era.[18] Cancer patients were now getting treatment that attacked their immune system and their bone marrow, the blood factory. So they were hematologically compromised but they weren't bleeding and didn't need whole blood. They needed parts of blood such as platelets and plasma. The technology to separate and fractionate blood now existed. So did a sterile plastic bag, developed in the 1950s by Professor Carl W. Walter. It took him five years, $1.5 million, and the understanding that thermoduric fungal spores persisted on labels made from southern pine but that paper made from Maine pines was spore-resistant and suitably sterile.[19] Now blood could be stored, transported, and transfused with much less risk. No more glass bottles and bungs. Fractionation technology could separate blood into red blood cells, plasma, platelets, cryoprecipitate, clotting factors. The era of component therapy (CT) began. CT meant choice: rather than having only whole blood, surgeons and doctors could pick and mix. It served chemotherapy patients—patients with chronic conditions including cancer still use nearly two-thirds of blood transfusions—and there was another issue. It could make more money. "You could break up this whole blood," says John Holcomb, a trauma surgeon in Houston and a military surgeon for twenty-three years. "And sell six or seven components instead of just selling one. Think of it: you go and donate blood for free, and the blood banks can break it into five or six components. The business model is incredible."

The switch to CT was swift and apparently irreversible, and not just for cancer patients. Within a decade, no one was using whole blood to treat bleeding anymore. In Holcomb's entire time at medical school, and even now, thirty-two years of medical career later, he has never seen whole blood used in a civilian setting. He has asked trauma surgeons who once happily used whole blood at the time why they abandoned it. "They don't have a good answer." It was like emergency medicine had flipped a switch, and no one protested. Holcomb calls it "unbelievable." The dogma changed, and it was comprehensive and unquestioned, at least anywhere that could afford the technology to separate blood. Ten years ago, at a conference, Holcomb met "folk from the less developed world" who apologized for transfusing with whole blood. Holcomb stood up and said, "Don't apologize. It's a better product."

Even a 1:1:1 ratio of blood components—1 unit of plasma, platelets, and red blood cells—is not the same as whole blood. It contains additives and anticoagulants that are not in blood. In a modern hospital ward, the red stuff dripping into veins looks like blood, but it is something different. Both Holcomb and Brohi think it is something that, at least for bleeding trauma patients, is an inferior product.

Mogadishu, on the night of October 3, 1993. Holcomb was two years out of medical school, a young surgeon working in a field hospital in Somalia. It was the night now mostly known as Black Hawk Down, though the more official name of events is the Battle of Mogadishu, and its story has been well told: a planned forty-five-minute operation that became a seventeen-hour battle; eighteen Americans killed and several hundred and possibly thousands of Somalis.[20] Seventy-three Americans were injured and arrived at Holcomb's field hospital. They were "a bunch of bleeding soldiers that were all blown up and shot up." They needed transfusions, so the medics got out bags of fresh frozen plasma to thaw, because that's what they had. PRBCs to bolster the blood, fresh frozen plasma to help it clot. A third of the plasma bags broke. They had no platelets, which have a shelf life of only five days and don't survive transit to most war zones. No plasma, no platelets, nothing to help the bleeding soldiers to clot. Nothing to stop the bleeding.

Holcomb did what he thought was revolutionary. "We were all young surgeons and we were doing what we thought was crazy." They drew blood from volunteers and used that. A third of the combat hospital personnel volunteered their blood and went straight back to work. The blood was matched using blood types inscribed on dog tags. As it turns out, this wasn't revolutionary at all. Later, he read about whole blood use in the Civil War, the two world wars, Korea, Vietnam. In most major conflicts except the one he was in, the United States had given its shocked and bleeding soldiers whole blood because it made sense.

The soldiers in Mogadishu seemed to do better with whole blood, and it was enough for the US military to begin to trial what it called Warm Fresh Whole Blood, a rare example of plain speaking from medicine, a profession with a fondness for hermetic Latin and Greek. (Why say "hospital-derived" when you can say "nosocomial" or "hiccup" when you've got "singultus"?) The warm fresh stuff was drawn from a human there and then and put into another. Or it could be stored for up to twenty-one days. In the conflicts in Iraq and Afghanistan, 10,300 units of whole blood had been transfused by November 2016.[21] In Iraq, a walking blood bank was used.[22] Troops were prescreened before departure, then if their blood was needed it could be collected on the spot, screened for infection (HIV and syphilis tests take the longest, but still only twenty minutes), and transfused. This suited military ships, where the use of blood components was difficult: platelets' shelf life was too short, and they were often days from the nearest blood bank. It made more sense to keep all of the components in human form and tap it when needed. In austere, far-forward situations or battlefields, there was another option, "buddy transfusion": emergency medics quickly find a suitable donor nearby, usually from the casualty's peers, and transfuse on the spot.[23]

The Norwegian special forces are the most far-forward with this: wanting to know whether turning their special forces into single trans-fusing units would affect their performance, they have bled special forces operatives, then sent them either walking up a steep mountain-side with a 20-kilogram (40-pound) backpack, or on a combination of a punishing treadmill test and push-ups or pull-ups, or on a round of pistol shooting. Tests were done before and ten minutes after donat-ing. All the operatives showed hardly any decrease in performance, although the study authors pointed out that the results should be taken with reservations, as they were done in nonstressed environments, and hardly anyone is as fit as a special forces soldier.[24] The debate is ongo-ing, and in most militaries, fluids are still the most usual treatment for a severely hemorrhaging soldier. Holcomb sees a future where hospi-tals will have a two-track system: one blood bank for people for whom component therapy is perfectly good—cancer patients or other people needing large transfusions—and the other for severely bleeding trauma patients, for whom whole blood is better. He calls his presentations on the topic "Back to the Future."

Opposition may come from blood bankers, committed to a logistic and financial model that sends out blood as components. But Karim Brohi believes that they will inch forward to a different way of doing things. There is more and more research showing that stored blood is inferior to fresh blood for certain patients, whether you divide it into components or not. First, red blood cells are always mixed with a storage medium (saline, adenine, glucose, or mannitol), which impedes their ability to deliver oxygen. Second, blood ages, as we do, and gets weaker, as we do. This is called "storage lesion." It may not affect cells' ability to carry oxygen but, says Brohi, "they are very bad at delivering it. They don't give it up to the tissues that they are supplying." Some research into storage lesions has concluded that nitric oxide is to blame. This, usually present in red blood cells, is what enables them to dilate capillaries so they can deliver oxygen. It is the pathfinder, the gate opener. But only three hours after being drawn, blood used in transfusions was found to have lost 70 percent of its nitric oxide. After a few days it was 90 percent.[25] Other studies have found that the age of red blood cells made no significant difference to outcome. The debate continues, with the only agreement being that storage definitely does something to blood. The storage lesion issue, Brohi translates for me, "really just means 'use fresher blood.'"

After four and a half hours, the operating room team at the Royal London closed her up. Her heart was beating, again. Brohi, who knew that hardly anyone survived that level of blood loss, still thought she might make it. It would be extraordinary but not impossible. Research done at the Centre for Trauma Science has flourished alongside the creation of the London Major Trauma System, a logistical program that means that critically injured trauma patients, rather than being sent to the nearest hospital, are now always sent to London's four dedicated major trauma centers.[26] The research, the logistics, the damage control resuscitation: something is working. In the last five years, London's trauma network has cut death rates in severely hemorrhaging patients by more than half.[27] In 2009, 34 percent of trauma victims in the London trauma system died of hemorrhage. In 2015, it was 18 percent. When three men decided to mow down then slash and stab people on a night out in London Bridge, in the summer of 2017, the major trauma centers received forty-eight casualties. Many had severe and complex injuries. Some, ten years ago, would have been judged unsurvivable. All of the casualties lived.[28]

In the early afternoon, after surgeons had closed their cuts and invasions, her heart stopped again. She died, too young. In the case of the woman, or anyone else who had that level and depth of injury, there is nothing that they could have done better. Many emergency services, says Brohi, would not have done a roadside thoracotomy on her "because the outcomes are dismal." In most places in the world, she would not have made it to hospital. "I don't think there are many services in the world that would put that amount of effort into a case like hers."

A few days after the Code Red, I tell Brohi that when I left the hospital that day and took a tube home, I was still so shaken I missed my stop twice. I tell him that since then, I've been scared at pedestrian crossings. I tell him it left me unsettled, uncertain. I lost the armor that protects me from the inevitable statistics, that allows us to walk and cycle on roads with buses and trucks and believe they won't kill us. I ask him what the clinical staff do to deal with cases like the cyclist, and that I heard them talking about it all day long. She was young, like they are. It was tragic. They are human. He says you decompress however you can. "Phone. Pub. Wife, husband. Alone." They were upset because "the thing about her is that she was a talk-and-die. She didn't have a big head injury. If you could get it right, she could be a survivor and she could have gone back to work. That's what hurts people most, that she was talking under the coach and then she died. She was salvageable, so why can't we save her?"

Knowledge of what happens in the body immediately after trauma— the "bad physiology"—is still opaque. This is not for want of trying. I lose count of the number of journals devoted to care that is critical, emergency, resuscitative, trauma. But it is difficult to carry out good, robust research on a just-injured body: How can you ask for patient consent to do randomized control studies on a person who has just hit a wall at eighty miles per hour? Trauma doctors in Maryland want to study whether inducing hypothermia in trauma patients, and lowering metabolism, can give doctors more time. A longer golden hour. To do this, they have to hand out pre-consent forms in shopping malls and hope that someone who takes one will then have a terrible accident and be a candidate.[29] Good research is usually retrospective (examining cases that have already happened), which is considered a lesser standard.

The Royal London now has funding to do a study into the efficacy of whole blood in transfusions. The University of Texas Health Science Center in Houston, where Holcomb works, has begun trialing whole blood in the hospital and on its air ambulances. Over the last decade, medicine has begun to notice that transfusion, one of the most common medical procedures, and one thought to do unquestionable good, needs better scrutiny, more money, more attention. "Trauma," Brohi wrote in a journal on traumatic injury, "is one of the greatest killers of humanity." Population growth and climate change are going to make trauma more common and more severe. Yet, "despite its high global impact, trauma remains low on governmental priorities, a side concern for funders and of only passing interest to most scientists." In the UK, where forty-six people are killed each day by injury, trauma science still gets only 1 percent of medical research funding.[30]

I ask Brohi how trauma care will be different in ten years. Perhaps hospitals will differentiate between different types of blood for different people: not just cross-matching, but oxygen-delivery matching. I'm expecting a technical answer about science and oxygenation and coagulation. But he says, "I hope they will have banned buses and trucks from London so they don't come into contact with cyclists and pedestrians. They obviously can't share the road." As for science, "in terms of treatment, we need a better way to protect the cells and to deliver them what they need while we're fixing people." He thinks trauma care will be more personalized, less one-red-bag-fits-all. As Londoners keep falling from windows or getting stabbed and shot and crushed and run over, trauma teams will keep fixing them, as best they can. Because each time it happens, each Code Red or Code Black or bull's-eye or one under, each catastrophically injured human being they encounter, "we just get that little bit better."

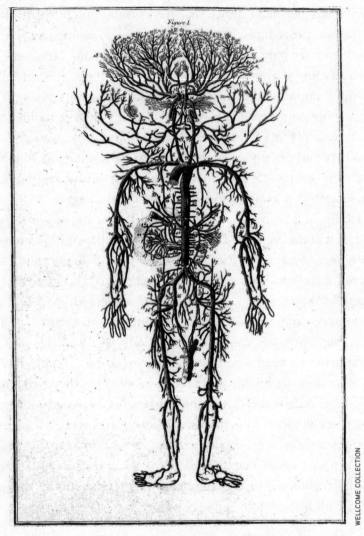

The circulation of blood, by A. J. Defehrt, 1762

NINE

BLOOD LIKE GUINNESS:
THE FUTURE

The Canadian border official has been taking demeanor lessons from the Americans. His is icy, unwelcoming, as solid as his shape. He begins his quiz: Why are you here? How long for? What do you do? How can you prove it? To look at a plasma clinic; not long; I am writing a book about blood; the Internet will vouch for me.

He looks at me, granite. I wait to be refused, expelled, taken to the special room where the undesired go (there is always a special room). I expect sanction, reproach, or worse.

He says, "Will there be vampires?"

My relief feels like cool water. I babble like cool water. Yes! Clinical vampirism is a thing. There's an underculture in New Orleans of people who drink blood.[1] A young man has changed his name to Darkness Vlad Tepes after the original prince Dracul (a vile tyrant, and a Transylvanian, but not a vampire) and says he is bullied for it. And it's been found that vampire bats feed other vampire bats who haven't fed, and isn't altruism in vampire bats fascinating?[2] But the border agent looks disappointed. He wanted Twilight, Buffy, True Blood, and Tom Cruise, white sharp teeth and fairy tales, not teenage goths and desmodonitae. A scan, a stamp, a stare, and he lets me go.

———

Puncture much of history and vampires will emerge. I recommend puncturing the historian Richard Sugg's vampirology work because what seeps from it is not Dracula's slithering fog, but a human fear of bloodsuckers that is ancient and enduring.[3] How terrifying the vampire, and how everywhere. In the mid-nineteenth century, two British army officers stayed for months in a Bulgarian village. In the neighboring village, a newborn baby seemed dead and was buried, but its mother later heard it crying and disinterred it. The child was seen, the council assembled, and the baby was declared a vampire and condemned to death by staking. This was done. This is horrific. But, as Sugg says, the question was not just how people could be so barbaric, but how people could be so terrified. People in Romania, Bulgaria, Greece, the Balkans, Poland, Russia, and elsewhere across Europe: for centuries, they were so petrified that "they staked and beheaded and burned and reburied the living and the dead, from friendless strangers to sons and mothers. They were so terrified that they had nervous breakdowns, became paralyzed, and in some cases, actually died of fear."[4]

What were they scared of? A vampire in the seventeenth century was not, despite the imagination of Bram Stoker, an elegant aristocrat dressed in a black cape. Open the door to one, and you would probably find, according to the historian Paul Barber, "a plump Slavic fellow with long fingernails and a stubbly beard, his mouth and left eye open, his face ruddy and swollen. He would wear informal attire—a linen shroud—and he would look for all the world like a disheveled peasant."[5] He may want to kill you or do some evil deed. But for most of vampire history, he would not in fact have been interested in drinking your blood.

For these old Europeans, death could be sudden but it wasn't immediate. Some people thought it took forty days for a dead person to properly die. For all that time, they were dead, but not quite. Slightly dead, as Sugg puts it. Others made holes in their walls, then bricked them up, to fool only slightly dead people wanting to make their way back in. The corpse had power after death, which is why mummy—powdered or preserved flesh—and skull moss were popular cures for centuries.

One of the first blood-sucking vampires was Arnold Paole, a Slavic peasant who died (for the first time) by falling off a hay cart. Paole's vampirism was documented by Johann Flückinger, an officer in the Austrian imperial army stationed in Serbia. His *Visum et Repertum* (usually translated as "Seen and Discovered," though I prefer "Been There, Seen That") was a bestseller. Before his death, Paole had reportedly told villagers that he had been visited by vampires. After it, villagers and cattle had their blood drunk, and four peasants died mysteriously. Paole's body was dug up. Oh dear. There was blood. His body was undecayed, and fresh blood flowed from his eyes, nose, mouth, and ears. His shroud was blood-soaked. His nails seemed new and his skin was flaky, both effects of decomposition that were understood by the ancient Egyptians, who sometimes attached thimbles to a corpse's nails to get them to stay put (Paole's "new" nails were probably nail beds), and both now better understood.[6] He was staked and burned, and he troubled neither man nor cow again.

Flückinger's account launched the era of the bloodsucking vampire as we now understand it, and a belief that has endured into modern times and in strange places. The ethnographer Luise White found convincing evidence in the 1950s that people in several African countries firmly believed that firefighters are vampires. Others thought it was police who were the secret blood drinkers. In 1952, as a villager interviewed by White recounts, a man returned to his home, much to the surprise of his neighbors. "He had been missing since 1927. We thought he had been slaughtered by the Nairobi Fire Brigade between 1930 and 1940 for his blood, which we believed was taken for use by the Medical Department for the treatment of Europeans with anemic diseases." The villager, a local politician named Anyango Mahondo, knew this was nonsense. People just thought firefighters were vampires because of the color of their equipment. He knew, instead, that it was the police who did the bloodsucking, capturing their victims and keeping them in pits under the police station.[7] Death, colonialism, the fear of turbulent change: vampires could be useful to account for a lot of anxieties. They were the most powerful, because what they stole was the most powerful substance and the most mysterious.

There were drinkers of blood in history, but they were at public executions, not police or fire stations. Two thousand years ago, the physician Celsus wrote of Romans who rushed to the side of dying gladiators to drink their hot blood from the men's cut throats.[8] Fresh warm blood was then, and for two thousand years since, considered a cure for the mystifying and terrifying "falling sickness" of epilepsy. (Before this, wrote two medical historians from the Institute of Medicine and Medical Ethics in Cologne, "there is no evidence that the sanguineous humor of a slain swordsman was regarded as a special drug for epileptics."[9])

All Romans writing about Romans drinking blood made sure to condemn it as barbarous and inhuman. But epileptics were desperate, and the understanding that their appalling affliction was electrical, not magical, was nineteen hundred years in the future. If falling down was because of weakness, then someone else's blood could provide the force to pull them up again. Blood-drinking epileptics were written about in Germany, Sweden, and Austria. Epileptics brought their own cups and beakers to scaffolds, or sometimes made do with white handkerchiefs. In 1851 in the southern Swedish town of Ystad, known to modern readers as the home of fictional detective Kurt Wallander, a rare judicial beheading caused a frenzy. Before dawn, the plain where the scaffold had been built was crowded with people who had brought "bowls, glasses, cups and saucepans." Saucepans! When the man was beheaded, a mob rushed the scaffold, but they were beaten off by soldiers; the corpse was swiftly removed by horseback, "and the ground was dug up so as to destroy the trace of blood."[10]

Consumption of fresh blood was thought to have more general benefits, the most powerful of which was to extend life. This concept should seem ordinary to millions of Christians, who symbolically drink blood every Sunday. It is also why the tale of Pope Innocent VIII has endured: that when he was dying, he was brought three young men, and they were opened for him and he drank all their blood. He wanted their youth and life. Throughout time, humans have sought to extend their own. If we are so clever, why can't we make life longer and stop death? There have been potions galore to achieve this, and theories and experiments. The notorious countess Elizabeth Báthory of Hungary was supposed to have bathed in the blood of young girls, although this was never mentioned in depositions at the time. The Ming emperor Jiajing, an unpleasant character who ruled in the sixteenth century, was persuaded that drinking secret potions containing lead, cinnabar, and the menstrual blood of young women would lengthen his life. He was almost wrong: this habit, and his awful cruelty, led to a concubine plot to kill him. Sadly it failed, and the concubines were executed by "slicing" or death by a thousand cuts.[11] Other rejuvenating substances were experimented with. *The Psychrolousia, or The History of Cold Bathing*, published in 1715 by Sir John Floyer, describes a man in the north of England, a cowherd "of an extreme age" (over sixty), who laid himself down at night next to his cows so he could drink in their breath. "The Breath of a Cow is a Cordial," he reported, "and much refreshes me when I am faint." Cowherds and shepherds who knew what was best breathed in the "salubrious Volatile Salts" in the breaths of their beasts, "early in a Morning before the Beams of Light and Heat exhale them, and rob them of the best Nose-gay in the World."[12] This cow halitosis cure didn't last. But that a dose of blood can give life and youth has never gone out of fashion.

It's 1905 in Moscow, but Leonid, a Russian mathematician, is on Mars. He has been politely kidnapped by Martians and taken on a ten-week journey on a nuclear photonic rocket to reach the red planet. There, he encounters a utopian society that is egalitarian, socialist, polyamorous, and long-living. He develops unsettling romantic feelings for Netti, a doctor he believes to be male (Martian genders cannot be told apart, making the polyamory even more interesting), who tells him that a fifty-year life span is very young for a Martian. Most live to twice that. Leonid is astonished. What is the secret?

We renew life, says Netti. "It is actually very simple." Just as simple cells feed and renew each other, fusing to extend their life spans, so can Martians, by performing mutual blood transfusions. This is not done to cure illness but to counter aging. The circulatory systems of two Martians are joined—it is not explained how, though Martians are very good at engineering and have impressive canals—and blood flows from one to the other and back again. "If all precautions are taken," says Netti, "it is a perfectly safe procedure. The blood of one person continues to live in the organism of the other, where it mixes with his own blood and thoroughly regenerates all his tissues." Leonid wonders why humans don't do the same. If Martians could shrug, Netti would have. "Perhaps it is due to your predominantly individualistic psychology, which isolates people from each other so completely that the thought of fusing them is almost incomprehensible to your scientists." We poor selfish humans, confined to our seventy or so years of life because we refuse these "regular comradely exchanges of life."[13]

Red Star, the tale of Leonid's trip to Mars and back, was written by Alexander Bogdanov, a doctor and revolutionary. He was close to Lenin and Stalin, a force in the select genre of Bolshevik science fiction and also someone who attempted to make his science fiction more scientific than fictional. Bogdanov was convinced that Martian "blood exchanges" could have rejuvenating properties for humans. This was not an implausible thought in the context of the 1920s in Russia and Europe, when rejuvenation and the conquering of the new concept of "senescence" was flourishing and fashionable. Most rejuvenation theorists focused on the sex hormones, as Nikolai Krementsov explores in his wonderful book on Bogdanov and Soviet science, *A Martian Stranded on Earth*. The use of animal gland extract became so popular that a newspaper cartoon showed monkeys delivering "an endocrine breakfast" to a patient at a noted Moscow hospital.[14]

In 1926, Stalin gave Bogdanov funds to set up a blood transfusion institute in Moscow. It was the first in the world, and an extraordinary achievement in a country whose health care system was otherwise not particularly impressive. There were other experiments in transfusion, but the most notorious was the "physiological collectivism," a comradely exchange of blood. Bogdanov, by now over fifty, took the blood of younger students but also gave his own. He swore it halted his balding and improved his eyesight.[15] A fellow revolutionary wrote that "forgetting [his age and health], he sometimes runs up four to five flights of stairs." His wife, Natalia Bogdanova, "also feels great: gout symptoms in her feet are gone; before she had to wear shoes made to order, now she wears regular ones."[16] Bogdanov looked seven, no, ten years younger.[17] Over two years, he took eleven liters of blood from the students, but the last was fatal. By then, blood groups were broadly understood, but not antigens and antibodies. Each dose of blood from a foreign body would have stimulated Bogdanov's immune system to prepare antibodies for the next dose. The eleventh liter of blood came from a student who had tuberculosis and malaria: in proper Martian style, Bogdanov hoped mixing their blood would cure the young man. Although he was the right blood group, Bogdanov's body reacted badly. The student survived. Despite another blood transfusion, Bogdanov died. But his belief that blood could rejuvenate did not.

The website is almost insultingly minimalist. Ambrosia, it reads. Young Blood Treatment. There is a list of locations: San Francisco and Tampa, and a contact form. That is all the written information available. The rest must be read from the imagery: a moving image of water, rippling harmoniously. A young man who is pictured sitting on a bench next to a vintage road racing bike. His hair is black and glossy, his skin is smooth. He has forgotten to wear socks. His moody beauty, his pale skin, his rosy lips: He looks like a vampire.

"Really?" says Jesse Karmazin. A vampire? He never intended such a thing. "I think we just thought athleticism, because of the bike. Maybe I'll look at it again." Karmazin and I talked late one night over Skype, transatlantically. It was past my bedtime, in the hours of witching and vampires. Karmazin was the name my "young blood" Google alert produced now and then and consistently for the past year or so. I researched him before we spoke and found images of a fit Paralympian rower (he was born missing half a leg). He doesn't look, in these pictures, far removed from his picturesque male model vampire. He followed the Paralympics with a medical degree at Stanford, then practiced psychiatry at Brigham and Women's Hospital in Boston, before he signed a voluntary agreement not to practice with the state of Massachusetts, a disciplinary step that is just short of a withdrawal of a license, and which is usually reserved for physicians whom the state believes to be "an immediate, serious threat to the public." This is important, because Karmazin's second career involves transfusing "young blood" to willing patients. He got the idea from Bogdanov, having read Krementsov's book about the comradely exchanges. He called his company Ambrosia, after the food of gods—who must be eating something right as they are immortal—but, he says, he could just as easily have called it after Bogdanov.

It was at Stanford that Karmazin starting wondering about blood. Clinicians were giving so many transfusions all the time, to all sorts of people. "It just sort of clicked." If blood could be given to heal sick people, why would its powerful effects not work on healthy people? A kind of superboost, but not involving wheatgrass or horrible green sludge. He considered red blood cells first, which he called little bags of hemoglobin, an essential protein. That would probably work. "But it turns out if you give someone excess red blood cells, the body gets overloaded with iron and there's no good way for the body to get rid of it." Plasma had proteins too: that was a better and safer option. Also, there was science to back him up.

In 1933, two anatomists named Eduardo Bunster and Roland K. Meyer published a paper called "An Improved Method of Parabiosis."[18] The name is as beautiful as the procedure is hideous, as it comes from the Greek for "beside" and "life." It is the splicing of two creatures, an artificial version of Siamese twins (conjoined humans) or freemartins (cattle). Parabiosis, the messing with the bodily integrity of creatures, had been tried before: I find an intriguing paper from 1912, concerned with the "Parabiosis in Brazilian Ants."[19] It was also tried on frogs and hydra, jellyfish-like sea creatures. But rats and mice were the most popular victims.

Bunster and Meyer's report is thorough, even providing the recipe for the depilatory used to remove the rats' fur (it's 35 grams of barium sulfide, 33 grams of talcum powder, 35 grams of flour, and 5 grams of soap chips, a formula I'm not sure will threaten Veet or Gillette anytime soon). They treated their stitched-together animals well, they wrote, despite shattering bones with needles and drilling holes to make the sutures stick better. They were wise to do this: when the biochemist Clive McCay at Cornell stitched together sixty-nine pairs of rats in 1956 (two by two), he learned two important things. The older and younger rats reacted differently to the same dose of barbiturate. And patience was essential. "If two rats are not adjusted to each other, one will chew the head of the other until it is destroyed."[20]

In Bunster and Meyer's illustrations, one parabiont rat stretches a foot to stroke his littermate's, as if they are intimate by choice not grotesqueness. The rats are cut, then sutured with No.1 black silk, and their systems, theoretically, are joined. Blended blood runs through two rats, two hearts, two circulations. That was surely wonder enough. But over the next twenty years other scientists explored a more specific question. What if young blood could transmit youth to old blood? In 1972, another study found that elderly rats transfused with young rat blood lived four to five months longer than normal.[21] But parabiosis for some reason was stopped. Too grotesque? A waste of soap chips?

Whatever the reason, parabiosis has been rejuvenated, along with many mice, rats, and Silicon Valley millionaires (probably). That the age of blood matters is known to anyone specialized in trauma or transfusion: younger, fresher blood is better, but most blood donations are a mixture of ages, of the blood cells themselves, and of the donation's shelf life. "Today we're in a weird situation," says Karmazin. "Most people who donate blood in the US are elderly. Sixty-five or older. So if you're young or get in an accident, you're likely to get blood from someone who is older and that probably has a detrimental effect even if it's lifesaving in the moment." He is wrong about the average blood donor: in the United States, it is a white college-educated man aged between thirty and fifty who is married and has an above-average income.[22] But his general point makes sense. Time does something to blood, so why can't blood do something to time?

In 2013, a team led by Amy Wagers of Harvard University reported on their experiments with stitching two mice together, one older than the other. (*Science* called this "Help the Aged.")[23] They found that stem cells in the older mouse began acting like younger ones, healing injuries better. They already thought a protein called GDF11 may be responsible. Later studies found that the older mice developed a stronger heart, better cells, shinier fur. The results were persuasive enough for human trials to be set up, including that of Ambrosia and its ilk (because by now it has ilk). A small trial explored the effects of giving young plasma to eighteen Alzheimer's patients. Its major finding was that the procedure was safe (no one died), though there were "hints," said the lead researcher, that recipients got better at remembering to take medication and being able to pay bills.[24]

The brain is aerobic and fueled by blood like every other part of us, and this fact may be why humans conquered the planet. A recent study of skulls throughout evolution concluded that brains have increased in size by 350 percent, but the volume of blood passing into the brain has increased by 600 percent. You can judge this from the holes in your head. The brain gets blood from carotid arteries, whose different sections—including cavernous, ophthalmic, and communicating—can be remembered with a helpful mnemonic of Please Let Children Consume Our Candy. The skull researchers noticed that the holes that the carotid arteries pass through increased in size too, to enable more blood to be delivered. "We believe," said project leader Professor Emeritus Roger Seymour, "this is possibly related to the brain's need to satisfy increasingly energetic connections between nerve cells that allowed the evolution of complex thinking and learning."[25] Intelligence is as bloodthirsty as everything else.

Is GDF11 the elixir of youth? Is the fountain of youth actually red and consisting of nine to twelve pints? Subsequent studies found that too much GDF11 caused damage not growth, impediment not immortality.[26] The debate is dynamic enough that the most common trait of scientists working in this field should be caution. Jesse Karmazin is different. He needs to be: although he claims he wanted to set up a trial in the usual way, funded by investors, he couldn't find funding. "You can't patent this, using young blood for aging. If you can't patent something, that scares off investors." Instead, he set up two clinics, employed two doctors to administer transfusions, and charged clients $8,000 to participate. In return, they had one hundred biomarkers in their blood analyzed before and after, and they were given plasma from a donor aged under thirty-five. A hundred people signed up. The results, says Karmazin, were so staggering, he cut short the trial. What results? "I have to preface this by saying that this has not yet been peer reviewed." Some patients report feeling younger. One of his patients, who had Alzheimer's, has now been assessed by a neurology team and can live independently again. "They no longer have Alzheimer's."

Sorry, are you saying the treatment eradicated dementia?

"Oh yes. This is the part where it breaks apart and you think, it can't be true."

But he continues. Two liters of plasma can cure cancer, heart disease, and diabetes. My skeptical alarm is sounding, and it gets louder when Karmazin says these results were found after only one infusion. Ambrosia has had plenty of criticism: a true scientific trial cannot be a cash cow. There is no control group (Karmazin says taking blood biomarkers before and after treatment means his cohort acts as its own control group). There are things to be concerned about, not least the grandiose claims. He wouldn't dare make them, he says, on TV or in print. "With the FDA, there's a very long process we have to go through to be able to say things like that and that happens. I can tell you because this is for a book." Still, Karmazin plans to publish his research, but in a "mid-tier journal."

For now, it is safest to say that the field of blood rejuvenation is in flux, an appropriate word meaning a discharge of blood. And I salute the mice and rats who continue to be stitched together, although now they are shaved not soap-chipped, and joined at knee and elbow not torso, and they are female usually, as female mice are less likely to chew each other's heads off. Thank you, mice, for perhaps enabling us to live a bit longer to experiment on more mice.

I know this sounds churlish. My dismay is directed at the vanity, at the rumors of the insultingly rich, people such as PayPal billionaire Peter Thiel, perhaps paying to have themselves injected with young blood or plasma. Though I do admire a satirical show named *Silicon Valley* that portrayed a mogul who employed a Blood Boy for regular rejuvenating transfusions. On *The Late Show*, wrote Tad Friend in a *New Yorker* exploration of the longevity industry, Stephen Colbert told young people that President Trump was going to replace Obamacare with mandatory parabiosis. "He's going to stick a straw in you like a Capri Sun."[27] I would prefer an antiaging emphasis that is not about extension but betterment, that can alleviate Alzheimer's, Parkinson's, and the other diseases that have stepped into our longer life spans to end them. I have seen how someone dies of dementia, and if that can be cured with blood, fetch me a straw.

The trouble started on the gurney. As she lay there prepped and ready for surgery, Hazel Jenkins was as nervous as any patient before a significant operation, but not unduly. She had been diagnosed with bowel cancer the year before and had an operation to cut out her bowel. Then the cancer had spread to her stomach and duodenum, so they had to be removed. There was no question of her refusing consent: she had faith that modern medicine could cure her. But she had a handicap. She was a Jehovah's Witness. This millenarian Christian sect with nine million members worldwide believes there will be a Second Coming, that birthdays are pagan and wrongful, and that they should never accept whole blood. This decision is based on a few Bible passages. Don't consume blood, it is written in Deuteronomy, because blood is life. "You must not eat the life along with the meat." The book of the Acts of the Apostles forbids food offered to idols, blood and the meat from strangled animals. Leviticus is the most severe: anyone who eats blood will bring severe opposition from God, and "must be cut off from their people."[28]

To me, all this reads as divine instructions for vegetarianism. But after the Second World War, after transfusion had been accepted as a powerful medical good, the church elders interpreted "eating" as "transfusion" and they judged it sinful.

"It's the life blood," Hazel tells me in a quiet hotel lounge in Leeds, where she has come after getting her latest scan at a nearby hospital. It should not be squandered. Witnesses are often confused with Christian Scientists (who consider medicine quackery), but actually they accept most that modern medicine can give them, with gratitude. "We believe in a loving God who wants us to live." But they will not have blood. Hazel's husband, Bob, is on the other side of the table: he accompanied her to the scan, and he also has cancer. They are both very Yorkshire, by which I mean plain speaking, no flummery. When Hazel describes how her bowel cancer was discovered, she says she got to the emergency department and had her temperature taken "to see if I was about to peg out or whether we could wait the four hours [to be seen by a doctor]." She and Bob became Witnesses in the 1960s. They researched the scriptures and at the same time read up on blood scripturally and medically. "Well, it's a core belief." They wanted to understand it properly.

Although they have five children and four grandchildren and though blood transfusions are often given in childbirth, the Jenkinses had no cause to test their belief until recently. First, Bob was diagnosed with myeloma, a cancer of certain white blood cells. Severe cases are often treated with a stem cell transplant, whereby the bone marrow is killed off, new stem cells are injected, and, if all goes to plan, new and undamaged blood cells begin to reproduce within weeks. Although the removed and replanted stem cells are hematopoietic—they produce blood—Witnesses will accept stem cell transfusions. But the weeks of having reduced immunity are perilous, and blood transfusions would normally be given. A patient with leukemia, a cancer of white blood cells that can also be treated with bone marrow transplants, might receive 30 units a year. In fact, although transfusions may be associated with severe bleeding and trauma, this is not their main use in the developed world. A recent report on red cell use in the UK found that only a quarter of transfusions were used in surgery. Most red cells were used for nonsurgical purposes such as hematology or oncology.[29] In developing countries, the typical recipient of a blood transfusion is an accident victim or woman in childbirth. Here, the typical recipient is an elderly person with a chronic disease. Someone like Bob, if he weren't a Witness.

The trouble on the gurney arose because science has raced ahead of biblical prohibitions. Richard Carter, a Witness, serves on the Hospital Liaison Committee, which acts as an information resource for Jehovah's Witnesses and clinicians. He speaks in measured tones and in a fluent medicalese learned from twenty-seven years of committee work and attending many meetings with hematologists and blood specialists. These hospital liaison committees began to be formed in the 1990s, after many years of trouble. For decades, Jehovah's Witnesses were a severe problem for medicine and medics. How could they have surgery without blood? How could anyone have surgery without blood? It was believed to be unassailable as a force for clinical good.

Yet the widespread adoption of blood as medicine after the Second World War had not been accompanied by any of the rigorous clinical trials that new drugs or therapeutics usually undergo. Countless transfused and resuscitated soldiers were judged evidence enough. "Before," says Dr. Dana Devine, editor of the blood journal *Vox Sanguinis*, "it was, any kind of blood is good for you. We topped it up like you top up oil in your car. Now we're understanding better how to do effective transfusion." Even by 1956 some blood specialists were wondering whether such liberal blood use was wise or safe. Dr. Theodore Zeltin, director of the South London Blood Transfusion Centre, wrote an editorial in the *Manchester Guardian* that warned of the "over-free use of blood transfusions." Transfusion had become so common "that nowadays in some hospitals pints of blood are dispensed more liberally than pints of beer."[30] Blood was so trusted, remembers Dr. Harvey Klein, chief of the Department of Transfusion Medicine at the National Institutes of Health, people were given a dose as a tonic. Half a liter of vim and zing.

Yet here were these strange religious people who were emphatic about not wanting blood but did not want to die because of it. They did die. Often, there were court cases, particularly when children were involved. There were angry clinicians, who must have found it hard to stomach such obstinacy when their job was to preserve life. Witnesses' views, wrote the BBC, "can spark immense upset, [and be] blamed for unnecessary deaths, as can their demands for special bloodless treatment, which can be both costly and ineffective."[31] They still die: just before Christmas 2016 in Canada, Éloise Dupuis, a young Witness, died of hemorrhagic shock after a cesarean section, having refused blood ten times. Her aunt accused members of the local hospital liaison committee, who were present in the hospital room, of putting her under duress, but a coroner found she had made her choice freely.[32] Several prominent doctors wondered publicly whether autonomy and patient choice had gone too far.[33]

For Richard Carter, committees like his are nothing more sinister than a resource, set up to link Witnesses—he has thirteen thousand in his catchment area—and the medical profession. They are meant to inform each about the other. There is no pressure exerted (though plenty of former Jehovah's Witnesses write copious blog posts suggesting otherwise). Carter is keen to inform me too, and does so thoroughly and carefully.

He gives me a copy of the medical directive that Witnesses carry around. It is simply a paper form that includes a cutout card that reads NO BLOOD next to an image of a blood bag crossed out. Witnesses carry these in their wallets or handbags, and it usually works, though not always. In an emergency, urgency can supersede going through a wallet. And emergency staff, says Carter, don't like to go through handbags. Maybe a bracelet would be better? Maybe. They're thinking of an app or digital alert.

The rest of the form is for details. A section lists what is forbidden. These are the "big fractions": red blood cells, white blood cells, plasma, and platelets. Below that, Carter says, "it is up to your conscience." Things are more fluid. Witnesses can make their minds up about whether they accept autologous blood transfusions, where their own blood is removed and re-transfused; cell salvage, where red blood cells are removed during an operation, filtered, and re-transfused; glues and sealants; recombinant erythropoietin (EPO), which can raise the level of oxygen in the blood to counter any possible anemia; and many other possibilities. They are happy to give blood samples, although studies have shown that patients in intensive care, for example, can be asked for 40 ml of blood a day. Some doctors call this "ICU vampirism" and wonder at the wisdom of taking all the blood from patients who are probably anemic and testing blood so often to see if someone needs a transfusion that they need a transfusion.[34]

These three Witnesses seem like kind and polite people, and we are having a nice cup of tea together, but I have to inquire about the lack of logic. Why refuse blood but not sizeable fractions of blood? Why refuse platelets—which are only 1 percent of blood—but accept albumin? Why reject blood but give it away for samples to be consumed in a pathology lab? Why stem cell transplants? You draw your line, says Hazel. She resorts to Noah to explain. "In the scriptures after the flood, Noah was given permission to start eating the animals. Of course they had to be bled. But when you buy your steak from the butcher's, there is a residue. If you ran it under the tap there would be a minute amount of blood. I don't think it's possible to remove every tiny amount. So it's a similar sort of thing, that's what it makes me think of." I think the essence of this is that you choose your limits. Some Witnesses won't accept even a "taint," she says. Some have, in extremis, accepted blood transfusions. Until 2000, when the church changed its policy, this transgression would mean Witnesses were "disfellowshipped" and expelled from their congregations.[35] In the all-encompassing culture of a cult, this meant losing friends, family, and church, all at once. Hazel answers carefully when I ask her what would happen now if anyone accepted blood. "It would be an indicator of whether you were actually a Witness. If you're going to pick and choose which laws you obey, you're not really a practicing Witness." You disqualify yourself by default.

Witnesses who have had accidents and been in a critical state have been given blood. That, says Carter, is forgivable. Sometimes the card isn't looked for. There is no family present to alert medics. It happens. Hazel carries her card and her form. She knew about fractions, and unlike with her urgent bowel surgery, this operation had been planned for and discussed. The chief surgeon had operated on her the first time, and on many other Witnesses. He was, says Hazel, "relaxed." But the anesthetist was the one who would administer drugs and products, and here was the anesthetist, who had not been present in the presurgical discussions, asking her if she consented to a blood fraction she had never heard of. Hazel was stumped. Bob had already left to get breakfast in the cafeteria and couldn't help. She said to the medical staff, I trust you, and she had reason: she was being cared for in a hospital that had performed a pioneering liver transplant without blood.

But the anesthetist needed her express consent. So there she was, horizontal and phoning Bob, then Carter. It was all a bit frantic. "My wife will tell you," says Carter, "that if the phone rings early on a Monday morning, it's an anesthetist who's got somebody already scrubbed up and waiting." Hazel asked her questions. Did they think she could have it? Was it allowed? Was it a small fraction or a big one? In the end, the surgeon came out of the operating theater to ask what the holdup was, and said, "Oh, then we'll just use something else," and that was that. Like she said, relaxed. "They obviously have so much in the armory."

This armory grew fastest after HIV infected the blood supply. Suddenly blood could be tainted and poisonous. Suddenly there was more scrutiny and there were more questions. At the same time, the concept of patient choice was beginning to percolate into paternalistic medicine. "We probably fit right into that," says Carter. "If you went back before that, you didn't ask a surgeon what he was going to do. Now things are very different." Questions began to be asked about blood in all sorts of ways. How much, what type, and, although it clearly did good, whether it also did bad. The Canadian doctor Paul Hébert caused a stir with a 1999 study of 823 critically ill patients. Critical care transfusions were given according to the transfusion trigger, a precise hemoglobin level in a patient's blood. When there is too little hemoglobin, the blood struggles to deliver enough oxygen to where it needs to go, and the patient is considered to be unsafely anemic. Since after the war, the trigger was 10 grams of hemoglobin per deciliter. Hébert's study split the patients into two groups. One set was transfused according to the regular trigger. This was the liberal transfusion group. The others were given restrictive transfusion: blood wasn't given until their hemoglobin had dropped to 7 grams per deciliter. The shocking result found that outcomes were no different in either group.[36] Patients younger than fifty-five actually did better on the restrictive regime. A big study called Transfusion Requirements in Critical Care (TRICC) backed this up.

The trigger dropped to 7 grams, although many doctors think any one-size-fits-all trigger is too restricting. And other innovations began, stimulated by this extraordinary new thought: What if blood wasn't an unquestioned good? What if there were alternatives? If they were going to use less blood or cut it out altogether, then they needed to prevent patients from bleeding. Early on, autologous blood transfusions were used. Blood could be removed from a patient before surgery, then reinfused into them if needed. These weren't popular with Witnesses, who thought storage was the same as pouring blood out on the ground. Once blood is mixed with a storage medium, says Carter, who says that their stance is spiritual, not clinical, but seems to prefer medical definitions over spiritual, then it is unacceptable. Also, perhaps you don't want to remove blood from an ill patient who is already at risk of anemia, even if you're going to put it back.

Carter hands me a booklet produced by the Royal College of Surgeons, meant as a guide for clinicians faced with people who refuse blood. It lists many possibilities.[37] Preoperatively: treating someone with EPO for several weeks can increase red cell blood production in the bone marrow sevenfold. More blood means more can be safely spilled. It also means muscles get more oxygen and effort is easier. That's why EPO will be forever associated with professional sporting cheats, and why if Witnesses weren't very sick patients awaiting treatment they could probably win marathons. Apart from their obvious surgical handicap, Witnesses are actually good patients. "We're quite a good clinical risk," says Carter. "We don't smoke. We've got good lungs, we can cope with low [blood] counts. We drink in moderation. We're not perfect but the nature of the Witness, being principled and disciplined is part of our culture. We'll take the drugs you give us, we'll take your dietary advice."

To conserve the loss of red cells, diluting agents such as hetastarch can be used before surgery. During procedures, clinicians can clamp off blood vessels to minimize bleeding. They can aim to do laparoscopic surgery, not open surgery. They can inject tiny beads into the bloodstream that block off the blood supply, a technique called radiological embolization. They can use drugs that aid clotting, such as tranexamic acid. There are tourniquets, vasoconstrictors, clamps. Apparently there are harmonic scalpels. I really should look up what they are but I don't want to ruin my mental image of a choir of steel blades. Marginal gains may be got using common sense. Raising a patient, says Carter, will lower their arterial pressure. Some interventions are now done using a "beach chair" position.

Restricting blood use may have been sparked by Witnesses, and fueled by HIV and hepatitis, but it has spread beyond both. Restricting blood use in surgery is now common enough to have a vocabulary. Bloodlessness. Dry surgery. Restrictive transfusions. There is a Society for the Advancement of Blood Management, founded in 2001, whose aim is "that blood management ought to be the standard of care, and that blood transfusion should be viewed as the alternative."[38] Blood transfusion, writes the society, "is costly to us all. While blood transfusion is safer than it has ever been, serious health risks remain." Acute and adverse reactions in a country with a good blood supply are highly unlikely, but they are not impossible. "Documented risks include potentially fatal transfusion reactions, acute lung injury, immune system changes that may lead to increased infection rate and circulatory overload." Transfused patients stay longer in the hospital. Some suffer immune reactions after receiving blood. Studies have shown "an association between liberal transfusion policies and increased incidence of health-care associated infections including surgical site infections and sepsis."[39]

Perhaps this occurs because of what red blood cells are mixed with. Perhaps it is the length of time they sit on the shelf. No cell is static; in thirty-seven days a lot can happen, and some of it might not be good. A new concept in blood is "appropriate." A transfusion will often be transformative, but should it always be done? A recent study had fifteen experts survey a huge number of cases of transfusion in medical, surgical, and trauma cases, then vote on the outcome. Transfusion, they concluded, had been appropriate in only 11.8 percent of the cases. They judged 28.9 percent to be "uncertain," and the highest fraction—nearly 60 percent—was inappropriate.[40] Also, bloodlessness is better for budgets. In the United States, transfusion of 1 unit of red blood cells costs $1,200. If, as the society contests, up to half of blood transfusions are prescribed for no justifiable reason, then the United States is spending $8.4 billion a year on pointlessness. Patient blood management has meant that the use of red cells has been declining. Over the last fifteen years in the UK, red blood cell use has dropped by 25 percent.[41] In the United States, the number of units collected and distributed by the American Red Cross dropped by 26.4 percent between 2009 and 2016. Transfusions have fallen by a quarter and may decrease by 40 percent by 2020 if the rate keeps up.[42]

Blood saves lives. It makes life better for millions of people. If you can, donate it. But perhaps we can use it better. Or what if we could replace it entirely?

Filling their empty veins with airy wine
That being concocted turns to crimson Blood

—Christopher Marlowe, *Tamburlaine the Great*

Dr. Kenji Igawa has invented synthetic blood. There is a powerful need for it in the world. Donors are increasingly hard to find and a safe blood supply must be expensively monitored and protected. How wonderful, then, to create something that can stand in for our blood; that can be carried to austere military environments more easily than the real fragile and perishable liquid; that can fill the gaps and fissures in the blood supply of developing nations. No more blood touts outside hospitals preying on desperate relatives. No more wastage. The WHO estimates that nearly two million blood donations globally are discarded because they could have transmitted infection.[43] But many red cells also go to waste because they exceed their use-by date. In the United States, 9 percent of red cells are thrown out for this reason. That doesn't sound like much, but 9 percent of 12.5 million, the number of red cells collected in 2015, is more than a million units.[44]

Attempting to mimic blood is an old human endeavor: as soon as William Harvey realized that blood was a circulating liquid, all manner of liquids that could circulate have been tried. Animal blood, but also wine and milk. In the late nineteenth century, two Canadian doctors, frustrated with the haphazard success rates of blood transfusion and desperate to deal with a cholera epidemic, decided to try cow and goat milk instead. The scientific basis for this was shaky: one doctor believed that the oil and fat particles of milk were transformed into white blood corpuscles, and that white blood became red blood. But a cow was brought to the hospital and milked, its milk filtered through gauze and injected. (Later, when the doctors applied to the city of Toronto for "a good cow" for their experiments and were refused, they resigned from their public posts.) Other doctors in the United States and the UK subsequently tried goat milk. Most patients reacted badly, by dying. One woman said her head was bursting. A Dr. Joseph Howe, not deterred by having failed to keep alive tuberculosis patients by giving them cow blood, decided human breast milk may be better.

> He attempted the infusion of three ounces of milk obtained from a healthy postpartum woman. The patient, a woman with suppurative lung disease, complained of pain in her chest and back shortly after

the injection began, and stopped breathing after two ounces had been given; however, she was resuscitated by artificial respiration and by "injections of morphine and whiskey."[45]

There's nothing old-fashioned about this concept or aspiration: we do something similar every time we give a bleeding patient a bag of a salty fluid called crystalloid. It's not blood but an understudy.

So Igawa's invention has been groundbreaking. Finally, a product that contains all the properties of blood, that can do the job of hemoglobin, and that has the benefits of blood's many enzymes and proteins. People who receive his product have shown no side effects. It works. Except Igawa's invention is called Tru Blood, and it is fake but not in the way we want. Igawa is a minor character in the vampire TV show *True Blood*, and Tru Blood is what vampires drink in place of humans. Synthetic blood is possible in film and fiction and, as yet, nowhere else.

The dizzying possibilities of a blood substitute are matched by the dizzying numbers of the sums so far spent on finding one. The possibility is so close and so far that "holy grail" is a phrase found even in sober scientific literature about artificial blood. Many blood substitutes have been tried. Some have got far enough along the research process to have been given brand names. PolyHeme. Hemopure. ErythroMer. Sanguinate. Most are hemoglobin carriers, or hemoglobin-based blood substitutes: they are designed to transport free hemoglobin from the lungs to tissues, as blood does, where it can deliver its oxygen, as blood does. Some use human hemoglobin and some use the bovine kind. But hemoglobin is contained within red cells for good reason. If freed from the confines of a cell, it scavenges nitric oxide from blood vessels, causing blood vessels to constrict, blood flow to drop, and strokes and heart attacks. Human or cow: same result. So far, the only thing that all these efforts have in common is that they are still limited. Hemopure, also known as HBOC-21, was licensed for clinical use in South Africa in 2001[46] but was never approved in the US or UK (though it can be used in trials).[47] PolyHeme failed to get FDA approval after it was given to nearly eight hundred trauma patients across the Midwest without their consent—as they were severely injured ambulance patients, they could hardly give it—and there was an outcry.[48]

Dr. Dana Devine, who is also chief medical and scientific officer at Canadian Blood Services, doesn't think a viable blood substitute will arrive in her lifetime or mine. "Mother Nature is a whole lot smarter than we are. We understand more than we used to, but we haven't overcome the problem of getting donor-derived blood." She is more optimistic about blood brewing in a lab. "We can take the cells, put them in a culture, tickle them with hormones and grow blood."

In a laboratory in Bristol, NHSBT scientists have done that. They took hematopoietic stem cells—the ones that grow blood—from an adult donor or from umbilical cords, which mothers can donate. Then they tried to mimic the bone marrow, brewing red cells in a laboratory. They weren't the first to attempt this.[49] A US team had done the same in 2008, and in 2011 in Paris volunteers received transfusions of ten billion artificially grown red blood cells. (That's only two milliliters of blood.) Twenty-six days after transfusion, the cells were still circulating.[50] And in the words of Dana Devine, "No one had keeled over." But the NHSBT team, explains Dr. Nick Watkins, assistant director of research, have done something different. They have immortalized cells. The word is as arresting as the achievement. "When you take a stem cell from an adult or [umbilical] cord blood and you produce red blood cells from those," says Watkins, "that's a linear process, you can only do it once." Instead, the NHSBT scientists have used proteins to create an erythroblast—an immature red cell—that will not become a red cell until it is given a signal. Millions of dormant cells, waiting to be activated, silent troops. The potential is alluring. "When you take a blood donation from a blood donor, the red blood cells you get vary in age from 0 days (i.e., just released from bone marrow) up to 120 days (the life span of a red blood cell). In a donation from a donor you have red blood cells that span that age gap. And the thinking behind the manufacturing process is that all the cells you produce are new and therefore you don't have these 120-day cells in your donation and therefore they may survive longer." When the cells are trialed on human volunteers, they will be tracked with a radioactive label, a small amount of radioactive material called chromium 51. I wonder at the chances of finding volunteers for that. "Will you accept fake blood with a dose of radiation, please? Sign here." With a fair wind, Watkins says, trials should be under way by the end of 2017. (But they weren't.)

Dr. Harvey Klein has sat on the NHSBT review board for many years. He is impressed, though cautiously. "These are major strides forward," he says. "And yet no one there has any illusions that this technology is going to provide millions of units of blood in the foreseeable future." It is far too expensive. "It isn't even close to providing as much blood as we can provide Guinness," says Klein, with one of the more unexpected analogies I've encountered while interviewing blood experts. "The research in the stem cell and hematopoiesis space is scientifically very satisfying," says Devine. "But the economic picture is just awful." It could be useful for giving blood to people with rare diseases or rare blood types. Watkins points to the famous case of a French teenager with sickle cell disease. Sickle cell is caused by a defect in the gene that governs hemoglobin, which causes hemoglobin to clump, and blockages and awful pain. It can be successfully treated with a stem cell transplant, but finding a donor is as easy as with any other transplant: it's not. When the boy was thirteen, stem cells were withdrawn and genetically manipulated to produce a functioning version of the hemoglobin gene. Two years later, they were transplanted back into him, and within three months his body was producing red cells with normal hemoglobin. He came off powerful opioids, he stopped getting painful episodes, and he is considered cured.[51] "I can imagine a scenario in the future in the Western world," says Watkins, "where patients with sickle cell disease might be faced with a choice of how they are treated. Do they receive donated blood, manufactured product, or do they have an autologous genetically engineered stem cell transplant? I think that's where the field will go in ten or twenty years. There will be more choice for the patient."

Watkins sounds like a scientist. Measured, cautious, but actually excited. There is possibility and promise. I glimpse this excitement in the names of companies founded to sound the depths of what blood can do for us. I read about Illumina, a massive DNA-sequencing company that has launched a start-up to work on a blood test that can detect cancer. It has raised $1 billion in funding from investors including Jeff Bezos and Bill Gates. Of course it is called Grail.[52] My database has hundreds of documents and articles about the widening and deepening abilities of blood to diagnose, defy death, defeat disease, with what has come to be called a liquid biopsy. I learn that blood will soon be able to diagnose manifold cancers, dementia, depression, with what is always called "a simple blood test." When I ask Dana Devine about the potential of liquid biopsies, she is as circumspect as other hematologists and transfusion specialists I speak to. "Liquid biopsy is not a magic bullet. If you read some of the accessible literature, you think it's going to be like a *Star Trek* tricorder." Take prostate-specific antigen, which can be detected in blood. "The great white hope years ago was PSA. But as we acquired more and more data, some people with elevated PSA did have prostate cancer, but some didn't and some with cancer didn't have elevated PSA. You have to be careful."

Harvey Klein is more optimistic. "We hear a lot of press today about precision medicine to the point where some people just think it's being overhyped, but the technology is advancing so rapidly. Will liquid biopsies replace the pathology laboratory? No. But they are going to make it a lot easier, faster, better than it is today where we have to go in and use an imaging procedure to find something and biopsy it."

There will be no tricorders for a while, but what we can do already is futuristic enough. To trap a stem cell and use it as the NHSBT team has done is called making it immortal. Immortality, elixirs, blood chimerism, for those equally trapped and tricked stitched-together rodents, that are "sacrificed" after use. The holy grail of that immensely powerful simple blood test sought by Grail. I wonder at this language of myth and fairy tale in sober science. I wonder if science reaches for words that belong to fairies and stories because they seem richer than ordinary language and life. So they suit blood better, because they convey better that despite all the simple blood tests there is nothing simple about blood.

Such possibility. It glitters like the gold and stardust inside us. But still we are the cowherds. The future holds people who will look back at us and think our achievements as limited as the belief that cows can breathe good health despite our editing DNA and growing stem cells and transforming lives by giving people blood. It's an extraordinary thing already, to achieve "the amending of bad blood by borrowing from a better body," as Samuel Pepys wrote five hundred years ago. But we will do better yet. Blood is not done teaching us what it can do. More wonder will come.

NOTES

ONE: MY PINT

1. A donation of blood in the UK is 470 ml (www.blood.co.uk/the-donation -process/what-happens-on-the-day/). There are precise ways to calculate a person's blood volume, but the general rule is that blood volume makes up 8 percent of a person's body weight. Roughly (www.blood.co.uk/the-donation -process/after-your-donation/how-your-body-replaces-blood/). Also www .hematology.org/Patients/Basics/.

2. I weigh 65 kilograms (143 pounds). Eight percent of 65 kilograms is 5.2 kilograms. Converting kilograms to pints (though it's mass to liquid) gets 9.15 pints. Dr. Harvey Klein, chief of transfusion medicine at the US National Institutes of Health, backed me up on this. "I've seen your TED talk. Yes, I'd say about nine pints."

3. When mixed with additives, red blood cells are allowed to be kept and used for twenty-one days in Japan, thirty-five days in the UK, forty-two days in the United States, Canada, China, and many other countries, and between forty-two and forty-nine days in Germany, depending on the additive used. Willy A. Flegel, Charles Natanson, and Harvey G. Klein, "Does Prolonged Storage of Red Blood Cells Cause Harm?," *British Journal of Haematology* 165, no. 1 (2014): 3–16.

4. "I remained there steadfastly until my mother came up and drank the dark blood. At once then she knew me, and with wailing she spoke to me winged words." Odyssey 11:155 from Homer, *The Odyssey*, trans. A. T. Murray, 2 vols. (Cambridge, MA: Harvard University Press, 1919), http://data.perseus .org/citations/urn:cts:greekLit:tlg0012.tlg002.perseus-eng1:11.138-11.179 (accessed January 2018).

5. The death of supernovas. https://spaceplace.nasa.gov/review/dr-marc-space /supernovas.html.

6. "That Lance Armstrong used banned blood transfusions to cheat. That Lance Armstrong would have his blood withdrawn and stored throughout the year and then receive banned blood transfusions in the team doctor's hotel room on nights during the Tour de France." *Report on Proceedings Under the World Anti-Doping Code and the USADA Protocol*, United States Anti-Doping Agency, Claimant, v. Lance Armstrong, Respondent. Reasoned Decision of the United States Anti-Doping Agency on Disqualification and Ineligibility, pp. 14, 61, 62.

7. The following are prohibited: "the administration or reintroduction of any quantity of autologous, allogenic (homologous) or heterologous blood, or red blood cell products of any origin into the circulatory system" (World Anti-Doping Agency, "Prohibited List," January 2017, p. 5).

8. E. M. Rose, *The Murder of William of Norwich: The Origins of Blood Libel in Medieval Europe* (New York: Oxford University Press, 2015).

9. "Hamas Revives Passover Blood Libel," *Times of Israel*, November 30, 2015.

10. Frank Capra, *Hemo the Magnificent*, 1957, www.youtube.com/watch?v =08QDu2pGtkc, 2.40–3.39.

11. *Proceedings of the International Seminar*, Royal College of Pathologists, November 13, 1998; "Altruism: Is It Alive and Well?," *Transfusion Medicine* 9, no. 4 (1999): 358.

12. The shape of a red blood cell is a biconcave disc with a flattened center. In other words, both faces of the disc have shallow bowl-like indentations. www.hematology.org/Patients/Basics/.

13. Matthew J. Loe and William D. Edwards, "A Light-Hearted Look at a Lion-Hearted Organ (or, a Perspective from Three Standard Deviations Beyond the Norm), Part 1 (of Two Parts)," *Cardiovascular Pathology*. 13, no. 5 (2004): 282–92.

14. "Blut ist ein ganz besondrer Saft," in *Faust*, Kapitel 7, http://gutenberg .spiegel.de/buch/-3664/7.

15. P. H. B. Bolton-Maggs, ed., Serious Hazards of Transfusion (SHOT) Steering Group, "The 2016 Annual SHOT Report," July 2017, https://www .shotuk.org/wp-content/uploads/SHOT-Report-2016_web_7th-July.pdf.

16. *2016 Global Status Report on Blood Supply and Availability* (Geneva: World Health Organization, 2017), 31.

17. Table of Blood Group Systems, International Society of Blood Transfusion, http://www.isbtweb.org/working-parties/red-cell-immunogenetics-and -blood-group-terminology/.

18. "Karl Landsteiner—Biographical," www.nobelprize.org/nobel_prizes/medicine /laureates/1930/landsteiner-bio.html. There is a short silent video of Land- steiner after he had arrived in Stockholm for the Nobel Prize ceremony in 1930, in which he looks no less severe than in still images. www.nobelprize .org/mediaplayer/index.php?id=1099.

19. Carl Zimmer, "Why Do We Have Blood Types?," *Mosaic*, July 14, 2014.

20. Jason B. Harris and Regina C. LaRocque, "Cholera and ABO Blood Group: Understanding an Ancient Association," *American Journal of Tropical Med- icine and Hygiene* 95, no. 2 (2016): 263–64.

21. F. Matthew Kuhlmann, Srikanth Santhanam, Pardeep Kumar, et al., "Blood Group O-Dependent Cellular Responses to Cholera Toxin: Parallel Clinical and Epidemiological Links to Severe Cholera," *American Journal of Tropi- cal Medicine and Hygiene* 95, no. 2 (2016): 440–43.

22. Erdal Benli, Abdullah Çırakoğlu, Ercan Öğreden, et al., "Are Erectile Func- tions Affected by ABO Blood Group?," *Archivio Italiano di Urologia e Andrologia* 88, no. 4 (2016): 270–73.

23. For an exhaustive examination of Nazis and blood, see chapter 5 of Doug- las Starr, *Blood: An Epic History of Medicine and Commerce* (New York: Perennial, 2002), 72.

24. Ruth Evans, "Japan and Blood Types: Does It Determine Personality?," BBC News, November 5, 2012, www.bbc.co.uk/news/magazine-20170787 (accessed February 10, 2017).

25. Erica Angyal's books on health and beauty by blood type, such as *Bijo no Ket- suekigata Book* ("Beautiful Women's Blood Type") and *Bijo no Ketsuekigata- bestu Obento Book* ("Beautiful Women's Lunch Box by Each Blood Type"), include type-specific diet and exercise tips. Type As should eat rice and grains, fruits and vegetables, as they descended from agricultural tribes. They're not good with dairy. Type Bs need a lot of protein or they tire easily. Noodles make Bs fat. ABs aren't as good at digesting meat as As: they should stick to soy beans. Yoga is good for all types: Bs should play golf, and ABs should do aero- bics "to let stress out." "The Importance of Blood Type in Japanese Culture," *Japan Today*, January 20, 2102, https://japantoday.com/category/features /lifestyle/the-importance-of-blood-type-in-japanese-culture (accessed Febru- ary 10, 2017).

26. Evans, "Japan and Blood Types: Does It Determine Personality?"

27. Elizabeth K. Wolf and Anne E. Laumann, "The Use of Blood-Type Tattoos During the Cold War," *Journal of the American Academy of Dermatology* 58, no. 3 (2008): 473.

28. "Physicians from the Atomic Energy Commission, the New York State Medical Board, and the Chicago Medical Society called for mass tattooing

of blood types either on the wrist or under the arm. The underarm area was chosen for the tattoo mark rather than the arms or legs, Chicago physician Andrew C. Ivy explained to a reporter from the *Chicago Tribune*, because arms and legs 'might be blown off by the atomic explosion.'" Susan E. Lederer, "Bloodlines: Blood Types, Identity, and Association in Twentieth-Century America," *Journal of the Royal Anthropological Institute* 19, no. S1 (2013): S118–29.

29. Wolf and Laumann, "The Use of Blood-Type Tattoos During the Cold War," 472.

30. Our View, *Logan City (UT) Herald Journal*, March 12, 1999.

31. Cells die and renew at different rates. Some—in the lens and heart—stay constant. But seven years for all cells to renew is roughly true. Adam Cole, "Does Your Body Really Refresh Itself Every Seven Years?" NPR, June 28, 2016, www.npr.org/sections/health-shots/2016/06/28/483732115/how-old -is-your-body-really (accessed October 10, 2017).

32. Robert S. Franco, "Measurement of Red Cell Lifespan and Aging," *Transfusion Medicine Hemotherapy* 39, no. 5 (2012): 302–7.

33. George Acton, *Physical Reflections upon a Letter Written by J. Denis, Professor of Philosophy and Mathematicks, to Monsieur de Montmor, Counsellor to the French King, and Master of Requests Concerning a New Way of Curing Sundry Diseases by Transfusion of Blood* (London: Printed by T. R. for J. Martyn, at the Bell without Temple Barr, 1668).

34. Statement of Bernice Steinhardt, Director Health Services Quality and Public Health Issues, Health, Education and Human Services Division, to the Subcommittee on Human Resources (Washington, DC: United States General Accounting Office), 1997, 3, http://www.gao.gov/archive/1997/he97143t .pdf (accessed May 2018).

35. The plan was for twelve pairs of pigeons to carry blood on their backs in a special container. "A taxi takes on average 12 minutes to arrive at Devonport hospital," said Hilary Sanders, whose idea it was, "and another 10 minutes to complete the journey. A pigeon would cover the two-and-a-half miles in less than five minutes." A hospital worker reported seeing only one pigeon take off with a sample. It was never seen again. Sarah Waddington, "Plymouth's Crazy Idea to Fly Blood Samples Between Hospitals—by Pigeon," *Plymouth Herald*, February 10, 2018.

36. Editorial, "Improving Blood Safety Worldwide," *Lancet* 370, no. 9585 (2007): 361.

37. Kara W. Swanson, *Banking on the Body: The Market in Blood, Milk, and Sperm in Modern America* (Cambridge, MA: Harvard University Press, 2014), 57.

38. A recent comprehensive report by the RAND corporation into the US blood supply called it "complex" but also "robust." Harvey Klein and colleagues think differently. The financial bullying might of huge hospital conglomerates,

they wrote in 2017, is forcing blood collection centers to lower their prices to unsustainable levels. There is fierce competition and the reduction of margins to the point where research is being cut. Andrew W. Mulcahy, Kandice A. Kapinos, Brian Briscombe, et al., *Toward a Sustainable Blood Supply in the United States: An Analysis of the Current System and Alternatives for the Future* (Santa Monica, CA: RAND Corporation, 2016), https://www .rand.org/pubs/research_reports/RR1575.html; Harvey G. Klein, J. Chris Hrouda, and Jay S. Epstein, "Crisis in the Sustainability of the U.S. Blood System," *New England Journal of Medicine* 377, no. 15 (2017): 1485–88.

39. Sadaguru Pandit, "10 Million Indians Made to Donate Blood Reveals NACO Data," *Hindustan Times*, July 12, 2017.
40. Jacob Copeman, *Veins of Devotion: Blood Donation and Religious Experience in North India* (New Brunswick, NJ: Rutgers University Press, 2009), loc. 1685, Kindle.
41. Jacob Copeman, "Religion, Risk and Excess in the Indian Blood Donation Encounter," in *Giving Blood: The Institutional Making of Altruism*, ed. Johanne Charbonneau and André Smith (Abingdon: Routledge, 2016), 130.
42. Ibid., 131.
43. Indian blood banks were given two years to comply. Sanjay Kumar, "Indian Supreme Court Demands Cleaner Blood Supply," *Lancet* 347, no. 8994 (1996): 114.
44. Sunil Raman, "Illicit India 'Blood Farm' Raided," BBC News, March 18, 2008, http://news.bbc.co.uk/1/hi/world/south_asia/7302649.stm (accessed October 3, 2010).
45. Rohit Singh, "Blood Racket: Arif, from Zardozi Artisan to Racketeer," *Hindustan Times*, June 7, 2017.
46. Nikhil M. Babu, "Inside India's Blood Black Market," *Business Standard* (India), December 24, 2016, www.business-standard.com/article/current -affairs/inside-india-s-blood-black-market-116122400708_1.html (accessed October 3, 2010).
47. Shuriah Niazi, "Weather-Weary Indian Farmers Resort to New Cash Crop— Blood," Reuters, February 17, 2016.
48. Vidya Krishnan, "Bad Blood: 2,234 Get HIV After Transfusion," *Hindu* (New Delhi, India), May 31, 2016.
49. India's National AIDS Control Organization, in response to right to information (RTI) requests, told reporters that "blood transfusion accounts for less than one per cent of total HIV infection."
50. HIV rates transmitted by transfusion range from 0.001 percent in high-income countries to just over 1 percent in low-income countries. "Blood Safety and Availability," World Health Organization, Fact Sheet, June 2017, www.who.int/mediacentre/factsheets/fs279/en/ (accessed June 4, 2018).
51. US Food and Drug Administration, "CPG Sec. 230.150 Blood Donor Classification Statement, Paid or Volunteer Donor," 2002, revised 2011,

available at https://www.fda.gov/ucm/groups/fdagov-public/@fdagov-afda-ice /documents/webcontent/ucm122798.pdf (accessed June 2018).

52. Melissa Lafsky, *How Much for That Pint of Blood?*, Freakonomics.com, blog, June 4, 2007, http://freakonomics.com/2007/06/04/how-much-for-that -pint-of-blood/ (accessed October 3, 2017).

53. Campbell Robertson, "For Offenders Who Can't Pay, It's a Pint of Blood or Jail Time," *New York Times*, October 19, 2015.

54. Susan E. Lederer, *Flesh and Blood: Organ Transplantation and Blood Transfusion in Twentieth-Century America* (New York: Oxford University Press, 2008), 93.

55. Kieran Healy, *Last Best Gifts: Altruism and the Market for Human Blood and Organs* (Chicago: University of Chicago Press, 2006), 73.

56. "Blood Connects Us All—Blood Donation Text Message Service in Sweden," World Health Organization, Health Topics, June 14, 2016, www.euro .who.int/en/health-topics/Health-systems/blood-safety/news/news/2016/06 /blood-connects-us-all-blood-donation-text-message-service-in-sweden (accessed October 3, 2017).

57. Interview with Mike Stredder.

58. Lederer, *Flesh and Blood*, 117.

59. William H. Schneider, "Blood Transfusion Between the Wars," *Journal of the History of Medicine and Allied Sciences* 58, no. 2 (2003): 187–224.

60. Lederer, *Flesh and Blood*, 135.

61. Interview with Mike Stredder.

62. *2016 Global Status Report on Blood Safety and Availability* (Geneva: World Health Organization, 2017), 6.

TWO: THAT MOST SINGULAR AND VALUABLE REPTILE

1. George Horn, *An Entire New Treatise on Leeches, Wherein the Nature, Properties and Use of That Most Singular and Valuable Reptile, Is Most Clearly Set Forth* (London: H. D. Symonds, 1798).

2. D. P. Thomas, "The Demise of Bloodletting," *Journal of the Royal College of Physicians of Edinburgh* 44, no. 1 (2014): 72.

3. Ibid., 73.

4. Audrey Davis and Toby Appel, *Bloodletting Instruments in the National Museum of History and Technology*, Smithsonian Studies in History and Technology, no. 41 (Washington, DC: Smithsonian Institution Press, 1979), 10.

5. Bloodletting was so popular it inspired the naming of the august medical journal the *Lancet*.

6. "Medicinal Leech (*Hirudo medicinalis*) in the Romney Marsh Natural Area," Romney Marsh Countryside Project, www.rmcp.co.uk/MedicinalLeech .html (accessed October 4, 2010); "Mixed Fortunes for New Forest

Bloodsuckers," Forestry Commission England, news release no. 16626, October 25, 2016.

7. Robert N. Mory, David A. Mindell, and David A. Bloom, "The Leech and the Physician: Biology, Etymology, and Medical Practice with *Hirudinea medicinalis*," *World Journal of Surgery* 24, no. 7 (2000): 878–83.

8. "Grandma Moses," Smithsonian National Museum of Natural History, Department of Invertebrate Zoology, http://invertebrates.si.edu/Features /stories/haementeria.html (accessed October 4, 2017).

9. "Leech," BBC Natural Histories, BBC Radio 4, August 8, 2016, www.bbc .co.uk/programmes/b07m5gwr (accessed October 10, 2017).

10. Until 2007, the European medicinal leech was thought to be *Hirudo medicinalis*. Then the leech specialist Mark E. Siddall and colleagues discovered there were three distinct genetic types of "European medicinal leech," and most commercially grown leeches were actually *Hirudo verbana*. Mark E. Siddall, Peter Trontelj, Serge Y. Utevsky, et al., "Diverse Molecular Data Demonstrate That Commercially Available Medicinal Leeches Are Not *Hirudo medicinalis*," *Proceedings of the Royal Society B* 274, no. 1617 (2007): 1481.

11. Haycraft experimented on dogs to prove the anticoagulant power of leech saliva, and that it continued long after the leech had detached. Apart from making the dogs "a bit sad," the leech saliva had no lasting effects. Hirudin was isolated by a German team and the patent licensed to Merck in 1905, which sold it as a drug. Robert G. W. Kirk and Neil Pemberton, *Leech* (London: Reaktion Books, 2013), 161.

12. J. Harsfalvi, J. M. Stassen, M. F. Hoylaerts, et al., "Calin from *Hirudo medicinalis*, an Inhibitor of von Willebrand Factor Binding to Collagen Under Static and Flow Conditions," *Blood* 85, no. 3 (1995): 705–11.

13. The calin binds to collagen, which would normally bind to platelets, aiding them to aggregate and form a clot. Biopharm researchers have likened this to "collagen-coating paint." Roy T. Sawyer, "Novel Cardiovascular Drugs from Bloodsucking Animals," *University of Wales Science and Technology Review*, no. 8 (1991): 3–12. Also Harsfalvi et al., "Calin from *Hirudo medicinalis*, an Inhibitor of von Willebrand Factor Binding to Collagen Under Static and Flow Conditions."

14. Results taken from Espacenet patent database, https://worldwide.espacenet .com/searchResults?submitted=true&locale=en_EP&DB=EPODOC&ST =advanced&TI=&AB=&PN=&AP=&PR=&PD=&PA=Biopharm&IN =Sawyer&CPC=&IC=&Submit=Search (accessed October 4, 2010).

15. The Old English *laece* meant "worm" and came from Middle Dutch. The other Old English *laece* came from Old Frisian (*letza*), Old Saxon (*laki*), and Old High German (*lakki*), according to a comprehensive history of the leech in the *World Journal of Surgery*, and meant a physician. Mory et al., "The

Leech and the Physician: Biology, Etymology, and Medical Practice with *Hirudinea medicinalis.*"

16. Roy T. Sawyer, "Scientific Rationale Behind the Medical Use of Leeches," chapter available on Researchgate, www.researchgate.net/profile/Roy_Sawyer/.

17. There may be no such poet, or several poets. Whoever he was, he does not always inspire praise. "Readers who chance upon Nicander's poetic oeuvre of nearly 1,600 lines, devoted almost entirely to snakes, spiders, and poisons and marked by an arcane style and recondite vocabulary, typically do not fall in love with their discovery. Professional classicists also sometimes run from Nicander as if from a venomous creature." Enrico Magnelli, "Nicander," in *A Companion to Hellenistic Literature*, ed. James J. Clauss and Maratine Cuypers (Malden, MA: Blackwell, 2010), 211; N. Papavramidou and H. Christopoulou-Aletra, "Medicinal Use of Leeches in the Texts of Ancient Greek, Roman and Early Byzantine Writers," *Internal Medicine Journal* 39, no. 9 (2009): 624–27.

18. Kirk and Pemberton, *Leech*, 47.

19. Avicenna, *The Canon of Medicine of Avicenna* (New York: AMS Press, 1973), 501.

20. Ibid., 502.

21. Hermann Samuel Glasscheib, *The March of Medicine: The Emergence and Triumph of Modern Medicine*, trans. Mervyn Savill (New York: Putnam, 1964), 156.

22. There is patchy written evidence, according to medieval historian Dr. Katherine Harvey, but Gilbert the Englishman (Gilbertus Anglicus) wrote of "gonorrhoea," a condition caused by the "flowing of man's seed against his will," that could be caused by "plenty of blood" and could be alleviated by letting blood. Correspondence with Dr. Harvey. Also Faye M. Getz, *Healing and Society in Medieval England: A Middle English Translation of the Pharmaceutical Writings of Gilbertus Angelicus* (Madison: University of Wisconsin Press, 1991), 272–73.

23. Thomas Dudley Fosbroke, *British Monachism; or, Manners and Customs of the Monks and Nuns of England* (London: M. A. Nattali, 1843), 234.

24. Worshipful Company of Barbers, "History of the Company," http://barbers company.org/history-of-the-company/ (accessed October 4, 2017).

25. City of London, *Liber Albus: The White Book of the City of London*, comp. John Carpenter, clerk Richard Whitington, trans. Henry Thomas Riley (London: Richard Griffin, 1861), 236.

26. Public Act, 32 Henry VIII, c. 40, "An Act Concerning the Privileges of Physicians"; Sidney Young, *The Annals of the Barber-Surgeons of London* (London: Blades, East and Blades, 1890).

27. Roy T. Sawyer, "History of the Leech Trade in Ireland, 1750–1915: Microcosm of a Global Commodity," *Medical History* 57, no. 3 (2013): 420–41.

28. James Webster, *Travels Through the Crimea, Turkey and Egypt: Performed During the Years 1825–1828* (London: Henry Colburn and Richard Bentley, 1830), 336–37.

29. I. S. Whitaker, J. Rao, D. Izadi, and P. E. Butler, "*Hirudo medicinalis*: Ancient Origins of, and Trends in the Use of Medicinal Leeches Throughout History," *British Journal of Oral and Maxillofacial Surgery* 42, no. 2 (2004): 133–37.

30. Kirk and Pemberton, *Leech*, 57.

31. "Ces sangsues ne doivent pas être épargnées, surtout lorsque le cas est traumatique, lorsque, par exemple, une roue à passé sur le corps." François-Joseph-Victor Broussais, *Cours de pathologie et de thérapeutique générales*, vol. 2 (Paris: J. B. Baillière, 1834), 226, accessed at http://gallica.bnf.fr/ark: /12148/bpt6k77300x?rk=64378;0.

32. "Vous enlevez en un instant une phlegmasie de six pouces à un pied d'étendue." Ibid., 156.

33. Kirk and Pemberton, *Leech*, 59.

34. *He told me that to these waters he had come*
To gather Leeches, being old and poor:
Employment hazardous and wearisome!
And he had many hardships to endure:
From Pond to Pond he roamed, from moor to moor,
Housing, with God's good help, by choice or chance:
And in this way he gain'd an honest maintenance.

William Wordsworth, *Poems, in Two Volumes* (London: Longman, Hurst, Rees, and Orme, 1807), 89–97.

35. Martine Hubert-Pellier, "La pêche à la sangsue," *Amis du Vieux Chinon* 11, no.1 (2007): 41.
Original text (my translation):
C'était ce qu'on nommait, au pays, "la pêche au sang". [. . .] On voyait soudain une fille s'amollir, vaciller, comme prise d'ivresse ou de vertige, quelquefois même s'avachir dans la barbotière, les fesses dans le bourbier mais l'esprit dans les nuages. Ses compagnes savaient ce qu'une telle défaillance signifiait: un affaiblissement de la volonté causé par le vampirisme insatiable des sangsues. Alors elles s'empressaient de hisser l'étourdie hors de la gadouille pour la libérer de ses parasites visqueux. Une franche rasade de pinard achevait de la requinquer.

36. "History," Ricarimpex, http://leeches-medicinalis.com/the-company/history/ (accessed October 6, 2017).

37. Hubert-Pellier, "La pêche à la sangsue," 42.

38. Ibid., 44–45.

39. Roy T. Sawyer, "The Portuguese Leech Trade in the 19th Century: The First Trans-Atlantic Commerce in Medicinal Leeches," *Anuário do Centro de Estudos de História do Atlântico* 7 (2015): 283–322.

40. Francis Bruno, Letter to the Editor, *Times*, January 28, 1825.

41. "Extract of *A Narrative of Lord Byron's Last Journey to Greece*," London *Literary Gazette*, no. 418, January 22, 1825.

42. A. R. Mills, "The Last Illness of Lord Byron," *Proceedings of the Royal College of Physicians Edinburgh* 28, no. 4 (1998): 76.

43. Joseph-Marie Audin-Rouvière, *Plus de sangsues!* (Paris: Le Normant Fils, 1827).

44. J. D. Rolleston, "F.J.V. Broussais (1772–1838): His Life and Doctrines," *Proceedings of the Royal Society of Medicine* 32, (January 11, 1939), 408.

45. *Times*, November 21, 1838.

46. The Centro Medico François Broussais is on Rome's Largo Antonio Sarti while the Hôpital Broussais is in the 14th arrondissement of Paris.

47. The common Italian and French names for leeches are pleasingly straightforward: both *sanguisuga* and *sangsue* translate as "bloodsucker."

48. Department of Health and Human Services, Food and Drug Administration, Letter to Brigitte Latrille, Ricarimpex SAS, June 21, 2004, www.accessdata .fda.gov/cdrh_docs/pdf4/k040187.pdf (accessed October 10, 2017).

49. Biopharm also sells HirudoSalt, to make up a saline solution to keep leeches in; HirudoMix, a moist matrix that dispenses with the need for frequent water changes; and HirudoGel, "a revolutionary material for keeping leeches healthy in hospital pharmacies"; www.biopharm-leeches.com/maintenance -products1.html (accessed October 10, 2017).

50. Keith L. Mutimer, Joseph C. Banis, and Joseph Upton, "Microsurgical Reattachment of Totally Amputated Ears," *Plastic and Reconstructive Surgery* 79, no. 4 (1987): 535–41.

51. Daniel Q. Haney, "Doctors Combine Modern Microsurgery and Ancient Leeching to Save Ear," Associated Press, September 24, 1985, www .apnewsarchive.com/1985/Doctors-Combine-Modern-Microsurgery-and -Ancient-Leeching-To-Save-Ear/id-f271f9b1c1cbd5dba4bbb17eeca88e83 (accessed October 10, 2017).

52. Ibid.

53. In fact, under the direction of a Professor Lavric, leeches had been in "constant use" at the Surgical Clinic in Ljubljana, used mainly to treat thrombosis and phlebitis. "This fact," wrote the two surgeons, "is of such importance that the chemist of the 2,000-bedded hospital in Ljubljana is never without a sufficient quantity of leeches to meet the demand." M. Derganc and F. Zdravic, "Venous Congestion of Flaps Treated by Application of Leeches," *British Journal of Plastic Surgery* 13 (1960): 187–92.

54. James Hamblin, "Please, Michael Phelps, Stop Cupping," *Atlantic*, August 9, 2016.

55. "According to Brynjolfsson and McAfee, such talk misses the point: trying to save jobs by tearing up trade deals is like applying leeches to a head

wound." Elizabeth Kolbert, "Our Automated Future," *New Yorker*, December 19 and 26, 2016.

56. "15 Most Bizarre Medical Treatments Ever," CBS News, www.cbsnews.com /pictures/15-most-bizarre-medical-treatments-ever/2/ (accessed May 3, 2018).

57. I. S. Whitaker, D. Izadi, D. W. Oliver, et al., "*Hirudo medicinalis* and the Plastic Surgeon," *British Journal of Plastic Surgery* 57, no. 4 (2004): 351.

58. Roy T. Sawyer, "A Sanguine Attachment: 2,000 Years of Leeches in Medicine," in *Medical and Health Annual* (London: Encyclopaedia Britannica, 1998), 97.

59. https://cites.org/eng/app/appendices.php (accessed January 10, 2018).

60. Whitaker et al., "*Hirudo medicinalis* and the Plastic Surgeon," 351.

61. Douglas B. Chepeha, Brian Nussenbaum, Carol R. Bradford, and Theodoros N. Teknos, "Leech Therapy for Patients with Surgically Unsalvageable Venous Obstruction After Revascularized Free Tissue Transfer," *Archives of Otolaryngology—Head and Neck Surgery* 128, no. 8 (2002): 961.

62. "Leech Therapy," Guy's and St. Thomas' NHS Foundation Trust, patient information leaflet, www.guysandstthomas.nhs.uk/resources/patient-information /surgery/Plastic-surgery/leech-therapy.pdf (accessed August 25, 2017).

63. Valerie Curtis, Nicole Voncken, and Shyamoli Singh, "Dirt and Disgust: A Darwinian Perspective on Hygiene," *Medische Antropologie* 11, no. 1 (1999): 148.

64. William Miller, quoted in ibid., 149.

65. Kirk and Pemberton, *Leech*, 137.

66. Mukund Jagannathan, Vipin Barthwal, and Maksud Devale, "Aesthetic and Effective Leech Application," *Plastic and Reconstructive Surgery* 124, no. 1 (2009): 338.

67. Anonymous, "Leeches Drunk Will Bite till Sober," letter, *Lancet* 2 (1849): 683.

68. James Rawlins Johnson, *A Treatise on the Medicinal Leech: Including Its Medical and Natural History, with a Description of Its Anatomical Structure: Also, Remarks upon the Diseases, Preservation and Management of Leeches* (London: Longman, Hurst, Rees, Orme, and Brown, 1816).

69. Alison Reynolds and Colm OBoyle, "Nurses' Experiences of Leech Therapy in Plastic and Reconstructive Surgery," *British Journal of Nursing* 25, no. 13 (2016): 729–33.

70. Claire Lomax, "How Leeches Could Save My Life," *Telegraph and Argus*, October 7, 2007.

71. "Blood-Sucking Leeches Save a Woman from Cancer," *Daily Mail* (London), October 9, 2007.

72. Mel Fairhurst, "Battling Michelle Beaten by Cancer," *Telegraph and Argus*, May 7, 2008.

73. Glynn Maples, "A Sucker's Born Every Minute on This Farm in Swansea, Wales," *Wall Street Journal*, September 21, 1989.

74. Derganc and Zdravic, "Venous Congestion of Flaps Treated by Application of Leeches," 189.

75. Thomas Moore, *The Journal of Thomas Moore*, ed. Wilfred S. Dowden (Newark: University of Delaware Press, 1983), 1450.

76. George Merryweather, "An Essay Explanatory of the Tempest Prognosticator in the Building of the Great Exhibition for the Works for the Industry of All Nations: Read Before the Whitby Philosophical Society, February 27, 1851," https://archive.org/stream/b2804163x/b2804163x_djvu.txt (accessed October 10, 2017).

77. Merryweather, in his published essay, expresses his obligations to "the Gentlemen of the Committee of Management of 'Lloyd's,' for the handsome manner in which I have been treated by them, and for giving publicity to a number of my experiments." For Lloyd's doing its own testing, see Kirk and Pemberton, *Leech*, 108.

78. Sawyer, "A Sanguine Attachment," 96.

79. "National Export Quotas," Convention on International Trade in Endangered Species of Wild Fauna and Flora (CITIES), n.d., www.cites.org/eng/resources/quotas/export_quotas (accessed February 10, 2018).

80. Richard G. Fiddian-Green, "Treating Heart Failure and Sepsis with Bloodletting and Leeches," responding to *British Medical Journal* 320, no. 7226 (2000): 39.

81. Jack McClintock and Elinor Carucci, "Bloodsuckers," *Discover*, December 1, 2001.

THREE: JANET AND PERCY

1. Dame Janet Vaughan, University of Oxford Medical Sciences Division, www.medsci.ox.ac.uk/about/history/women-in-oxford-medical-sciences/dame-janet-vaughan-1899-1993 (accessed October 11, 2017).

2. In 2015, 12,591,000 units of red blood cells were collected, and 11,349,000 were transfused. Add that to 1,983,000 transfusions of platelets, plus 2,727,000 of plasma, and divide by 31,536,000 seconds in a year and the figure comes to 1.018455099. Calculations done by American Association of Blood Banks (personal communication), using figures available in Katherine D. Ellingson, Matthew R. P. Sapiano, Kathryn A. Haass, et al., "Continued Decline in Blood Collection and Transfusion in the United States—2015," *Transfusion* 57, no. S2 (2017): S1588–98. Also, "Significant Shortages Impact U.S. Blood Supply," American Association of Blood Banks, press release, July 11, 2016.

3. "70 Years of Life Saving Blood Donations," NHSBT, September 26, 2016, www.blood.co.uk/news-and-campaigns/news-and-statements/70-years-of-life-saving-blood-donations/ (accessed May 2018).

4. NHSBT, "What We Do," www.nhsbt.nhs.uk/what-we-do/blood-services/blood-donation/ (accessed February 2018); American Red Cross, "Blood

Facts and Statistics," n.d., www.redcrossblood.org/learn-about-blood/blood -facts-and-statistics (accessed February 2018).

5. "Blood Safety and Availability Fact Sheet," World Health Organization, reviewed June 2017, www.who.int/mediacentre/factsheets/fs279/en/ (accessed October 12, 2017).

6. Janet Vaughan, "Jogging Along, or, A Doctor Writes," unpublished manuscript, collection of Somerville College, University of Oxford.

7. Virginia Woolf, *Mrs. Dalloway* (Adelaide, South Australia: eBooks@Adelaide, 2015), e-book, https://ebooks.adelaide.edu.au/w/woolf/virginia/w91md/ (accessed May 2018).

8. E. Cobham Brewer, *Dictionary of Phrase and Fable* (Philadelphia: H. Altemus, 1898).

9. Vaughan, "Jogging Along."

10. Ibid.

11. In 1920, the university decided to admit women as full members and allowed them to graduate (before that, plenty had studied at Oxford and got the ferry to Dublin to graduate, earning them the name "Steamboat Ladies"). "Women at Oxford," University of Oxford, www.ox.ac.uk/about/oxford-people/women -at-oxford (accessed February 2018); "Only in Britain," *Telegraph* (London), October 7, 2016.

12. *Women of Our Century*, BBC, first broadcast on BBC Two, August 3, 1984.

13. Ibid.

14. Vaughan, "Jogging Along."

15. Minot won the Nobel Prize in Physiology or Medicine in 1934, along with George H. Whipple and William P. Murphy, for "their discoveries concerning liver therapy in cases of anaemia," www.nobelprize.org/nobel_prizes/medicine /laureates/1934/minot-facts.html (accessed October 12, 2017).

16. Vaughan, "Jogging Along." *The Edinburgh Book of Plain Cookery Recipes*, published in 1932, includes raw liver recipes "for the use of patients suffering with pernicious anaemia," prepared by Miss Pybus, Sister Dietitian to the Royal Infirmary, Edinburgh. Miss Pybus recommends ox or calf liver, chicken or pig liver at a push, and kidney if no liver is available. First, create raw liver pulp by mincing liver, weighing out five ounces, then adding water and beating until it is the consistency of thick cream. For raw liver sandwiches, cut very thin brown bread and butter, spread thinly with Marmite or Bovril. Add some of the liver pulp flavored with a little lemon juice, pepper, and salt. Whenever serving liver, "do not remove the lid of the dish in front of the patient, as the smell is sometimes nauseating." *The Edinburgh Book of Plain Cookery Recipes*, rev. enl. ed. (London: Thomas Nelson and Sons, 1932), 308.

17. Janet Vaughan interviewed by Dr. Max Blythe, Royal College of Physicians/ Oxford Brookes University Medical Sciences Interview Archive, 1987.

18. This mincing apparently transformed Janet into a much more interesting character in the eyes of her cousin Virginia, who wrote in her diary, when

Janet was twenty-six, that "Good dull Janet Vaughan [. . .] joined us." Virginia Woolf, *A Room of One's Own* (Peterborough: Broadview Press, 2001), 98. Also, Virginia Woolf, *Diary of Virginia Woolf*, entry for May 13, 1926, http://www.woolfonline.com/?node=content/contextual/transcriptions &project=1&parent=41&taxa=42&content=6317&pos=21 (accessed May 2018).

19. Janet Vaughan interviewed by Dr. Max Blythe.

20. "There was a problem because Harvard didn't have any women. I was, you see, a Rockefeller fellow, they couldn't say no to me but I was a woman so I couldn't work with patients. I decided I should work with mice. I ordered some mice from these stewards, the mice didn't come. I went to the stewards, why haven't my mice come? Well, there aren't many Boston mice available. I said, well there are some excellent mice in Philadelphia; there was a very famous strain of mouse in Philadelphia. They were well known these Philadelphian mice. So there I was, no mice. So I had to work with pigeons." Janet Vaughan interviewed by Dr. Max Blythe.

21. Vaughan, "Jogging Along."

22. Ibid.

23. Janet Vaughan interviewed by Dr. Max Blythe.

24. Vaughan, "Jogging Along."

25. "I was never a good party member however. I did not care for being instructed in the Communist Manifesto, it seemed like a waste of time, and I was far too concerned with my Blood Transfusion Service, my air raid casualties and my medicine to carry on beyond a few months as a party member." Vaughan, "Jogging Along."

26. The exact predictions were six hundred thousand killed and 1.2 million casualties. Richard M. Titmuss, *Problems of Social Policy* (London: Her Majesty's Stationery Office, 1950), 13, www.ibiblio.org/hyperwar/UN/UK /UK-Civil-Social/UK-Civil-Social-2.html#fn2 (accessed May 2018).

27. Vaughan says they were told to expect thirty-seven thousand in her interview with Polly Toynbee and fifty-seven thousand in Vaughan, "Jogging Along."

28. Janet Vaughan interviewed by Dr. Max Blythe.

29. Ovid, *Metamorphoses*, Bk. VII:234–93, http://ovid.lib.virginia.edu/trans /Metamorph7.htm (accessed November 7, 2017).

30. Harvey probably reached his theory of circulation in 1616 when he was a Lumleian lecturer at St. Bartholomew's Hospital. Geoffrey Keynes, *Blood Transfusion* (London: Henry Frowde and Holder & Stoughton, 1922; republished by Leopold Classics Library, 2015).

31. Holly Tucker, *Blood Work: A Tale of Medicine and Murder in the Scientific Revolution* (New York: W. W. Norton, 2011), 22.

32. William Harvey, English physician, in *Encyclopædia Britannica*, www .britannica.com/biography/William-Harvey (accessed November 7, 2017).

33. John Aubrey, *Aubrey's "Brief Lives,"* ed. Andrew Clark (London: Henry Frowde, 1898), 300.

34. Keynes, *Blood Transfusion*, 3.

35. Richard Lower, "The Method Observed in Transfusing the Bloud out of One Animal into Another," in Royal Society (Great Britain), *Philosophical Transactions* (1665–1678), vol. 1 (1665–1666) (London: Royal Society of London), 353–58.

36. H. F. Brewer et al., *Blood Transfusion*, ed. Geoffrey Keynes (London: John Wright & Sons, 1949), 9.

37. J. Denis, "A letter concerning a new way of curing sundry diseases by transfusion of blood, written to Monsieur de Montmor, Councellor to the French King, and Master of Requests," *Philosophical Transactions* 2, no. 27 (1666): 489–504.

38. J. Denis, "An Extract of a Letter Written by J. Denis, Doctor of Physick, and Professor of Philosophy and the Mathematicks at Paris, touching a late Cure of an Inveterate Phrensy by the Transfusion of Bloud," *Philosophical Transactions* 2, no. 32 (1666): 621.

39. "He had but 20s. for his suffering it, and is to have the same again tried upon him: the first sound man that ever had it tried on him in England." Samuel Pepys, *Diary*, November 30, 1667, www.pepysdiary.com/diary/1667 /11/30/ (accessed May 2018).

40. Brewer et al., *Blood Transfusion*, 16.

41. Keynes, *Blood Transfusion*, 9.

42. James Blundell, *The Principles and Practice of Obstetricy, as at Present Taught* (London: E. Cox, 1834), 420.

43. "Heavy Bleeding After Birth (Postpartum Haemorrhage)," Royal College of Obstetricians and Gynaecologists, Fact Sheet, 2013, www.rcog.org.uk /globalassets/documents/ . . . /heavy-bleeding-after-birth.pdf (accessed January 22, 2018).

44. Harold W. Jones and Gulden Mackmull, "The Influence of James Blundell on the Development of Blood Transfusion," *Annals of Medical History* 10 (1928); 242–48.

45. "Isidore Colas, le brave petit Breton," *Le Progrès du Morbihan*, November 14, 1914, www.bannalec.fr/isidore-colas-lhistoire-de-la-transfusion-sanguine/ (accessed May 2018).

46. Ibid.

47. Ministry of Information, *Life Blood: The Official Account of the Transfusion Services* (London: HMSO, 1945), 4.

48. Hugh J. McCurrich, Correspondence, *British Medical Journal* 2, no. 3960 (1936): 1110.

49. Ministry of Information, *Life Blood*, 3.

50. Ibid., 15.

51. *Blood Transfusion*, directed by H. M. Nieter, Ministry of Information (UK), propaganda film, 1941.

52. Oliver used the phrase "a pint of the best" frequently in his lectures, for example, during a cinematograph lecture titled "The Romance of Blood Transfusion," in Horley, England, reported by the *Surrey Mirror*, September 30, 1932.

53. Ritchie Calder, "They Also Served with Their Blood: Official Story of the Transfusion Service," *Lincolnshire Echo*, March 22, 1945.

54. "A full ten months earlier, in December 1920, the Camberwell Red Cross Division minutes mentioned a similar request by the hospital. In fact, 'a number of members had already volunteered' to give their blood." Kim Pelis, "'A Band of Lunatics Down Camberwell Way': Percy Lane Oliver and Voluntary Blood Donation in Interwar Britain," in *Medicine, Madness and Social History: Essays in Honour of Roy Porter*, ed. Roberta Bivins and John V. Pickstone (London: Palgrave Macmillan, 2007), 151.

55. *Medical Services, Surgery of the War*, ed. Sir W. G. MacPherson et al. (London: HMSO, 1922), 111.

56. Ibid., chapter 1, p. 1.

57. Harvey Cushing, *From a Surgeon's Journal: 1915–1918* (Boston: Little, Brown, 1936), 259.

58. Nicholas Whitfield, "Who Is My Stranger?: Origins of the Gift in Wartime London, 1939–45," *Journal of the Royal Anthropological Institute* 9, no. 51 (2013): S95–S117.

59. Alastair Masson, *Report of Proceedings, The Scottish Society of the History of Medicine, Session 1988–89* (Edinburgh: The Scottish Society of the History of Medicine, 1989), 13.

60. Francis Hanley, *The Honour Is Due: Personal Memoir of the Blood Transfusion Service—Now Known as the Great London Red Cross Blood Transfusion Service (1921–86)* (Surbiton, Surrey, UK: JRP, 1998), 26.

61. Lederer, *Flesh and Blood*, 82.

62. Geoffrey Keynes, "Blood Donors," *British Medical Journal* 2, no. 3327 (1924): 613–15.

63. "Blood Transfusion Congress in Paris," *British Medical Journal* 2, no. 4009 (1937): 924.

64. Editorial, *British Medical Journal* 2, no. 4263 (1942): 342–43.

65. Schneider, "Blood Transfusion Between the Wars."

66. Lederer, *Flesh and Blood*, 88.

67. Charles V. Nemo, "I Sell Blood," *American Mercury*, February 1934, 194–203.

68. Lederer, *Flesh and Blood*, 84.

69. Letter to the *British Medical Journal*. from J. Bagot Oldham, September 10, 1932.

70. Tom Richards, "Percy Lane Oliver, O.B.E. (1878–1944)," article hosted on the British Blood Transfusion Society website, www.bbts.org.uk/downloads /03_article_written_by_tom_richards_in_1994.pdf/ (accessed October 18, 2017).

71. "The Demand on the Blood Donor," *British Medical Journal* 2, no. 4262, (1942): 342–43.

72. Pelis, "'A Band of Lunatics Down Camberwell Way.'"

73. Frederick Walter Mills, "The London Blood Service and the Psychology of Donors," in Brewer et al., *Blood Transfusion*, 353.

74. "Problems of Blood Transfusion," *British Medical Journal* 2, no. 3959 (1936): 1035.

75. "Voluntary Blood Donors' Association First Annual Dinner," *Nottingham Evening Post*, March 5, 1934.

76. "Voluntary Blood Donors' Association," *British Medical Journal* 2, no. 3907 (1935): 1014.

77. Pelis, "'A Band of Lunatics Down Camberwell Way,'" 152.

78. Mills, "The London Blood Service and the Psychology of Donors," 354.

79. Editorial, *British Medical Journal*.

80. "Women Blood Donors Best. A World Meeting," *Gloucester Citizen*, August 7, 1937.

81. Pelis, "'A Band of Lunatics Down Camberwell Way,'" 156.

82. "A Derby Man's Diary," *Derby Daily Telegraph*, October 7, 1932.

83. Hanley, *The Honour Is Due*, 12.

84. Ibid., 45.

85. Mills, "The London Blood Service and the Psychology of Donors," 358.

86. W. Addison, "Blood Transfusion," *Saturday Review*, February 20, 1932.

87. Alastair H. B. Masson, *History of the Blood Transfusion Service in Edinburgh* (Edinburgh: Edinburgh and South-East Scotland Blood Transfusion Association, 1993), 14.

88. Hanley, *The Honour Is Due*, 22.

89. "Blood Transfusion Congress in Paris," *British Medical Journal* 2, no. 4009 (1937): 924.

90. *Blood Transfusion*, film by the Ministry of Information, 1941.

91. T. H. O'Brien, *History of the Second World War: Civil Defence* (London: Her Majesty's Stationery Office, Longman's, Green, 1955), 165.

92. Ibid., 330.

93. Peter Lewis, *A People's War* (London: Thames Methuen, 1986), 8.

94. Mollie Panter-Downes, "Letter from London," *New Yorker*, September 9, 1939.

95. Hansard, HC Deb 27 April 1937, vol 323 col.154.

96. The 1938 Annual Report of the London Blood Transfusion Service reported that a large blood supply depot had been set up in a "bomb-proof" building in Cheam with a capacity of 1,000 pints of blood. It would be able to get up

and running within seven days of hostilities. Apart from that, there were two pints of blood each in four London County Council hospitals for emergency maternity requirements. Masson, *Report of Proceedings*, 25.

97. "There was even a glass-blower, Juan Torrero, and his assistant Salvador Fuentes, enabling the tailor-production of glass ampoules for the manipulation and storage of blood." Linda Palfreeman, *Spain Bleeds* (Eastbourne, UK: Sussex Academic Press, 2015), 41.

98. Ibid., 49.

99. *Blood Transfusion*, directed by H. M. Nieter.

100. Minutes of the Emergency Blood Transfusion Service Scheme, Wellcome Library, London.

101. "It is at the Sklifassovski [Hospital] that the method of procuring blood for transfusion popularized by Professor Yudin is in use. Within six hours of death from suicide or by street accidents, or from angina, blood is drawn off, typed, subjected to the Wasserman test, collected in jars, and preserved in ice chests." George Sacks, "A Note on Surgery in Moscow," *British Medical Journal* 2, no. 3909 (December 7, 1935): 1118.

102. "Stored Blood in War," *Times* (London), June 15, 1939.

103. Michael Gearin-Tosh, *Living Proof: A Medical Mutiny* (New York: Scribner, 2010), ebook.

104. E. T. Burke, "Blood Transfusion Service for War," *British Medical Journal* 2, no. 4099 (1939): 247–48.

105. Editorial, "Problems of Blood Transfusion," *British Medical Journal* 2, no. 3959 (1936): 1035.

106. P. L. Oliver, "A Plea for a National Blood Transfusion Conference," *British Medical Journal* 2, no. 3959 (1936): 1032.

107. Janet Vaughan interviewed by Dr. Max Blythe.

108. *Medical Research in War: Report of the Medical Research Council for the Years 1939–1945* (London: HM Stationery Office, 1947), 183.

109. http://withlovefromgraz.blogspot.co.uk/2013/11/5-april-1940.html; reproduced with permission of Loraine Fergusson.

110. Janet Vaughan, "The London Blood Supply Depots, 1939–1945," draft chapter for *Medical History of the War*, Wellcome Library archives, GC/186/2.

111. O. M. Solandt, "The Work of a London Emergency Blood Supply Depot," *Canadian Medical Association Journal* 44, no. 2 (1941): 189–91.

112. US Government Accountability Office, "Maintaining an Adequate Blood Supply Is Key to Emergency Preparedness," September 10, 2002, pp. 7–8.

113. Vaughan, "The London Blood Supply Depots, 1939–1945."

114. History module, "Remote Damage Control Resuscitation," http://rdcr.org /wp-content/uploads/2017/02/A-RDCR-HISTORY-MODULE-1-edited -sm.compressed.pdf (accessed October 19, 2017).

115. "Beverley Women's Blood Saves Lives in Europe," *Driffield Times*, January 27, 1945.

116. Ministry of Information, *Life Blood*.

117. "Blood Donor Romances," *Dundee Courier*, August 5, 1943.

118. "Blood Donor Used to Pay to Be Bled," *Gloucestershire Echo*, May 3, 1943.

119. "White Feathers Sent to Girls: A Pernicious Practice," *Sevenoaks Chronicle and Kentish Advertiser*, February 6, 1942.

120. "Blood Donor Plea," *Lancashire Evening Post*, December 29, 1942.

121. The donor who gave this reply was a married man, aged forty-seven, three children, sales representative, £20–30 ($80–120) a week, 10 donations. Richard Titmuss, *The Gift Relationship* (London: George Allen & Unwin, 1970), 231.

122. Harriet Proudfoot, "The Doctor," in *Janet Maria Vaughan 1899–1993: A Memorial Tribute,* ed. Pauline Adams, privately published (Oxford: Somerville College, 1993), 18.

123. Janet Vaughan interviewed by Dr. Max Blythe.

124. Vaughan, "The London Blood Supply Depots, 1939–1945."

125. W. H. Ogilvie, "Surgical Advances During the War," *Journal of the Royal Army Medical Corps* 85, no. 6 (1945): 259–65.

126. "The Army Blood Transfusion Service," *British Medical Journal* 1, no. 4297 (1943): 610–11. The Silver Thimble Fund was founded by Miss Hope Elizabeth Hope-Clarke in 1915. Every embroiderer had a broken thimble, she reasoned, so why not melt down the broken thimbles to make medical equipment and raise funds? Between 1915 and 1919, sixty thousand thimbles, and other trinkets, had raised money to buy fifteen ambulances, five hospital motor launches, two dental surgery cars, and a disinfector. Silver Thimble Fund: https://historicengland.org.uk/listing/what-is-designation/heritage-highlights/how-did-thimbles-help-thousands-of-servicemen-in-the-first-world-war (accessed January 2018).

127. Shipping whole blood over the Atlantic—aside from the hazards caused by enemy attacks—was a nonstarter. It would take at least five days, much of the useful shelf life of blood. Plasma was easier to transport, would survive the journey better, and, because mismatched plasma does not cause the reactions that mismatched blood does, was a much more efficient and useful donation: Starr, *Blood*, 93–98.

128. Vaughan, "Jogging Along."

129. *Medical Research in War*, 183–84.

130. Vaughan, "Jogging Along."

131. "South West London Blood Supply Depot (memorandum)," author unknown, Ministry of Health, 1945, https://wdc.contentdm.oclc.org/digital/collection/health/id/1896 (accessed at University of Warwick digital collection on October 24, 2017).

132. Vaughan, "Jogging Along."
133. Letter from Janet Vaughan to George Minot, May 12, 1945, Wellcome Collection.
134. Letter from Janet Vaughan to Molly Hoyle, Somerville College archives.
135. Barbara Harvey and Louise Johnson, "Obituary: Dame Janet Vaughan," *Independent* (London), January 12, 1993.
136. "Orations Delivered at a Congregation for the Conferment of Honorary Degrees," University of Liverpool, July 14, 1973, Somerville College archives.
137. Sheila Callender, obituary of Janet Vaughan, Somerville College archives.
138. Janet Adam Smith, obituary of Janet Vaughan, Somerville College archives.
139. Barbara Harvey, "Memorial Address," in *Janet Maria Vaughan 1899–1993: A Memorial Tribute.*
140. Christian Parham (Fitzherbert), in ibid.
141. Hanley, *The Honour Is Due*, 1.
142. www.kch.nhs.uk/patientsvisitors/wards/m-o (accessed October 24, 2017).
143. www.english-heritage.org.uk/visit/blue-plaques/oliver-percy-lane-1878 -1944 (accessed October 24, 2017).
144. James Park, "The Family," in *Janet Maria Vaughan 1899-1993: A Memorial Tribute.*
145. Harvey, "Memorial Address."
146. Dame Alice Prochaska, *The Principal's Diary*, July 14, 2011, https:// principal2010.wordpress.com/2011/07/14/departures-and-returns/ (accessed October 24, 2017).
147. Janet Vaughan interview with Polly Toynbee.
148. http://withlovefromgraz.blogspot.co.uk/2013/11/5-april-1940.html; reproduced with permission of Loraine Fergusson.

FOUR: BLOOD BORNE

1. South Africa's most recent census data put the population at 391,749, www.statssa.gov.za/?page_id=4286&id=328. Government statistics concluded that 44.6 percent of people live in formal dwellings.
2. "We Can Beat Escalating Crime in Khayelitsha," Western Cape Government, press release, www.westerncape.gov.za/khayelitsha (accessed March 23, 2018).
3. Interview with Dr. Genine Josias of Thuthuzela.
4. International Partnership for Microbicides, Fact Sheet, July 2017, https://www.ipmglobal.org/sites/default/files/attachments/publication /ipm_general_fact_sheet_121217.pdf. Also "Women's Health," World Health Organization, Fact Sheet no. 334, updated September 2013, www.who.int /mediacentre/factsheets/fs334/en/.

5. *How AIDS Changed Everything*, UNAIDS (Geneva: United Nations Programme on HIV/AIDS, 2015), p. 436, http://www.unaids.org/sites/default/files/media_asset/MDG6Report_en.pdf (accessed May 2010).

6. Amy Maxmen, "Older Men and Young Women Drive South African HIV Epidemic," *Nature* 535, no. 7612 (2016): 335.

7. Maggie Norris and Donna Rae Siegfried, *Anatomy & Physiology for Dummies* (Hoboken, NJ: John Wiley & Sons, 2011), 258.

8. Tim Jonze, "'It Was a Life-and-Death Situation. Wards Were Full of Young Men Dying': How We Made the Don't Die of Ignorance AIDS Campaign," *Guardian*, September 4, 2017.

9. Some states and countries have HIV-specific legislation. In Texas, people have been arrested for spitting and prosecuted under an aggravated assault statute. In 2008, a homeless man named Willie Campbell was jailed for thirty-five years for spitting in the mouth and eye of a police officer. "HIV-positive Man Sentenced to 35 Years for Spitting at Officer," *Associated Press*, May 15, 2008.

10. Interview with Eric Goemaere.

11. "AIDS by the Numbers—AIDS Is Not Over, But It Can Be," UNAIDS (Geneva: United Nations Programme on HIV/AIDS, 2016), p. 3.

12. Global Database on HIV-Specific Travel and Residence Restrictions, www.hivtravel.org (accessed May 2018).

13. Centers for Disease Control and Prevention, "Final Rule Removing HIV Infection from U.S. Immigration Screening," www.cdc.gov/immigrantrefugeehealth/laws-regs/hiv-ban-removal/final-rule.html (accessed May 2018). Other diseases that will deny you entry to the USA: syphilis, infectious leprosy, gonorrhea, and active tuberculosis.

14. "AIDS by the Numbers," 7.

15. "The Scales Have Tipped: UNAIDS Announces 19.5 Million People on Life-Saving Treatment and AIDS-Related Deaths Halved Since 2005," UNAIDS, press release, July 20, 2017.

16. Avert, "HIV and AIDS in South Africa," www.avert.org/professionals/hiv-around-world/sub-saharan-africa/south-africa (accessed March 23, 2018).

17. UNAIDS country profile of South Africa, www.unaids.org/en/regionscountries/countries/southafrica (accessed March 23, 2018).

18. Maxmen, "Older Men and Young Women Drive South African HIV Epidemic."

19. "Violence Against Women and HIV," UK Consortium on AIDS and International Development, Fact Sheet, July 2013.

20. Interview with Dr. Sam Wilson.

21. Marta Darder, Liz McGregor, Carol Devine, et al., *No Valley Without Shadows: MSF and the Fight for Affordable ARVs in South Africa* (Brussels: Médecins Sans Frontières, 2014), 17.

22. Ibid., 18.

23. Interview with Nozizwe Madlala-Routledge, former deputy minister of health.

24. Rose George, "A 19th-Century Epidemic," *Mosaic Science*, July 6, 2015, https://mosaicscience.com/story/19th-century-epidemic/ (accessed May 2018).
    S. J. Connolly, "The 'Blessed Turf': Cholera and Popular Panic in Ireland, June 1832," *Irish Historical Studies* 23, no. 91 (1983): 214–32.

25. John Donnelly, "Activists Wonder if Life Imitates Television in U.S. Policy on AIDS," *Boston Globe*, June 18, 2001.

26. In their meta-analysis, researchers found that 55 percent of Americans and 77 percent of sub-Saharan Africans adhered properly to their treatment regime. E. J. Mills, J. B. Nachega, I. Buchanan, et al., "Adherence to Antiretroviral Therapy in Sub-Saharan Africa and North America: A Meta-analysis," *JAMA* 296, no. 6 (2006): 679–90.

27. Treatment Action Campaign, *Fighting for Our Lives: The History of the Treatment Action Campaign 1998–2010* (Cape Town, South Africa: Treatment Action Campaign, 2010).

28. "Hidden in the Mealie Meal," *Human Rights Watch* 19, no. 18(A) (2007): 26.

29. In 2016, more than two-thirds of people living with HIV knew their HIV status. Of the people who knew their status, 77 percent were accessing treatment, and of the people accessing treatment, 82 percent were virally suppressed. However, the 90-90-90 targets are off track in the Middle East, eastern Europe, and central Asia, where AIDS deaths have risen by 48 percent and 38 percent. *Ending AIDS: Progress Towards the 90-90-90 Targets*, UNAIDS, July 10, 2017.

30. Interview with Eric Goemaere.

31. Jon Cohen, "Large Study Spotlights Limits of HIV Treatment as Prevention," *Science*, July 25, 2016.

32. She was known as the Mississippi Baby, and for two years no virus could be detected in her body. Sarah Boseley, "Kick and Kill: Is This the Best New Hope for an AIDS Cure?" *Guardian*, July 15, 2014; "'Mississippi Baby' Now Has Detectable HIV, Researchers Find," US National Institute of Allergy and Infectious Diseases, press release, July 10, 2014.

33. StopAIDS, "It Ain't Over," https://stopaids.org.uk/our-work/why-hiv-matters /it-aint-over/ (accessed March 23, 2017).

34. Teresa Welsh, "World at Risk of Losing Control of HIV and AIDS Epidemic, PEPFAR Architect Says," *Devex*, April 25, 2018.

35. Sarah Boseley, "Think the AIDS Epidemic Is Over? Far from It—It Could Be Getting Worse," *Guardian*, July 31, 2016.

36. Edward C. Green, Daniel T. Halperin, Vinand Nantulya, and Janice A. Hogle, "Uganda's HIV Prevention Success: the Role of Sexual Behavior Change and the National Response," *AIDS and Behavior* 10, no. 4 (2006): 335–46.

FIVE: THE YELLOW STUFF

1. NHS, "Why Give Blood?," www.blood.co.uk/why-give-blood/how-blood-is -used/blood-components/plasma/ (accessed April 3, 2018).

2. NHS Blood and Transplant Price list 2017/18, hospital.blood.co.uk/media /29056/price_list_bc_nhs_2017-18.pdf (accessed March 23, 2018).

3. IVIG costs 39 euros a gram, according to Gwendal Le Masson, Guilhem Solé, Claude Desnuelle, et al., "Home Versus Hospital Immunoglobulin Treatment for Autoimmune Neuropathies: A Cost Minimization Analysis," *Brain and Behavior* 8, no. 2 (2018): e00923. The *Immunoglobulin Database Report 2015/2016* produced by Medical Data Solutions and Services found an average cost in the UK of £35 ($50) a gram, igd.mdsas.com/wp-content /uploads/ImmunoglobulinDatabaseReport201516.pdf (accessed April 3, 2018). Current gold spot price per gram: $42.20, per Money Metals Exchange, www.moneymetals.com/precious-metals-charts/gold-price (accessed May 2018).

4. These facts and countless others have been built into visualizations from data from the BACI International Trade Database, at the Observatory for Economic Complexity run out of MIT. Prepare to lose many hours there. https://atlas.media.mit.edu/en/visualize/tree_map/hs92/export/usa/all/show /2016/ (accessed March 23, 2018).

5. Jim MacPherson is not the only person to use the phrase. Andrew Pollack, "Is Money Tainting the Plasma Supply?," *New York Times*, December 5, 2009.

6. Creativ-Ceutical/EAHC-EU Commission-EU, *An EU-wide Overview of the Market of Blood, Blood Components and Plasma Derivatives Focusing on Their Availability for Patients/Creativ-Ceutical Report*, April 8, 2015.

7. National Hemophilia Foundation, Fast Facts, www.hemophilia.org/About -Us/Fast-Facts (accessed March 23, 2018).

8. World Federation of Hemophilia, "What Is Hemophilia?," www.wfh.org/en /page.aspx?pid=646 (accessed June 8, 2018).

9. World Federation of Hemophilia, "Severity of Hemophilia," www.wfh.org /en/page.aspx?pid=643 (accessed April 3, 2018).

10. The 4,689 figure comes from the UK Haemophilia Centres Doctors' Organ-isation. The number of survivors was revealed in a parliamentary written question on January 29, 2018, and added to a figure of 220 deaths from Scotland. Simon Hattenstone, "Britain's Contaminated Blood Scandal: 'I Need Them to Admit They Killed My Son,'" *Guardian*, March 3, 2018; "UKHCDO Bleeding Disorder Statistics for April 2010 to March 2011," p. 41, www.ukhcdo.org/docs/AnnualReports/2011/UKHCDO%20Bleeding%20 Disorder%20Statistics%20for%202010-2011.pdf (accessed June 8, 2018).

11. The All-Parliamentary Group (APPG) on Haemophilia and Contaminated Blood, "Inquiry into the Current Support of Those Affected by the Con-taminated Blood Scandal in the UK," January 2015, 9.

12. *The Penrose Inquiry, Final Report*, vol. 1: *Patients' Experiences*, March 2015, p. 28.

13. *Panorama*, "Contaminated Blood: The Search for the Truth," broadcast July 17, 2017.

14. *Final Report/Commissioner: Horace Krever*, Government of Canada, Commission of Inquiry on the Blood System in Canada, 3 vols. (Ottawa, Ontario: Commission of Inquiry on the Blood System in Canada, 1997), vol. 2, p. 370.

15. Julian Miller interviewed on *Good Health*, available at www.youtube.com /watch?v=mvOHWRxuBYM.

16. Starr, *Blood*, 240.

17. FDA, CPG Sec. 230.150, "Blood Donor Classification Statement, Paid or Volunteer Donor," https://www.fda.gov/ucm/groups/fdagov-public/@fdagov -afda-ice/documents/webcontent/ucm122798.pdf (accessed June 2018).

18. Allen M. Hornblum, "They Were Cheap and They Were Available: Prisoners as Research Subjects in Early Twentieth Century America," *BMJ* 315, no. 7120 (1997): 1437.

19. By 1945, 71,350 felons had given 100,000 pints of blood. Lederer, *Flesh and Blood*, 93.

20. Ibid.

21. Mara Leveritt, "Bloody Awful: How Money and Politics Contaminated Arkansas's Prison Plasma Program," *Arkansas Times*, August 16, 2007.

22. Jeffrey St. Clair, "Arkansas Bloodsuckers: The Clintons, Prisoners and the Blood Trade," *Counterpunch*, September 4, 2015. Also James Ridgeway, *It's All for Sale: The Control of Global Resources* (Durham, NC: Duke University Press, 2004), 184.

23. J. Garrott Allen, letter to W. D'A. Maycock, January 6, 1975, downloaded from www.taintedblood.info. This and many other documents were released to the Tainted Blood campaign after a Freedom of Information request but are currently unavailable due to an impending public inquiry and a lawsuit.

24. *World in Action*, "Blood Money," transcript, *Penrose Inquiry, Final Report*, March 2015, www.penroseinquiry.org.uk/downloads/transcripts/PEN0131400 .pdf (accessed January 18, 2018).

25. *Week In, Week Out*, originally broadcast on BBC Wales on January 26, 1988, www.youtube.com/watch?v=Ir0qLl3n94o (accessed April 4, 2018).

26. Rupert Harry Miller, *Life of a Salesman: A True Story* (self-pub., Kindle, 2014), p. 35, loc. 462.

27. Press release, Coalition Against Bayer Dangers (Germany), January 24, 2010.

28. H. Krever, *Final Report: Commission of Inquiry on the Blood System in Canada* (Ottawa: The Commission, 1997), vol. 3, p. 758.

29. Ibid., 760.

30. Editorial, *Lancet* 324, nos. 8417–8418 (December 29, 1984): 1433.

31. Walt Bogdanich and Eric Koli, "2 Paths of Bayer Drug in 80's: Riskier One Steered Overseas," *New York Times*, May 22, 2003.

32. Ibid.

33. Carol Anne Grayson, "Blood Flows Not Just Through Our Veins but Through Our Minds. How Has the Global Politics of Blood Impacted on the UK Haemophilia Community?" master's diss., University of Sunderland, 2007, http:// haemophilia.org.uk/support/day-day-living/patient-support/contaminated -blood/dissertation-carol-grayson-contaminated-blood-products/ (accessed May 2018).

34. Marianne Barriaux, "Iraqi Father Who Lost Five Sons to AIDS Still Struggling for Justice," *Middle East Eye/AFP*, December 18, 2014.

35. Letter from Arthur Bloom to all hemophilia center directors, January 11, 1982.

36. Simon Hattenstone, "Britain's Contaminated Blood Scandal," *Guardian*, March 3, 2018. Jan and Colin Smith were also interviewed for BBC's *Panorama*, "Contaminated Blood: The Search for the Truth," first broadcast July 13, 2017.

37. Armour did not modify its process until two years later. Krever, *Final Report: Commission of Inquiry on the Blood System in Canada*, vol. 2, p. 506.

38. Department of Health and Social Security Finance Department circular, March 5, 1985.

39. Geoff Leo, "HIV Rates on Sask. Reserves Higher Than Some African Nations," CBC, June 3, 2015.

40. Mark Lemstra and Cory Neudorf, "Health Disparity in Saskatoon: Analysis to Intervention," Saskatoon Health Region, 2008.

41. Canadian Hemophilia Society, "Commemoration of the Tainted Blood Tragedy," www.hemophilia.ca/en/commemoration-of-the-tainted-blood-tragedy/ (accessed April 2018).

42. Health Canada, "Plasma Donation in Canada," www.canada.ca/en/health -canada/services/drugs-health-products/biologics-radiopharmaceuticals -genetic-therapies/activities/fact-sheets/plasma-donation-canada.html (accessed April 2018).

43. Angela Kocherga, "Plasma Is Big Business Along the Border," ABC7 KVIA, February 27, 2012, www.kvia.com/news/plasma-is-big-business-along-the -border/53247859 (accessed April 2018).

44. Robert C. James and Cameron A. Mustard, "Geographic Location of Commercial Plasma Donation Clinics in the United States, 1980–1995," *American Journal of Public Health* 94, no. 7 (2004): 1224–29.

45. Analidis Ochoa-Bendaña, *The Big Business of Blood Plasma*, twodollarsaday .com, blog, June 16, 2016, www.twodollarsaday.com/blog/2016/6/16/the-big -business-of-blood-plasma (accessed April 2018).

46. Ibid.

47. The FDA allows Americans to sell plasma up to twice a week. "The volume of plasma collected during an automated plasmapheresis collection procedure shall be consistent with the volumes specifically approved by the

Director, Center for Biologics Evaluation and Research, and collection shall not occur less than 2 days apart or more frequently than twice in a 7-day period." US Food and Drug Administration, *Code of Federal Regulations*, Title 21, vol. 7, pt. 640: Additional Standards for Human Blood and Blood Products, Sec. 640.65, www.accessdata.fda.gov/scripts/cdrh/cfdocs/cfcfr /CFRSearch.cfm?CFRPart=640&showFR=1&subpartNode=21:7.0.1.1.7.7 (accessed May 2018).

48. Derek Norfolk, ed., *Handbook of Transfusion Medicine*, 5th edition (London: TSO, 2013), 16.

49. Kathryn J. Edin and H. Luke Shaefer, *$2.00 a Day: Living on Almost Nothing in America* (Boston: Houghton Mifflin Harcourt, 2015), 93.

50. Darryl Lorenzo Wellington, "The Twisted Business of Donating Plasma," *Atlantic*, May 28, 2014. Wellington concluded that the poor health effects of donating plasma are linked to the sodium citrate used to preserve blood products.

51. R. Laub, S. Baurin, D. Timmerman, et al., "Specific Protein Content of Pools of Plasma for Fractionation from Different Sources: Impact of Frequency of Donations," *Vox Sanguinis* 99, no. 3 (2010): 220–31.

52. Legislative Assembly of Ontario, Standing Committee on Social Policy, Bill 21, Safeguarding Health Care Integrity Act, 2014, Hearing, Monday December 1, 2014, http://www.ontla.on.ca/web/committee-proceedings /committee_transcripts_details.do?locale=en&Date=2014-12 -01&ParlCommID=9003&BillID=3015&Business=&DocumentID=28419 (accessed May 2018).

53. Ibid.

54. Lucy Reynolds, "Selling Our Safety to the Highest Bidder: The Privatisation of Plasma Resources UK," *openDemocracy UK*, April 24, 2013, https:// www.opendemocracy.net/ournhs/lucy-reynolds/selling-our-safety-to-highest -bidder-privatisation-of-plasma-resources-uk.

55. Suzanne Shelley, "Immunoglobulin (IG) Drives the Blood-Plasma Therapeutics Market," *Pharmaceutical Commerce*, April 4, 2016.

56. Although early results were promising, the third stage of the Baxter trial showed that "IVIg infusions performed every 2 weeks do not improve cognition or function at 18 months in patients with mild to moderate AD." Norman R. Relkin, Ronald G. Thomas, Robert I. Rissman, et al., "A Phase 3 Trial of IV Immunoglobulin for Alzheimer Disease," *Neurology* 88, no. 18 (2017): 1768–75; Madolyn Bowman Rogers, "Bon Appétit: Endogenous Antibodies Prod Microglia to Eat Aβ Deposits," *Alzforum*, February 25, 2016, www.alzforum.org/news/research-news/bon-appetit-endogenous-antibodies -prod-microglia-eat-av-deposits (accessed April 6, 2018).

57. S. Kile, W. Au, C. Parise, et al., "IVIG Treatment of Mild Cognitive Impairment Due to Alzheimer's Disease: A Randomised Double-Blinded Exploratory Study of the Effect on Brain Atrophy, Cognition and Conversion to Dementia,"

Journal of Neurology, Neurosurgery, and Psychiatry 88, no. 2 (2017): 106–11.

58. Anne Kingston, "What a Blood Plasma-for-Profit Clinic Means for Public Health Care," *Macleans*, January 14, 2017.

59. Sophia Chase, "The Bloody Truth: Examining America's Blood Industry and Its Tort Liability Through the Arkansas Prison Plasma Scandal," *William & Mary Business Law Review* 3, no. 2 (2012): 597.

60. European Blood Alliance, *Blood, Tissues and Cells from Human Origin: The European Blood Alliance Perspective* (Amsterdam: European Blood Alliance, 2013), 74.

61. CBC Radio, "The Debate over Paying Canadians for Plasma," February 25, 2018.

62. "Department of Health Secures Guaranteed Long-Term Supplies of Plasma for NHS Patients," Department of Health, press release, December 17, 2002.

63. It was sold to Bain for £230 million ($329 million), with the UK government retaining 20 percent. Four years later, Bain and the UK government sold BPL (UK and US branches both) to Creat for "a total cash consideration" of £820 million ($1.17 billion). Paul Gallagher, "'Is There No Limit to What This Government Will Privatise?': UK Plasma Supplier Sold to US Private Equity Firm Bain Capital," *Independent* (London), June 11, 2014; "Creat Group Corporation Agrees to Acquire Bio Products Laboratory Ltd.," BainCapital .com, press release, May 18, 2016.

64. Isabel Teotonio, "From American Vein to Canadian Vein," *Toronto Star*, July 18, 2014.

65. "Contaminated Blood and Blood Products," Parliament.uk, House of Commons, Hansard, November 24, 2016, vol. 617, debate, https://hansard .parliament.uk/commons/2016-11-24/debates/9369C591-D01B-4479 -B78A-E74243142B88/ContaminatedBloodAndBloodProducts (accessed April 5, 2018).

66. David Watters, testimony to the Archer Inquiry, January 19, 2012.

67. Plasma Protein Therapeutics Association, "PPTA Statement on 'How Blood-Plasma Companies Target the Poorest Americans,'" March 16, 2018.

68. Editorial, "The Big Business of Blood Plasma," *Lancet Haematology* 4, no. 10 (2017): e452.

SIX: ROTTING PICKLES

1. Shanti Kadariya and Arja R. Aro, "Chhaupadi Practice in Nepal—Analysis of Ethical Aspects," *Dovepress Journal: Medicolegal and Bioethics* 5 (2015): 53–58. *Padi* is usually translated as "woman," but also as "being." Nepal Fertility Care Center, *Assessment Study on Chhaupadi in Nepal: Towards a Harm Reduction Strategy*, March 2015, p. 3, nhsp.org.np/wp-content/uploads /formidable/7/Chhaupadi-FINAL.pdf (accessed May 2018).

2. Gopal Sharma, "Nepali Girl Suffocates to Death After Being Banished for Menstruating," Reuters, December 20, 2016.

3. Rajneesh Bhandari and Nida Najar, "Shunned During Her Period, Nepali Woman Dies of Snakebite," *New York Times*, July 9, 2017.

4. Nepal Fertility Care Center, *Assessment Study on Chhaupadi in Nepal: Towards a Harm Reduction Strategy*, p. 4.

5. Government of Nepal Central Bureau of Statistics/UNICEF/MICS, *Nepal Multiple Indicator Cluster Survey 2010, Final Report May 2012* (Kathmandu: 2012), 108.

6. I started my period aged thirteen. I'm now forty-eight. The average amount of menstrual blood lost is 30–40 milliliters, according to the NHS; 60–80 milliliters counts as heavy bleeding. I've never been pregnant so have menstruated every month for thirty-five years. As I have endometriosis, I'll factor in many episodes of heavy bleeding. So, 420 months/periods, at an average of 50 milliliters, comes to 21,000 milliliters, or 21 liters. NHS Choices, "Heavy Periods," www.nhs.uk/conditions/heavy-periods/.

7. Sara L. Read, "'Those Sweet and Benign Humours that Nature Sends Monthly': Accounting for Menstruation in Early-Modern England" (PhD thesis, Loughborough University, 2010), 4, available from Loughborough University Institutional Repository at https://dspace.lboro.ac.uk/2134/6542.

8. The Clue survey was carried out in 2015 and had ninety thousand responses, https://helloclue.com/survey.html (accessed March 31, 2018).

9. D. Emera, R. Romero, and G. Wagner, "The Evolution of Menstruation: A New Model for Genetic Assimilation: Explaining Molecular Origins of Maternal Responses to Fetal Invasiveness," *Bioessays* 34, no. 1 (2012): 26–35.

10. Ibid.

11. Suzanne Sadedin, "How and Why Did Women Evolve Periods?," *Forbes*, May 6, 2016, www.forbes.com/sites/quora/2016/05/06/how-and-why-did-women-evolve-periods/#6418868c57a3 (accessed May 2018).

12. Dyani Lewis, "Explainer: Why Do Women Menstruate?," *Conversation*, June 17, 2013, https://theconversation.com/explainer-why-do-women-menstruate-13744 (accessed March 31, 2018).

13. M. F. Ashley-Montagu, "Physiology and the Origins of the Menstrual Prohibitions," *Quarterly Review of Biology* 15, no. 2 (1940): 211–20.

14. Pliny the Elder, *The Natural History*, Book VII, Chapter 13 (15), ed. John Bostock (London: Taylor & Francis, 1855), www.perseus.tufts.edu/hopper/text?doc=Perseus%3Atext%3A1999.02.0137%3Abook%3D7%3Achapter%3D13.

15. Ibid., Book XXVIII, Chapter 23, www.perseus.tufts.edu/hopper/text?doc=urn:cts:latinLit:phi0978.phi001.perseus-eng1:28.23.

16. Janice Delaney, Mary Jane Lupton, and Emily Toth, *The Curse*, rev. ed. (Urbana: University of Illinois Press, 1988), 3.

17. Leviticus 15:19, New American Standard Version.

18. Christopher Cooper, *Blood: A Very Short Introduction* (Oxford: Oxford University Press, 2016), 5.

19. Herbert Ian Hogbin, *The Island of Menstruating Men: Religion in Wogeo, New Guinea* (Long Grove, IL: Waveland Press, 1996), 88–89.

20. Thomas Buckley and Alma Gottlieb, eds., *Blood Magic: The Anthropology of Menstruation* (Berkeley: University of California Press, 1988), 279.

21. Wynne Maggi, *Our Women Are Free: Gender and Ethnicity in the Hindu-kush* (Ann Arbor: University of Michigan Press, 2001).

22. Virginia Smith, *Clean: A History of Personal Hygiene and Purity* (New York: Oxford University Press, 2007), 32.

23. Ibid., 35.

24. United Nations Resident and Humanitarian Coordinator's Office, Field Bulletin, "Chaupadi in the Far-West," issue no. 1, April 2011, p. 3.

25. Festivals of India, www.festivalsofindia.in/rishipanchmi/rishi-panchmi.aspx.

26. United Nations Human Rights Office of the Commissioner, *Nepal Conflict Report* (Geneva: 2012), 3.

27. Rose George, "Celebrating Womanhood: How Better Menstrual Hygiene Management Is the Path to Better Health, Hygiene and Business," Water Supply and Sanitation Collaborative Council, 2013, p. 6.

28. Sarah House, Thérèse Mahon, and Sue Cavill, *Menstrual Hygiene Matters* (London: WaterAid, 2012), 216.

29. Pazhman Pazhohish, "Afghanistan: Breaking Taboos Around Menstruation," Institute for War and Peace Reporting, September 27, 2016.

30. Catherine S. Dolan, Caitlin R. Ryus, Sue Dopson, et al., "A Blind Spot in Girls' Education: Menarche and Its Webs of Exclusion in Ghana," *Journal of International Development* 26, issue 5 (2014): 648.

31. Marie Lathers, *Space Oddities: Women and Outer Space in Popular Film and Culture, 1960–2000* (New York: Continuum, 2010), 39.

32. Sally K. Ride interviewed by Rebecca Wright, NASA Johnson Space Center Oral History Project, Edited Oral History Transcript, October 22, 2002, www.jsc.nasa.gov/history/oral_histories/RideSK/RideSK_10-22-02.htm (accessed March 31, 2018).

33. Sarah Kaplan, "What Happens When a Woman Gets Her Period in Space?" *Washington Post*, April 22, 2016.

34. Jack Olsen, *Night of the Grizzlies* (Crime Rant Classics, 2014), loc. 200, e-book.

35. Caroline P. Byrd, "Of Bears and Women: Investigating the Hypothesis That Menstruation Attracts Bears" (master's thesis, University of Montana, 1988), Paper 7720, p. 2.

36. Ibid.

37. "Grizzly Grizzly Grizzly Grizzly" (Washington, DC: US Department of Agriculture, Forest Service, 1981), https://ia800908.us.archive.org/2/items /grizzlygrizzlygr239unit/grizzlygrizzlygr239unit.pdf (accessed May 2018).

38. Bruce S. Cushing, "The Effects of Human Menstrual Odors, Other Scents, and Ringed Seal Vocalizations on the Polar Bear" (master's thesis, University of Montana, 1980), Paper 7257.

39. Translation from Béla Schick's 1920 publication on menotoxin at the marvelous online Museum of Menstruation (sadly now closed in real life), www .mum.org/menotox.htm.

40. In *Orange Is the New Black*, the main character, Piper, is smacking a wall in the bathroom of the prison where she is incarcerated. A guard threatens to write her a "shot," or disciplinary note. "I'm so sorry," says Piper. "It's menses, it's menses badness." The guard says, "Ewww, just go." He doesn't write the shot.

41. A letter from V. R. Pickles pointed out that menotoxin was not toxic to rats (they had had a bacterial infection), nor was it particularly harmful to plants. Proof was still needed that it was related to premenstrual depression. He would be glad, he concluded, to hear from anyone interested in aspects of this problem. *Lancet* 303, no. 7869 (1974): 1292.

42. The son of Jiro Ono, who founded the famous sushi restaurant Sukiyabashi Jiro, told the *Wall Street Journal* in 2011 that there were no women working in their restaurant "because women menstruate. To be a professional means to have a steady taste in your food, but because of the menstrual cycle, women have an imbalance in their taste, and that's why women can't be sushi chefs." Mary M. Lane, "Why Can't Women Be Sushi Masters?," *Wall Street Journal* blog, February 18, 2011, https://blogs.wsj.com/scene /2011/02/18/why-cant-women-be-sushi-masters/ (accessed April 2018).

43. Tomi-Ann Roberts, Jamie L. Goldenberg, Cathleen Power, and Tom Pyszzynski, "'Feminine Protection': The Effect of Menstruation on Attitudes Towards Women," *Psychology of Women Quarterly* 26, no. 2 (2002): 131–39.

44. Ibid., 132.

45. A YouGov poll asked 2,140 men and women in the UK about their attitudes to periods. Although nearly half the women said that they would feel embarrassed discussing periods with their dads, only 9 percent of men said they would feel uncomfortable discussing periods with their daughters. Action-Aid blog, "1 in 4 Women Don't Understand Their Menstrual Cycle," May 23, 2017, www.actionaid.org.uk/blog/news/2017/05/24/1-in-4-uk-women-dont -understand-their-menstrual-cycle.

46. WaterAid Australia, WaterAid blog, "Leaks, Cramps and Cravings: Majority of Women Adapt Their Lifestyle Because of a Fear of 'Period Dramas,'" May 25, 2016, www.wateraid.org/au/articles/leaks-cramps-and-cravings -majority-of-women-adapt-their-lifestyle-because-of-a-fear-of. Also Rose George, "My Gold Medal Goes to Fu Yuanhui for Talking Openly About Her Period," *Guardian*, August 16, 2016.

47. Frank Bures, *The Geography of Madness: Penis Thieves, Voodoo Death, and the Search for the Meaning of the World's Strangest Syndromes* (Brooklyn, NY: Melville House, 2016), 39.

48. Thwe T. Htay, "Premenstrual Dysphoric Disorder Clinical Presentation," MedScape.com, https://emedicine.medscape.com/article/293257-clinical (accessed April 1, 2018).

49. Meredith Bland, "Man Asks, 'Is PMS Real?' Women Answer, 'Yes, Motherf**cker, It Is,'" http://www.scarymommy.com/man-asks-is-pms-real/ (undated), (accessed May 2018).

50. Erin Beresini, "The Myth of the Falling Uterus," *Outside*, March 25, 2013, www.outsideonline.com/1783996/myth-falling-uterus.

51. Travis Saunders, "Olympic Ski Jumping Competition Completed Without a Single Uterus Explosion," *PLoS* blogs, February 14, 2012, http://blogs.plos.org/obesitypanacea/2014/02/14/olympic-ski-jumping-competition-completed-without-a-single-uterus-explosion/ (accessed April 1, 2018).

52. "Sex Hormone-Sensitive Gene Complex Linked to Premenstrual Mood Disorder," National Institutes of Health, news release, January 3, 2017, www.nih.gov/news-events/news-releases/sex-hormone-sensitive-gene-complex-linked-premenstrual-mood-disorder (accessed April 1, 2018).

53. Colin Sumpter and Belen Torondel, "A Systematic Review of the Health and Social Effects of Menstrual Hygiene Management," *PLoS ONE* 8, no. 4 (2013): e62004.

54. World Bank, "Higher Education," October 5, 2017, www.worldbank.org/en/topic/tertiaryeducation.

55. UNESCO, "Education in Pakistan," Education for All Global Monitoring Report, Fact Sheet, October 2012, https://en.unesco.org/gem-report/sites/gem-report/files/EDUCATION_IN_PAKISTAN__A_FACT_SHEET.pdf.

56. Diana E. Hoffmann and Anita J. Tarzian, "The Girl Who Cried Pain: A Bias Against Women in the Treatment of Pain," *Journal of Law, Medicine & Ethics* 29, no. 1 (2001): 13–27.

57. Kounteya Sinhal, "70% Can't Afford Sanitary Napkins, Reveals Study," *Times of India*, January 23, 2011.

58. Amy Keegan, "Zombie Statistics: To Make Progress We Need to Kill Them Off for Good," https://washmatters.wateraid.org/blog/zombie-statistics-to-make-progress-we-need-to-kill-them-off-for-good (accessed May 2018).

59. Sarah Jewitt and Harriet Ryley, "It's a Girl Thing: Menstruation, School Attendance, Spatial Mobility and Wider Gender Inequalities in Kenya," *Geoforum* 56 (2014): 137–47.

60. House et al., *Menstrual Hygiene Matters*, 31.

61. UNICEF, *Menstrual Hygiene in Schools in 2 Countries of Francophone West Africa: Burkina Faso and Niger Case Studies in 2013*, 2013.

62. Community-Led Total Sanitation, "SHARE Policy Brief on Menstrual Hygiene Management," February 8, 2017, p. 4.

63. Sally Piper Pillitteri, *School Menstrual Hygiene Management in Malawi: More Than Toilets*, SHARE/WaterAID, 2012.

64. Paul Montgomery, Julie Hennegan, Catherine Dolan, et al., "Menstruation and the Cycle of Poverty: A Cluster Quasi-Randomised Control Trial of Sanitary Pad and Puberty Education Provision in Uganda." *PLoS ONE* 11, no. 12 (2016): e0166122.

SEVEN: NASTY CLOTHS

1. National Family Health Survey-4, Government of India, Ministry of Health and Family Welfare, 2015–16, p. 6.

2. The piece in *Quartz* was filed under "taboo." Isabella Steger and Soo Kyung Jung, "An Outcry over DIY Period Pads Made from Shoe Insoles Has Sparked a National Menstruation Conversation in Korea," *Quartz*, June 11, 2017.

3. Vibeke Venema, "The Indian Sanitary Pad Revolutionary," BBC News, March 4, 2014.

4. Muruga was listed under "pioneers." Ruchira Gupta, "Arunachalam Muruganantham," *Time*, April 23, 2014.

5. Jayaashree Industries, http://newinventions.in/project-overview/.

6. World Cancer Research Fund International, Cervical Cancer Statistics, www.wcrf.org/int/cancer-facts-figures/data-specific-cancers/cervical-cancer-statistics.

7. Community-Led Total Sanitation, "SHARE Policy Brief on Menstrual Hygiene Management," 3.

8. I treat market research predictions with caution. This one, from Research and Markets, valued the industry at $23 billion in 2016 and predicts it will grow to $32 billion by 2022. "Feminine Hygiene Products Market: Global Industry Trends & Forecasts to 2022—Research and Markets," BusinessWire.com, December 7, 2017, www.businesswire.com/news/home/20171207005496/en/Feminine-Hygiene-Products-Market-Global-Industry-Trends. A competing estimate from Allied Market Research predicted it would grow to $42.7 billion by 2022. "World Feminine Hygiene Products Market Is Expected to Reach $42.7 Billion by 2022," PRNewswire.com, Allied Market Research, April 13, 2016, www.prnewswire.com/news-releases/world-feminine-hygiene-products-market-is-expected-to-reach-427-billion-by-2022-575532151.html.

9. The 1560 Geneva Bible version of Isaiah 30:22 reads, "And ye shall pollute covering of the images of silver, and the riche ornament of thine images of golde, & cast them away as a menstruous cloth, and thou shalt say unto it, Get thee hence." Sara Read, "Thy Righteousness Is But a Menstrual Clout:

Sanitary Practices and Prejudice in Early Modern England," *Early Modern Women: An Interdisciplinary Journal* 3 (2008): 14.

10. John Bunyan, *The Practical Works of John Bunyan: With a Preliminary Essay on His Character and Writings, by the Rev. Alexander Philip*, vol. 4 (London: G. King, 1841).

11. Read, "Thy Righteousness Is But a Menstrual Clout," 6.

12. Patent, 1902, https://patents.google.com/patent/US737258.

13. Vern L. Bullough, "Merchandising the Sanitary Napkin: Lillian Gilbreth's 1927 Survey," *Signs: Journal of Women in Culture and Society* 10, no. 3 (1985): 615.

14. Lehman Brothers Collection, "Kimberly-Clark Corporation," www.library.hbs.edu/hc/lehman/company.html?company=kimberly_clark_corporation.

15. Read, "'Those Sweet and Benign Humours that Nature Sends Monthly': Accounting for Menstruation in Early-Modern England."

16. "Dr. Aveling's Vaginal Tampon-Tube," *British Medical Journal* 1, no. 956 (1879): 633.

17. Keynes, *Blood Transfusion*, 26.

18. "Tampon Inventor Certain His Creation Is Safe to Use," *Chicago Tribune*, May 5, 1981.

19. Earle C. Haas, "Catamenial Device," Patent 1964911A, patented July 3, 1934, by the United States Patent Office.

20. www.reddit.com/r/IAmA/comments/jmthl/i_design_tampons_ama/.

21. Tampax official history puts the date of the sale at 1936. History of Tampax, available at https://tampax.com/en-us/history-of-tampax. *Small Wonder*, an official history of Tambrands, Tampax's parent company, says the deal was concluded on October 16, 1933. Ronald H. Bailey, *Small Wonder: How Tambrands Began, Prospered, and Grew*, published by Tambrands in 1946, available at the Museum of Menstruation, www.mum.org/smallw.htm.

22. Ibid., 9.

23. Mona Chalabi, "How Many Women Don't Use Tampons?" *FiveThirtyEight*, October 1, 2015, https://fivethirtyeight.com/features/how-many-women-dont-use-tampons/ (accessed April 1, 2018).

24. Delaney et al., *The Curse*, 137.

25. Vintage Kotex ad images available from Duke University Libraries digital collection, https://library.duke.edu/.

26. Mary G. Cardwell, "Tampons in Menstruation," letter, *British Medical Journal* 1, no. 4242 (1942): 537.

27. Communication with the Advertising Standards Authority. Also see www.asa.org.uk/type/broadcast/code_section/32.html.

28. T. L. Stanley, "Pad Ad Takes the Bold Step of Showing Periods Are Actually Red," *Adweek*, July 7, 2011.

29. Ibid.

30. Wendee Nicole, "A Question for Women's Health: Chemicals in Feminine Hygiene Products and Personal Lubricants," *Environmental Health Perspectives* 122, no. 3 (2014): A70–A75.

31. Ibid.

32. Ed Yong, "You're Probably Not Mostly Microbes," *Atlantic*, January 8, 2016.

33. Researchers who surveyed 3,012 menstruating women in North America found that while more black women than white women carried *S. aureus* (8 percent versus 14 percent); of those, white women were more likely to be carrying the toxicogenic strains of staph (15 percent of white subjects; 6 percent of black subjects). Jeffrey Parsonnet, Melanie A. Hansmann, Mary L. Delaney, et al., "Prevalence of Toxic Shock Syndrome Toxin 1-Producing *Staphylococcus aureus* and the Presence of Antibodies to This Superantigen in Menstruating Women," *Journal of Clinical Microbiology* 43, no. 9 (2005): 4628–34.

34. J. Todd, M. Fishaut, F. Kapral, and T. Welch, "Toxic-Shock Syndrome Associated with Phage-Group-I Staphylococci," *Lancet* 2, no. 8100 (1978): 1116–18.

35. CDC, "Historical Perspectives Reduced Incidence of Menstrual Toxic-Shock Syndrome—United States, 1980–1990," www.cdc.gov/mmwr/preview /mmwrhtml/00001651.htm (accessed April 1, 2018).

36. Philip M. Tierno, *The Secret Life of Germs* (New York: Pocket Books, 2001), 73.

37. Sharma Louise Vostral, *Under Wraps: A History of Menstrual Hygiene Technology* (Lanham, MD: Lexington Books, 2008), 158.

38. *Chem Fatale Report: Potential Health Effects of Toxic Chemicals in Feminine Care Products*, Women's Voices for the Earth, 2013.

39. "Dioxins and Their Effects on Human Health," World Health Organization, Fact Sheet, October 2016, www.who.int/mediacentre/factsheets/fs225/en/ (accessed April 11, 2018).

40. Michael J. DeVito and Arnold Schecter, "Exposure Assessments to Dioxins from the Use of Tampons and Diapers," *Environmental Health Perspectives* 110, no. 1 (2002): 23.

41. Tampax, "What Are Tampons Made Of?," https://tampax.com/en-us/tips -and-advice/period-health/whats-in-a-tampax-tampon (accessed April 2, 2018).

42. US Food & Drug Administration, "Dioxin in Tampons," www.fda.gov /scienceresearch/specialtopics/womenshealthresearch/ucm134825.htm (accessed April 2, 2018).

43. *Chem Fatale Report*, 10.

44. US Food & Drug Administration, "Guidance for Industry and FDA Staff: Menstrual Tampons and Pads: Information for Premarket Notification Submissions (510(k)s)," July 27, 2005, www.fda.gov/downloads/MedicalDevices /DeviceRegulationandGuidance/GuidanceDocuments/ucm071799.pdf (accessed April 2, 2018).

45. "Hallaron glifosato en algodón, gasas, hisopos, toallitas y tampones de La Plata," Infobae.com, October 20, 2015, www.infobae.com/2015/10/20 /1763672-hallaron-glifosato-algodon-gasas-hisopos-toallitas-y-tampones-la -plata/ (accessed April 2, 2018). An earlier study found glyphosate in the urine of 90 percent of residents of General Pueyrredón province, both urban and rural. None of the residents had come into direct contact with glyphosate.

46. Dr. Jen Gunter, October 24, 2015, https://drjengunter.wordpress.com/2015 /10/24/no-your-tampon-still-isnt-a-gmo-impregnated-toxin-filled-cancer -stick/ (accessed April 2, 2018).

47. "Tampons and Sanitary Pads in Switzerland Declared Safe," Swissinfo.ch, March 2, 2017, www.swissinfo.ch/eng/toxic-chemicals_tampons-and-sanitary -pads-in-switzerland-declared-safe-/43000250 (accessed April 2, 2018).

48. The FDA's Code of Federal Regulations, Part 801: Labeling, requires manufacturers of scented and unscented tampons to list information regarding absorbency and the risks of TSS. www.accessdata.fda.gov/scripts/cdrh/cfdocs /cfCFR/CFRSearch.cfm?fr=801.430 (accessed April 2, 2018).

49. The Tampon Safety and Research Act of 1999, available at www.govtrack .us/congress/bills/106/hr890.

50. The Robin Danielson Feminine Hygiene Product Safety Act of 2017, available at www.govtrack.us/congress/bills/115/hr2379.

51. Carolyn Maloney, "You Know Where Your Tampon Goes. It's Time You Knew What Goes into It, Too."

52. "World Feminine Hygiene Products Market Is Expected to Reach $42.7 Billion by 2022."

53. Euromonitor International, *Sanitary Protection Report 2017* (cost: £875/ $1,325), www.euromonitor.com/category-update-sanitary-protection/report.

54. Anasuya Basu, "Compostable Sanitary Pads: A Sustainable Solution in Menstrual Hygiene?," *The Wire*, January 30, 2018.

55. Water Supply and Sanitation Collaborative Council, *Celebrating Womanhood* (Geneva: WSSCC, 2013), 11.

56. Ghebre E. Tzeghai, Funmilayo O. Ajayi, Kenneth W. Miller, et al., "A Feminine Care Clinical Research Program Transforms Women's Lives," *Global Journal of Health Science* 7, no. 4 (2015): 45–59.

57. Jill Craig, "In Slums of Nairobi, Sex for Sanitation," *Voice of America*, February 14, 2012.

58. Penelope A. Phillips-Howard, George Otieno, Barbara Burmen, et al., "Menstrual Needs and Associations with Sexual and Reproductive Risks in Rural Kenyan Females: A Cross-Sectional Behavioral Survey Linked with HIV Prevalence," *Journal of Women's Health* 24, no. 10 (2015): 801–11.

59. "Pupils Exchange Sex for Sanitary Pad—GES Reveals," GhanaWeb.com, October 7, 2016.

60. Linda Mason, Elizabeth Nyothach, Kelly Alexander, et al., "'We Keep It Secret So No One Should Know'—A Qualitative Study to Explore Young

Schoolgirls' Attitudes and Experiences with Menstruation in Rural Western Kenya," *PLoS ONE* 8, no. 11 (2013): e79132.

61. Hannah Parry, "Customs Officials Seize Half a Ton of Radioactive Sanitary Pads at Lebanon Airport," *Mail Online*, March 21, 2015.

62. Dominic Jackson, "Radioactive Sanitary Pads from China Seized by Lebanon," *Shanghaiist*, March 23, 2015 (accessed September 2017; site not currently available).

63. "Counterfeit Sanitary-Napkin Ring Smashed," *China Daily*, May 15, 2013.

64. Indian Standard: Specification for Sanitary Napkins, March 1993, Bureau of Indian Standards, first published May 1980.

65. Tzeghai et al., "A Feminine Care Clinical Research Program Transforms Women's Lives."

66. "Standard Test Method for Performing Behind-the-Knee (BTK) Test for Evaluating Skin Irritation Response to Products and Materials That Come into Repeated or Extended Contact with Skin," ASTM International, www.astm.org/Standards/F2808.htm.

67. HERproject, https://herproject.org/.

68. Lydia DePillis, "Two Years Ago, 1,129 People Died in a Bangladesh Factory Collapse. The Problems Still Haven't Been Fixed," *Washington Post*, April 23, 2015.

69. Business Case Studies, "Beyond Corporate Social Responsibility: A Primark Case Study," https://businesscasestudies.co.uk/primark/beyond-corporate-social-responsibility/the-value-of-the-herproject.html (accessed April 2, 2018).

70. Ibid.

71. https://herproject.org/impact.

72. George, "My Gold Medal Goes to Fu Yuanhui for Talking Openly About Her Period."

73. Clare O'Connor, "Why 2016 Was the Year of the Women-Led Period Startup," *Forbes*, December 22, 2016.

74. Lily Kuo, "Kenya Is Promising Free Sanitary Napkins to Help Keep Girls in School," *Quartz*, June 23, 2017, https://qz.com/1012976/uhuru-kenyatta-promises-free-sanitary-napkins-for-kenyan-school-girls/ (accessed April 2, 2018).

75. "Uganda: No Money for Sanitary Pads in Schools to Fulfil Museveni's Promise," *Monitor*, February 15, 2017 (accessed September 2017).

76. Alon Mwesigwa, "Jailed for Calling the Ugandan President a 'Pair of Buttocks,' Activist Vows to Fight On," *Guardian*, June 19, 2017.

77. Emily Wax, "Virginity Becomes a Commodity in Uganda's War Against AIDS," *Washington Post*, October 7, 2005.

78. Radhika Sanghani, "Indian Women Protest Temple That Wants to Scan Them for 'Impure' Periods," *Telegraph* (London), November 23, 2015.

79. Nikita Azad, "An Open Letter to You: From the 'Young, Bleeding Woman' Who Shook Sabarimala," Youth Ki Awaaz, August 30, 2016, www

.youthkiawaaz.com/2016/08/happy-to-bleed-open-letter-nikita-azad/ (accessed May 2018).

80. Class action filed by ACLU against Muskegon County et al. in 2014, p.13, accessed at www.aclumich.org/sites/default/files/2014_MuskegonComplaint.pdf.

81. Brooklyn Defender Services, testimony of Andrea Nieves before the New York City Council Committee on Women's Issues, June 2, 2016, http://bds .org/andrea-nieves-testifies-in-support-of-council-bill-requiring-doc-to -provide-free-feminine-hygiene-products-in-city-jails/.

82. Amy Fettig, "Arizona Needs Laws That Protect Women Prisoners' Menstrual Health," ACLU, February 9, 2018, www.aclu.org/blog/prisoners -rights/women-prison/arizona-needs-laws-protect-women-prisoners -menstrual-health (accessed March 14, 2018).

83. Jimmy Jenkins, "'Pads and Tampons and the Problems with Periods': All-Male Committee Hears Arizona Bill on Feminine Hygiene Products in Prison," kjzz.org, February 12, 2018, http://kjzz.org/content/602963/%E2%80%98pads -and-tampons-and-problems-periods-all-male-committee-hears-arizona-bill -feminine.

84. Under the revised tax rate, women's sanitary products taxed at 5 percent include sanitary towels, sanitary pads, panty liners not designed as incontinence products, tampons, sanitary belts for use with looped towels or pads, keepers (internal devices for the collection of menstrual flow), and maternity pads for the collection of lochia (vaginal discharge after birth containing blood, mucus, and uterine discharge). www.gov.uk/government/publications /vat-notice-70118-womens-sanitary-protection-products/vat-notice-70118 -womens-sanitary-protection-products (accessed April 3, 2018).

85. Damian McBride, *Power Trip: A Decade of Policy, Plots and Spin* (London: Biteback, 2013).

86. Ema Sagner, "More States Move to End 'Tampon Tax' That's Seen as Discriminating Against Women," npr.org, March 25, 2018, www.npr.org/2018 /03/25/564580736/more-states-move-to-end-tampon-tax-that-s-seen-as -discriminating-against-women.

87. BBC, "Why Is the US 'Tampon Tax' So Hated?," September 14, 2016, www .bbc.co.uk/news/world-us-canada-37365286 (accessed April 3, 2018).

88. "Tampon-Tax Protest Turns Zurich Fountains Red," *Local*, October 4, 2016, www.thelocal.ch/20161004/tampon-tax-protest-turns-zurich-fountains-red (accessed April 3, 2018).

89. The UK government has since pledged to use £15 million ($20 million) of tampon tax revenue to fund "projects that tackle sexual violence, address social exclusion and improve mental health and wellbeing." My mental health and well-being would be improved by free sanitary products. Ben Quinn, "Anti-abortion Life Charity Will Get Cash from UK Tampon Tax," *Guardian*, October 28, 2017. "Women and Girls Set to Benefit from £15 Million Tampon Tax Fund," UK government, press release, March 26, 2018,

www.gov.uk/government/news/women-and-girls-set-to-benefit-from-15
-million-tampon-tax-fund.

90. Sara Austin, "How These 3 Women Are Working to Make 'Menstrual Equity' a Reality," *Cosmopolitan*, November 17, 2016.

91. Gloria Steinem, "If Men Could Menstruate," *Ms.*, October 1978.

EIGHT: CODE RED

1. John 11:43–44, Bible, New International Version, http://biblehub.com/niv/john/11.htm (accessed October 26, 2017).

2. Hazen Burton, "The 'Blood Trinity': Robertson, Archibald and MacLean—The Canadian Contribution to Blood Transfusion in World War I," *Dalhousie Medical Journal* 35, 1, (2008): 21.

3. The WHO's latest figure is 5.8 million. "Injuries and Violence: The Facts," World Health Organization, Fact Sheet, www.who.int/violence_injury_prevention/key_facts/en/ (accessed October 26, 2017).

4. Ibid.

5. The WHO's *Global Health Estimates 2015* reported that 760,073 Africans died from HIV/AIDS, while 930,178 died from trauma. This category included self-inflicted injury. www.who.int/healthinfo/global_burden_disease/estimates/en/index1.html (accessed October 26, 2017).

6. B. J. Eastridge, M. Hardin, J. Cantrell, et al., "Died of Wounds on the Battlefield: Causation and Implications for Improving Combat Casualty Care," *Journal of Trauma and Acute Care Surgery* 71, Suppl. 1 (July 2011): S4–8.

7. Jennifer M. Gurney and John B. Holcomb, "Blood Transfusion from the Military's Standpoint: Making Last Century's Standard Possible Today," *Current Trauma Reports* 3, no. 2 (2017): 144.

8. J. B. Holcomb, "Major Scientific Lessons Learned in the Trauma Field over the Last Two Decades," *PLoS Medicine* 14, no. 7 (2017): e1002339.

9. John B. Holcomb, "Transport Time and Preoperating Room Hemostatic Interventions Are Important: Improving Outcomes After Severe Truncal Injury," *Critical Care Medicine* 46, no. 3 (2018): 447.

10. Interview with Karim Brohi.

11. J. F. LePage, "On Transfusion," *Lancet* 20, 3092 (1882): 970.

12. Brigadier General Douglas B. Kendrick, *Medical Department United States Army in World War II: Blood Program in World War II*, US Army Medical Department, Office of Medical History, http://history.amedd.army.mil/booksdocs/wwii/blood/chapter3.htm (accessed April 6, 2018).

13. Corinne S. Wood, "A Short History of Blood Transfusion," *Transfusion* 7, no. 4 (1967): 302.

14. Martin D. Zielinski, Donald H. Jenkins, Joy D. Hughes, et al., "Back to the Future: The Renaissance of Whole-Blood Transfusions for Massively Hemorrhaging Patients," *Surgery* 155, no. 5 (2014): 883–86. The paper gets its

name from the film of the same name (Part II) and is inspired by this quote from Dr. Emmet Lathrop "Doc" Brown: "I went to a rejuvenation clinic and got a whole natural overhaul. They took out some wrinkles, did hair repair, changed the blood, added a good 30 to 40 years to my life. They also replaced my spleen and colon. What do you think?"

15. London's Air Ambulance, "UK's First Air Ambulance to Carry Blood on Board," press release, March 4, 2012, https://londonsairambulance.co.uk/our-service/news/2012/03/uks-first-air-ambulance-to-carry-blood-on-board (accessed May 2018).

16. Matthew A. Borgman, Philip C. Spinella, Jeremy G. Perkins, et al., "The Ratio of Blood Products Transfused Affects Mortality of Patients Receiving Massive Transfusions at Combat Support Hospital," *Journal of Trauma* 63, no. 4 (2007): 805–13.

17. It was first used by HEMS in 2014. London's Air Ambulance, "World's First Pre-hospital REBOA Performed," June 16, 2014, https://londonsairambulance.co.uk/our-service/news/2014/06/we-perform-worlds-first-pre-hospital-reboa. The technique was pioneered in the Korean War by Colonel C. W. Hughes. C. W. Hughes, "Use of an Intra-aortic Balloon Catheter Tamponade for Controlling Intra-abdominal Hemorrhage in Man," *Surgery* 36, no. 1 (1954): 65–68.

18. National Institutes of Health, National Cancer Institute, "National Cancer Act of 1971," www.cancer.gov/about-nci/legislative/history/national-cancer-act-1971 (accessed April 4, 2018).

19. Carl W. Walter, "Invention and Development of the Blood Bag," *Vox Sanguinis* 47, no. 4 (1984): 318–24.

20. Patrick J. Kiger, "Behind the Battle of Mogadishu," *National Geographic*, n.d., http://channel.nationalgeographic.com/no-man-left-behind/articles/behind-the-battle-of-mogadishu/ (accessed January 2018).

21. Gurney and Holcomb, "Blood Transfusion from the Military's Standpoint," 149.

22. R. L. Davies, "Should Whole Blood Replace the Shock Pack?," *Journal of the Royal Army Medical Corps* 162, no. 1 (2016): 5–7.

23. G. Strandenes, A. P. Cap, D. Cacid, et al., "Blood Far Forward—a Whole Blood Research and Training Program for Austere Environments," *Transfusion* 53, no. S1 (2013): 124S–130S.

24. G. Strandenes, H. Skogrand, P. C. Spinella, et al., "Donor Performance of Combat Readiness Skills of Special Forces Soldiers Are Maintained Immediately After Blood Donation: A Study to Support the Development of a Pre-hospital Fresh Whole Blood Transfusion Program," *Transfusion* 53, no. 3 (2013): 526–30.

25. "Blood Simple," *Economist*, October 11, 2007. Also James D. Reynolds, Gregory S. Ahearn, Michael Angelo, et al., "S-Nitrosohemoglobin Deficiency: A Mechanism for Loss of Physiological Activity in Banked Blood,"

Proceedings of the National Academy of Sciences of the United States of America 104, no. 43 (2007): 17058–62.

26. The network reaches beyond London, to the tip of Kent, and includes the four Major Trauma Centres, plus thirty-five other hospitals. Map available from www.c4ts.qmul.ac.uk/london-trauma-system/london-trauma-system-map.

27. "London's Trauma System Dramatically Reduces Mortality Rates," news release, Queen Mary, University of London, October 27, 2015.

28. Ross Lydall, "Medical Miracle as All 48 London Bridge Terror Victims Who Made It to Hospital Survive Their Injuries," *Evening Standard*, June 9, 2017. A blog by Karim Brohi referred to thirty-six injured. http://blogs.bmj .com/bmj/2017/08/04/trauma-networks-and-terrorist-events/.

29. Nicola Twilley, "Can Hypothermia Save Gunshot Victims?" *New Yorker*, November 28, 2016.

30. "Transform Trauma," Barts Charity, http://bartscharity.org.uk/get-involved /appeals/transform-trauma/ (accessed April 5, 2018).

NINE: BLOOD LIKE GUINNESS: THE FUTURE

1. John Edgar Browning, "The Real Vampires of New Orleans and Buffalo: A Research Note Towards Comparative Ethnography," *Palgrave Communications* 1, no. 15006 (2015), online only at https://www.nature.com/articles /palcomms20156 (accessed May 2018). Also John Edgar Browning, "What They Do in the Shadows: My Encounters with the Real Vampires of New Orleans," theconversation.com, March 25, 2015, http://theconversation .com/what-they-do-in-the-shadows-my-encounters-with-the-real-vampires -of-new-orleans-39208 (accessed May 2018). This piece includes the photo caption "Before you ask, they probably haven't seen Twilight."

2. G. G. Carter and G. S. Wilkinson, "Food Sharing in Vampire Bats: Reciprocal Help Predicts Donations More Than Relatedness or Harassment," *Proceedings of the Royal Society B* 280, no. 1753 (2013): 20122573.

3. Richard Sugg, *The Faces of the Vampire* (forthcoming). See also Richard Sugg, *Mummies, Cannibals and Vampires: The History of Corpse Medicine from the Renaissance to the Victorians* (New York: Routledge, 2016).

4. Sugg, *The Faces of the Vampire* (forthcoming).

5. Paul Barber, "The Real Vampire," *Natural History* 99, no. 10 (1990): 74.

6. Ibid.

7. Luise White, *Speaking with Vampires: Rumor and History in Colonial Africa* (Berkeley: University of California Press, 2000), 4.

8. Richard Sugg, "The Art of Medicine: Corpse Medicine: Mummies, Cannibals, and Vampires," *Lancet* 371, no. 9630 (2008): 2078–79.

9. Ferdinand Peter Moog and Alex Karenberg, "Between Horror and Hope: Gladiator's Blood as a Cure for Epileptics in Ancient Medicine," *Journal of the History of the Neurosciences* 12, no. 2 (2003): 138.

10. Sugg, *Mummies, Cannibals and Vampires*, 124.

11. Hsieh Bao Hua, *Concubinage and Servitude in Late Imperial China* (Lanham, MD: Lexington Books, 2014), 197.

12. Sir John Floyer and Dr. Edward Baynard, *Psychrolousia. Or, The History of Cold Bathing: Both Ancient and Modern* (London: William Innys, 1715), 409–10.

13. Alexander Bogdanov, *Red Star: The First Bolshevik Utopia*, ed. Loren R. Graham and Richard Stiles, trans. Charles Rougle (Bloomington: Indiana University Press, 1984), 83–86.

14. Nikolai Krementsov, *A Martian Stranded on Earth: Alexander Bogdanov, Blood Transfusions, and Proletarian Science* (Chicago: University of Chicago Press, 2011), 81.

15. *New World Encyclopaedia*, "Alexander Bogdanov," http://www.newworld encyclopedia.org/entry/Alexander_Bogdanov (accessed May 2018).

16. Krementsov, *A Martian Stranded on Earth*, 59–60.

17. Ibid., 60.

18. E. Bunster and R. K. Meyer, "An Improved Method of Parabiosis," *The Anatomical Record* 57, 4 (1933): 339–43.

19. William M. Mann, "The Stanford Expedition to Brazil: Parabiosis in Brazilian Ants," *Psyche* 19, no. 2 (1912): 36–41.

20. C. M. McCay, F. Pope, and W. Lunsford, "Experimental Prolongation of the Life Span," *Bulletin of the New York Academy of Medicine* 32, no. 2 (1956): 91–101. Also C. M. McCay, F. Pope, W. Lunsford, et al., "Parabiosis Between Old and Young Rats," *Gerontologia* 1, no. 1 (1957): 7–17.

21. "Blood from Young Animals Can Revitalise Old Ones," *Economist*, July 15, 2017.

22. AABB, "Blood FAQ," www.aabb.org/tm/Pages/bloodfaq.aspx (accessed April 5, 2018).

23. M. Sinha, Y. C. Jang, J. Oh, et al., "Restoring System GDF11 Levels Reverses Age-Related Dysfunction in Mouse Skeletal Muscle," *Science* 344, no. 6184 (2014): 649–52.

24. Stanford Medicine News Center, "Clinical Trial Finds Blood-Plasma Infusions for Alzheimer's Safe, Promising," press release, November 4, 2017.

25. University of Adelaide, "Smarter Brains Are Blood-Thirsty Brains," press release, August 30, 2016.

26. The debate is ongoing. Jocelyn Kaiser, "Antiaging Protein Is the Real Deal, Harvard Team Claims," *Science*, October 21, 2015, http://www.sciencemag .org/news/2015/10/antiaging-protein-real-deal-harvard-team-claims (accessed May 2018).

27. Tad Friend, "Silicon Valley's Quest to Live Forever," *New Yorker*, April 3, 2017.

28. "But be sure you do not eat the blood, because the blood is the life, and you must not eat the life with the meat" (Deuteronomy 12:23). "Instead we

should write to them, telling them to abstain from food polluted by idols, from sexual immorality, from the meat of strangled animals and from blood" (Acts 15:20). "Anyone who eats blood must be cut off from their people" (Leviticus 7:27). All from International Standard Version.

29. Royal College of Physicians, National Comparative Audit of Blood Transfusion and NHSBT, *National Comparative Audit of Blood Transfusion: 2015 Audit of Patient Blood Management in Adults Undergoing Elective, Scheduled Surgery*, 2015, 7.

30. Richard Daniel, *Managing Without Blood* (Daniel Medico-Legal, 2012), 4.

31. Clare Murphy, "The Right to Die for Jehovah," BBC News, November 5, 2007.

32. Sidhartha Banerjee, "Quebec Coroner Says Jehovah's Witnesses Had Right to Refuse Blood Transfusions," *Canadian Press*, November 14, 2017.

33. Guylaine LaRose, an emergency pediatrician, and Antoine Payot, director of the ethics unit at Montreal's Sainte-Justine Hospital, appealed for Quebec's Civil Code to be amended so that doctors could override refusals of transfusion in life-threatening situations. Stephen Smith, "Calls to Amend Quebec Civil Code in Wake of Jehovah's Witness Death," CBC News, November 26, 2016.

34. Tamra Ramasinghe and William D. Freeman, "'ICU Vampirism': Time for Judicious Blood Draws in Critically Ill Patients," *British Journal of Haematology* 164, no. 2 (2013): 302–3.

35. Osamu Muramoto, "Bioethical Aspects of the Recent Changes in the Policy of Refusal of Blood by Jehovah's Witnesses," *British Medical Journal,* 322, no. 7277 (2001): 37–39.

36. Paul C. Hébert, George Wells, Morris A. Blajchman, et al., "A Multicenter, Randomized, Controlled Clinical Trial of Transfusion Requirements in Clinical Care," *New England Journal of Medicine* 340, no. 6 (1999): 409–17; V. Sim, L. S. Kao, S. Frangos, et al., "Can Old Dogs Learn New 'Transfusion Requirements in Critical Care': A Survey of Packed Red Blood Cell Transfusion Practices Among Members of the American Association for the Surgery of Trauma," *American Journal of Surgery* 210, no. 1 (2015): 45–51.

37. Royal College of Surgeons, *Caring for Patients Who Refuse Blood: A Guide to Good Practice for the Surgical Management of Jehovah's Witnesses and Other Patients Who Decline Transfusion*, November 2016.

38. Society for the Advancement of Blood Management, "Mission," www.sabm .org/mission (accessed April 6, 2108).

39. K. D. Ellingson, M. R. P. Sapiano, K. A. Haass, et al., "Continued Decline in Blood Collection and Transfusion in the United States—2015," *Transfusion* 57, no. S2 (2017): 1588–98.

40. A. Shander, A. Fink, M. Javidroozi, et al., "Appropriateness of Allogeneic Red Blood Cell Transfusion: The International Consensus Conference on Transfusion Outcomes," *Transfusion Medicine Reviews* 25, no. 3 (2011): 232–246.e53.

41. Dr. Megan Rowley, "Where Does Blood Go?," NHSBT presentation, https://www.transfusionguidelines.org/uk-transfusion-committees/regional-transfusion-committees/london/education (accessed May 2018).

42. Harvey G. Klein, Chris Hrouda, and Jay S. Epstein, "Crisis in the Sustainability of the U.S. Blood System," *New England Journal of Medicine* 377, no. 15 (2017): 1485–88.

43. World Health Organization, "Blood Donor Selection and Counselling," www.who.int/bloodsafety/voluntary_donation/blood_donor_selection_counselling/en/ (accessed April 2018).

44. From the National Blood Collection and Utilization Survey 2015. The actual figure is 12,591,000. Ellingson et al., "Continued Decline in Blood Collection and Transfusion in the United States—2015."

45. I am grateful to Drs. James Bovell and Edwin Hodder for enabling me to create the tag "milk transfusion." H. A. Oberman, "Early History of Blood Substitutes, Transfusion of Milk," *Transfusion* 9, no. 2 (1969): 74–77.

46. Ravi Nessman, "South Africa Approves Blood Substitute," *Los Angeles Times*, April 16, 2001.

47. See, for example, "Expanded Access Protocol Using HBOC-201," Johns Hopkins University study, at https://clinicaltrials.gov/ct2/show/NCT02684474. Also, "Hemoglobin Oxygen Therapeutics LLC Announces the World's First Human Liver Transplantation after Ex-situ Normothermic Machine Perfusion Using Hemopure," PR Newswire, October 18, 2017.

48. Brandon Keim, "Controversial Blood Substitute May Be a Killer," *Wired*, May 24, 2007.

49. University of Bristol, "Major Breakthrough in the Manufacture of Red Blood Cells," press release, March 23, 2017, www.bristol.ac.uk/news/2017/march/blood-cells.html (accessed April 5, 2018). Also, Nick Watkins, "Research, Development and Innovation—The Generation Game Changer," blog post for NHSBT, February 23, 2018.

50. M.-C. Giarratana, H. Douard, A. Dumont, et al., "Proof of Principle for Transfusion of In Vitro–Generated Red Blood Cells," *Blood* 118, no. 19 (2011): 5071–79 (accessed April 5, 2018). Colin Barras, "What Is Artificial Blood and Why Is the UK Going to Trial It?" *New Scientist*, June 25, 2015.

51. J. A. Ribeil, S. Hacein-Bey-Abina, E. Payen, et al., "Gene Therapy in a Patient with Sickle Cell Disease," *New England Journal of Medicine* 376, no. 9 (2017): 848–55. Andy Coghlan, "Gene Therapy 'Cures' Boy of Blood Disease That Affects Millions," *New Scientist*, March 1, 2017.

52. Antonio Regalado, "Grail's $1 Billion Bet on the Perfect Cancer Test," *MIT Technology Review*, June 5, 2017, www.technologyreview.com/s/607944/grails-1-billion-bet-on-the-perfect-cancer-test/ (accessed April 6, 2018).

FURTHER READING

Bivins, Roberta, and John V. Pickston, eds. *Medicine, Madness and Social History: Essays in Honour of Roy Porter*. London: Palgrave Macmillan, 2007.

Bogdanov, Georgi. *Red Star: The First Bolshevik Utopia*. Bloomington: Indiana University Press, 1984.

Carney, Scott. *The Red Market: On the Trail of the World's Organ Brokers, Bone Thieves, Blood Farmers, and Child Traffickers*. New York: William Morrow, 2011.

Charbonneau, Johanne, and André Smith, eds. *Giving Blood: The Institutional Making of Altruism*. New York: Routledge, 2016.

Coles, K. A., R. Bauer, Z. Nunes, and C. L. Peterson, eds. *The Cultural Politics of Blood, 1500–1900*. Basingstoke: Palgrave Macmillan, 2015.

Cooper, Christopher. *Blood: A Very Short Introduction*. Oxford: Oxford University Press, 2016.

Edin, Kathryn J., and H. Luke Shaefer. *$2.00 a Day: Living on Almost Nothing in America*. Boston: Mariner Books, 2016.

Hayes, Bill. *Five Quarts: A Personal and Natural History of Blood*. New York: Ballantine Books, 2005.

Healey, Kieran. *Last Best Gifts: Altruism and the Market for Human Blood and Organs*. Chicago: Chicago University Press, 2006.

Hill, Lawrence. *Blood: A Biography of the Stuff of Life*. Toronto: Oneworld, 2014.

Houppert, Karen. *The Curse: Confronting the Last Unmentionable Taboo: Menstruation*. London: Profile Books, 2012.

Keynes, Geoffrey. *Oxford Medical Publications: Blood Transfusion*. South Yarra, Victoria: Leopold Classic Library, 2015.

Kirk, Robert G. W., and Neil Pemberton. *Leech*. London: Reaktion Books, 2013.

Krementsov, Nikolai. *A Martian Stranded on Earth: Alexander Bogdanov, Blood Transfusions, and Proletarian Science*. Chicago: University of Chicago Press, 2011.

Lederer, Susan E. *Flesh and Blood: Organ Transplantation and Blood Transfusion in Twentieth-Century America*. New York: Oxford University Press, 2008.

Miller, Jonathan. *The Body in Question*. London: Jonathan Cape, 1979.

Palfreeman, Linda. *Spain Bleeds: The Development of Battlefield Blood Transfusion During the Civil War*. Eastbourne: Sussex Academic Press, 2016.

Parsons, Vic. *Bad Blood: The Tragedy of the Canadian Tainted Blood Scandal*. Toronto: Lester Publishing, 1995.

Picard, André. *The Gift of Death: Confronting Canada's Tainted Blood Tragedy*. Toronto: HarperCollins Canada, 1995.

Rose, E. M. *The Murder of William of Norwich: The Origins of the Blood Libel in Medieval Europe*. Oxford: Oxford University Press, 2015.

Seeman, Bernard. *The River of Life*. London: Lowe & Brydone, 1962.

Starr, Douglas. *Blood: An Epic History of Medicine and Commerce*. New York: Harper Perennial, 2002.

Stein, Elissa, and Susan Kim. *Flow: The Cultural Story of Menstruation*. New York: St. Martin's Griffin, 2009.

Steinberg, Jonny. *Three Letter Plague: A Young Man's Journey Through a Great Epidemic*. London: Vintage Books, 2008.

Sugg, Richard. *Mummies, Cannibals and Vampires: The History of Corpse Medicine from the Renaissance to the Victorians*. London: Routledge, 2016.

Swanson, Kara W. *Banking on the Body: The Market in Blood, Milk and Sperm in Modern America*. Cambridge, MA: Harvard University Press, 2014.

Tierno, Philip M. *The Secret Life of Germs: What They Are, Why We Need Them, and How We Can Protect Ourselves Against Them*. New York: Atria Books, 2003.

Titmuss, Richard. *The Gift Relationship: From Human Blood to Social Policy*. London: Allen & Unwin, 1970.

Tucker, Holly. *Blood Work: A Tale of Medicine and Murder in the Scientific Revolution*. London: W.W. Norton, 2012.

ACKNOWLEDGMENTS

This book grew out of *The Big Necessity*, my book on sanitation. After it was published, I began to be asked to write about periods, decided I wanted to write about them in more depth, and then broadened that idea into a book about all sorts of aspects of blood, because how could I not, once I started looking into it? This made sense, except that I found myself writing about a topic that required me to understand medicine, science, history, culture, religion, philosophy, and much more, while having barely any background in medicine or science. I have needed a lot of help and have received it from many quarters. I hope I remember to acknowledge everyone I should, but, if not, my apologies and thanks.

For answering endless questions and providing tours of blood labs, I thank David Bowen and Julian Barth of Leeds Teaching Hospitals NHS Trust; Anne Weaver, Karim Brohi, and others at the Royal London Hospital trauma department; and London's Air Ambulance, including the three HEMS staff who allowed me to accompany their night shift around London in a fast car. This episode didn't make it into the book because, thankfully, no one was profoundly injured. It was a pleasure sharing a near-midnight feast with you while alarming other diners with urgently orange uniforms.

I am grateful to Anurag Maloo, who interpreted for me in the corridors of Delhi hospitals (and also gave a pint of his blood for the first time); Laurence Hamburger and Stink TV for providing a last-minute home in Cape Town; and Colin Clay, for his obliging hospitality and research skills in Saskatoon. Many people in many press departments have answered my questions with patience. Particular thanks to Stephen Bailey of NHSBT; the London office of MSF, who arranged my Khayelitsha trip; and all the MSF staff in Khayelitsha who hosted me; as well as Laura Crowley and colleagues of WaterAid, who have answered many questions about periods. Thanks also to AABB for promptly answering such queries as "But how do you know it's every two seconds?" with prompt yet authoritative numerical breakdowns.

For enabling various parts of my research, I thank Professor Sophie Scott, Natalie Cooper, Hannah Newman, and Claire Bromley. Anne Manuel, librarian of Somerville College, Oxford, obligingly granted me access to the college archives to research Janet Vaughan.

I detest transcribing—who doesn't?—so am grateful to Jane Duffus, who transcribed endless interviews that often included difficult medical terminology with speed and skill. Dr. Margaret McCartney generously passed on my request to find doctors who might read through the manuscript, and Dr. Sarah Worboys, Dr. Pete Lowe, and Dr. Diana Wetherill read through drafts. If there are still any medical errors in the book, it is my fault not theirs. Other trusted readers were Molly Mackey (also known as Percy Oliver's biggest fan), Thomas Ridgway, and Ruth Metzstein.

My agents Erin Malone and Siobhan O'Neill at William Morris Endeavor are unfailingly superb at what they do: attentive, caring, and as talented at managing yet another bout of panic as they are at sourcing inspirational Playmobile operating theatres. I'm very lucky. At Portobello Books, my book has been capably edited and cared for by Laura Barber and Ka Bradley, and copyedited by Mandy Woods with hawk-eye accuracy. The smart cover design is by James Paul Jones. Years after we began working together, I am still grateful to have landed in the peerless editing hands of Riva Hocherman at Metropolitan Books, who is as incisive and sharp as she is patient and understanding, all qualities that any author should seek in an editor. Thank you also to Grigory Tovbis, Christopher O'Connell, and others at Metropolitan for their editing, copyediting, and production skills. The excellent bloody cover of the Metropolitan edition was designed by Nicolette Seeback.

Two years ago, I was diagnosed as peri-menopausal. I've got my medication figured out now, more or less, but it has been a rocky road and I have lost many days to debilitating depression. Authors writing books are tricky enough; authors going through the menopause writing books are trickier. So I thank again all my editors and agents, who— once I dared to tell the truth about why I couldn't sometimes work—were extremely kind and compassionate. My best coping mechanism has been running in the fells so I'm grateful for the company and encouragement of all my fell-running friends and club mates. But it's my friends, family, and partner who have dealt with the rockiness at close quarters and they have been exceptional. Thank you to my mother, Sheila Wainwright; to my siblings; nephew and nieces; and to Neil "Braveshorts" Wallace for his love, support, and company off and on the hills. Finally, I thank every phlebotomist, health-care assistant, trauma surgeon, GP, and the millions of others who work for our beloved but besieged National Health Service, seventy years old this year. It is under attack so let us borrow from Janet Vaughan's operating system. Organize, and fight.

INDEX

ABOUT THE AUTHOR

ROSE GEORGE is the author of *The Big Necessity* and *Ninety Percent of Everything*. Her journalism has been published in *The New York Times*, *Scientific American*, *The Guardian*, and the *New Statesman*, among other publications. She lives in Yorkshire, England.